26 Springer Series in Solid-State Sciences

Edited by Ekkehart Kröner

Springer Series in Solid-State Sciences

Editors: M. Cardona P. Fulde H.-J. Queisser

Volumes 1 - 39 are listed on the back inside cover

Isaak A. Kunin

Elastic Media with Microstructure I

One-Dimensional Models

With 45 Figures

Springer-Verlag Berlin Heidelberg New York 1982

Professor Isaak A. Kunin

Department of Mechanical Engineering, University of Houston
Houston, TX 77004, USA

Guest Editor:

Professor Dr. Ekkehart Kröner

Institut für Physik, Universität Stuttgart
D-7000 Stuttgart 80, Fed. Rep. of Germany

Series Editors:

Professor Dr. Manuel Cardona
Professor Dr. Peter Fulde
Professor Dr. Hans-Joachim Queisser

Max-Planck-Institut für Festkörperforschung, Büsnauer Strasse 171,
D-7000 Stuttgart 80, Fed. Rep. of Germany

Title of the original Russian edition:
Teoriya uprugikh sred s mikrostrukturoi

ISBN-13:978-3-642-81750-2 e-ISBN-13:978-3-642-81748-9
DOI: 10.1007/978-3-642-81748-9

Library of Congress Cataloging in Publication Data. Kunin, I. A. (Isaak Abramovich), 1924- Elastic media with microstructure. (Springer series in solid-state sciences ; 26) Rev. and updated translation of: Teoriia uprugikh sred s mikrostrukturoĭ. Bibliography: v. 1, p. Includes index. Contents: 1. One-dimensional models -- 1. Elasticity. I. Title. II. Series. QA931.K8913 531'.3823 81-18268 AACR2

Softcover reprint of the hardcover 1st edition 1982

2153/3130-543210

Preface

Crystals and polycrystals,composites and polymers, grids and multibar systems can be considered as examples of media with microstructure. A characteristic feature of all such models is the existence of scale parameters which are connected with microgeometry or long-range interacting forces. As a result the corresponding theory must essentially be a nonlocal one.

The book is devoted to a systematic investigation of effects of microstructure, inner degrees of freedom and nonlocality in elastic media. The propagation of linear and nonlinear waves in dispersive media, static problems, and the theory of defects are considered in detail. Much attention is paid to approximate models and limiting transitions to classical elasticity.

The book can be considered as a revised and updated edition of the author's book under the same title published in Russian in 1975. The first volume presents a self-contained theory of one-dimensional models. The theory of three-dimensional models will be considered in a forthcoming volume.

The author would like to thank H. Lotsch and H. Zorsky who read the manuscript and offered many suggestions.

Houston, May 1982

ISAAK A. KUNIN

Contents

1. Introduction

In recent years, new physical and mathematical models of material media, which can be considered far-reaching generalizations of classical theories of elasticity, plasticity, and ideal and viscous liquids, have been intensively developed. Such models have appeared for a number of reasons. Primary among them are the use of new construction materials in extreme conditions and the intensification of technological processes. The increasing tendency toward rapproachement of mechanics with physics is closely connected with these factors. The internal logic of the development of continuum mechanics as a science is also important.

This treatment is devoted to the study of models of elastic media with microstructure and to the development of the nonlocal theory of elasticity. Starting from such models as a crystal lattice and simple discrete mechanical systems, we develop the theory and its applications in a systematic way.

The Cosserat continuum was historically one of the first models of elastic media which could not be described within the scope of classical elasticity [1.1] However, the memoirs of *E.* and *F. Cosserat* (1909) remained unnoticed for a long time, and only around 1960 did the generalized models of the Cosserat continuum start to be developed intensively. They are known as oriented media, asymmetric, multipolar, micromorphic, couple-stress, etc., theories. For short we shall call them couple-stress theories. Essential contributions to the development of couple-stress theories were made, for example, by Aero, Eringen, Green, Grioli, Günther, Herrmann, Koiter, Kuvshinsky, Mindlin, Naghdi, Nowacki, Palmov, Rivlin, Sternberg, Toupin, and Wozniak; their fundamental works are listed in the Bibliography. The Bibliography is far from being complete. A survey of works before 1960 can be found in the fundamental treatment of *Truesdell* and *Toupin* [1.2]; later ones are quoted in papers by *Wozniak* [1.3], *Savin* and *Nemish* [1.4], *Iliushin* and *Lomakin* [1.5], as well as in monographs by *Misicu* [1.6] and *Nowacki* [1.7].

From the very beginning of the development of the generalized Cosserat models, attention was turned to their connections with the continuum theory of dislocations. In 1967 a symposium was organized by the International Union of Theoretical and Applied Mechanics which had great significance in summing up the tenyear period of development [1.8]. In the symposium a new trend closely connected with the theory of the crystal lattice was also presented which contained the above-indicated models as a longwave approximation, namely, a nonlocal theory of elasticity. The nonlocal theory of elasticity was also developed in works of Edelen, Eringen, Green, Kröner, Kunin, Laws and others, given in the Bibliography. A rather complete listing on media with microstructure is contained in [1.9]. It is worth mentioning here that the very term "nonlocal elasticity" seems to have been introduced by *Kröner* in 1963 [1.10], and the first monograph on the subject was published by this author in 1975 [1.11].

We start out with a brief classification of the theories of elastic media with microstructure. Explicit or implicit nonlocality is the characteristic feature of all such theories. The latter, in its turn, displays itself in that the theories contain parameters which have the dimension of length. These scale parameters can have different physical meanings: a distance between particles in discrete structures, the dimension of a grain or a cell, a characteristic radius of correlation or action-at-a-distance forces, etc. However, we shall always assume that the scale parameters are small in comparison with dimensions of the body.

One has to distinguish the cases of strong and weak nonlocality. If the "resolving power" of the model has the order of the scale parameter, i.e., if, in the corresponding theory, it is physically acceptable to consider wavelengths comparable with the scale parameter, then we shall call the theory nonlocal or strongly nonlocal (when intending to emphasize this). In such models, one can consider elements of the medium of the order of the scale parameter, but, as a rule, distances much smaller than the parameter have no physical meaning. The equations of motion of a consistently nonlocal theory necessarily contain integral, integrodifferential, or finitedifference operators in the spatial variables. In nonlocal models, the velocity of wave propagation depends on wavelength; therefore, the term "medium with spatial dispersion" is also used, frequently.

Let us emphasize that nonlocality or spatial dispersion can have different origins. They can be caused by a microstructure of the medium (in particular, by the discreteness of the micromodel) or by approximate consideration of such parameters as thickness of a rod or plate. One can speak therefore about the physical or geometrical nature of nonlocality. In the latter case, the nonlocal model is, as a rule, one or two dimensional and serves as an effective approximate description of a local three-dimensional medium.

If the scale parameter is small in comparison with the wavelengths considered, but the effects of nonlocality cannot be neglected completely, then a transition is possible to approximate models, for which integral and finite-difference operators are replaced by differential operators with small parameters attached to their highest derivatives. In such a case, one can speak about the model of the medium with weak spatial dispersion. The corresponding theory will be called weakly nonlocal. All above-mentioned couple-stress theories belong to this type, although they are usually constructed on a purely phenomenological basis.

Finally, the consideration of sufficiently long waves (zeroth long-wave approximation) leads to a transition to a local theory in the limit, already containing no scale parameters. This property of locality, i.e., the possibility of considering "infinitesimally small" elements of the medium, is inherent in all the classical models of the mechanics of continuous media.

Let us return to nonlocal models. They can be divided into two classes: discrete and continuous. Discrete structure of a medium could be taken into account in the usual way, for example, as is done in the theory of the crystal lattice. However, the apparatus of discrete mathematics is most cumbersome; therefore, we shall also use the mathematical model of quasicontinuum for an adequate description of the discrete medium. Its essence is an interpolation of functions of discrete argument by a special class of analytical functions in such a way that the condition of one-to-one correspondence between quasicontinuum and the discrete medium is fulfilled. The advantages of such an approach consist in an ability to describe discrete and continuous media within the scope of a unified formalism and, in particular, to generalize correctly such concepts of continuum mechanics as strain and stress. It is to be emphasized that the model of quasicontinuum is applicable not only to crystal lattices but also to macrosystems.

We shall also distinguish media of simple and complex structures. In the first case, the displacement vector is the only kinematic variable and it determines a state of the medium completely. Body forces are the corresponding force variable. To describe a medium of complex structure, a set of microrotations and microdeformations of different orders characterizing the internal degrees of freedom and the corresponding force micromoments is additionally introduced.

The difference between media of simple and complex structures, generally speaking, is conserved in the approximation of weak nonlocality, but this is dispalayed only for high enough frequencies of the order of the natural frequencies of the internal degrees of freedom. At low frequencies, the internal degrees of freedom can be excluded from the euqations of motion so that they will contribute to the effective characteristics of the medium only. The difference between the quasicontinuum and the continuous medium completely disappears in the approximation of weak nonlocality.

In the zeroth long-wave approximation, at not very high frequencies, a complete identification of different models of a medium with microstructure takes place: all of them are equivalent to the classical model of elastic continua which was obtained on the basis of general phenomenological postulates. Only effective elastic moduli "know" about the structure of the initial micromodel, but this information cannot, of course, be derived from them. It follows that an explicit consideration of microstructure effects and, in particular, of the internal degrees of freedom is possible only with the simultaneous consideration of nonlocality, i.e., a consistent theory of elastic medium with microstructure must necessarily be nonlocal.

Schematically, the connections between different theories are shown in Fig. 1.1.

Our main purpose is the investigation of the effects of microstructure and nonlocality. In addition, we wish to elucidate the domain of applicability of different theories of media with microstructure. Such theories are considerably more complex than the usual theory of elasticity, although they reduce to this in

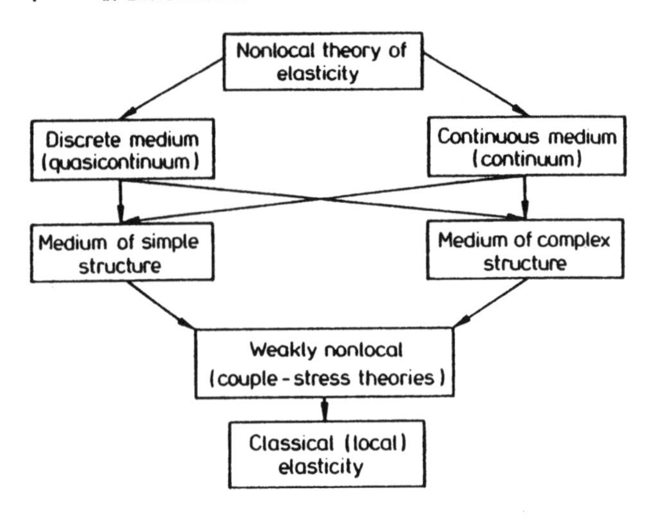

Fig. 1.1.

certain limits; their application is, as a rule, reasonable only when they describe qualitatively new effects which are not derivable from the local theory.

We find it advisable to consider a number of simple models of media with microstructure in order to acquire some nonlocal intuition before proceeding to generalizations.

For these reasons, the treatment is divided into two parts. One-dimensional models, for which cumbersome tensor algebra is not needed, are studied in this first volume. At the same time, one can trace a number of distinctions of the nonlocal theory from the local one already in one-dimensional models. These distinctions have both physical and methodological character. Particularly, one has to analyze critically the possibility of using in nonlocal theories such habitual notions as stress, strain, energy density and flux.

In the second volume [1.1], three-dimensional models are considered, the main attention being payed to specific three-dimensional effects. The last chapters are devoted to point defects and dislocations in the nonlocal theory of elasticity.

The author has tried to avoid complex mathematical methods in the first part. It is assumed that the reader is acquainted with the Fourier transform and has some skill in working with δ-functions (though it is easy to acquire it in the process of reading the book).

The necessity of simultaneous use of space-time and frequency-wave representations is one of the peculiarities of the formalism used in the nonlocal theory. This way of thinking is quite habitual for physicists, but could be, to some extent, unusual for engineers. Because of this, the rate of presenting the material is slow in the beginning and then speeds up gradually. For the same purpose, a number of results are presented in the form of problems, which are considered a part of the text and they are referred to in the later text.

In conclusion, let us note that a number of important questions are omitted. In particular, this is related to the thermodynamics of the nonlocal models, which was contained in the original plan of the book. Unfortunately, the author has not succeeded in representing this problem in a sufficiently simple and physically motivated form. An axiomatic approach to the nonlocal thermodynamics was developed by *Edelen* [1.13].

2. Medium of Simple Structure

By definition, the only kinematic variable of the medium of simple structure is displacement; it determines the state of the medium completely. The simplest model of such a medium is a linear chain of point masses connected by elastic bonds. Similar continuous models of microstructure are also of interest.

In this chapter we develop the theory of such media or, which is the same, the one-dimensional nonlocal elasticity. We introduce the notion of a quasi-continuum which permits one to treat discrete and continuous models in the scope of the same formalism. Much attention is paid to the analysis of the physical meaning of such basic notions as strain, stress, energy density and flux, and boundary conditions for nonlocal models. We investigate in detail new physical phenomena specific to nonlocal elasticity, approximate models and the transition to classical elasticity in the limit of long waves.

2.1 Simple Chain

Let us start with the simplest model of a medium with microstructure, namley, a linear chain of point masses connected by elastic bonds. It can be considered either as a mechanical system with given characteristics of weihtless bonds (for example, springs) or as a one-dimensional model of a crystal. In the latter case, effective characteristics of the bonds can be found from the interaction potential of the atoms.

In this section, we shall study the infinite, simple, homogeneous chain. More complicated models will be investigated later.

The chain is called homogeneous if a periodically repeating cell can be distinguished. If the cell contains only one particle, the chain is called simple.

Let the particles of mass m be located along the x axis and have in equilibrium coordinates na where a is the distance between particles and n runs through all integers. It is supposed that displacements of the particles $u(na)$ can be only longitudinal, i.e., along the x axis. Let us set temporairly, for convenience in writing, $a = 1$ (which can always be achieved by the proper choice of the unit of length).

The potential energy Φ of the chain is a functional of the displacement field $u(n)$. When displacements are small, Φ can be expanded in a series in $u(n)$ and be represented in the form

$$\begin{aligned}\Phi &= \Phi_0 + \Phi_1 + \Phi_2 + \Phi_3 + \cdots \\ &= \Phi_0 + \sum_n \Phi(n)u(n) + \frac{1}{2}\sum_{nn'} \Phi(n, n')\, u(n)\, u(n') \\ &\quad + \frac{1}{3!}\sum_{nn'n''} \Phi(n, n', n'')u(n)\, u(n')\, u(n'') + \cdots . \end{aligned} \quad (2.1.1)$$

Henceforth, the summation is carried out over all n. The coefficients of the expansion, obviously, are the derivatives of Φ in displacements (or coordinates) in the equilibrium state.

The first term of the expansion, i.e., the constant Φ_0, is the energy of the chain at equilibrium, which is of no further interest. The term Φ_1 must be equal to zero because the expansion is made about equilibrium. Thus, as the first approximation for the potential energy, Φ can be expressed as

$$\Phi = \frac{1}{2}\sum_{nn'} \Phi(n, n')\, u(n)\, u(n'), \quad (2.1.2)$$

which is completely defined by the constants $\Phi(n, n')$, constituting the parameters of the model. Usually, they are called the force constants.

This approximation is known in the crystal-lattice theory as harmonic and obviously leads to linear equations of motion. Taking into account the next terms of the expansion (anharmonic model) corresponds to an approximate considera-tion of nonlinearlity. At first we shall restrict ourselves to the study of linear models. Specific nonlinear effects will be considered in Chaps. 5 and 6.

The kinetic energy of the chain with displacements $u(n, t)$ depdending on time is

$$T = \frac{m}{2}\sum_n \dot{u}^2(n, t). \quad (2.1.3)$$

The equation of motion can be obtained, as usual, from the condition of sta-tionarity of action

$$\mathscr{L} = \int_{t_1}^{t_2} L(u, \dot{u})\, dt, \quad (2.1.4)$$

where $L(u, \dot{u}) = T - \Phi$ is the Lagrangian. If external forces $q(n, t)$ act on the particles of the chain, then it is necessary to include in the expresssion for L a term which takes into account the contribution of these forces. For the model under consideration we have in the harmonic approximation,

$$L = \frac{m}{2}\sum_n \dot{u}^2(n, t) - \frac{1}{2}\sum_{nn'} \Phi(n, n')\, u(n, t)\, u(n'\, t) + \sum_n q(n, t)\, u(n, t)\, . \quad (2.1.5)$$

The Largrange equations

$$\frac{d}{dt}\frac{\partial L}{\partial \dot{u}} - \frac{\partial L}{\partial u} = 0$$

now give the euqations of motion

$$m\ddot{u}(n, t) + \sum_{n'} \Phi(n, n')\, u(n', t) = q(n, t)\,. \tag{2.1.6}$$

The physical meaning of the force constants $\Phi(n, n')$, $n \neq n'$, is found easily from the equations. In fact, let a particle with the coordinate n' have a unit static displacement, displacements of others being zero. Then, the external force, applied at a point n and compensating the reaction of the elastic bond, is equal to $\Phi(n, n')$. Thus the force constants, which determine properties of the model, are the parameters of the elastic bonds between the particles of the mechanical system or charactensitics of interaction between atoms in a crystal. The physical meaning of the self-action constants $\Phi(n, n)$ will be made clear below.

Let us now consider general properties of the force constants. It follows from (2.1.2) that for a unique definition of $\Phi(n, n')$ it is necessary to demand that the symmetry condition[1]

$$\Phi(n, n') = \Phi(n', n)\,. \tag{2.1.7}$$

be fulfilled. It follows, from the homogeneity of the chain, that

$$\Phi(n + n'', n' + n'') = \Phi(n, n')$$

for any n, n', n''. This is possible only if

$$\Phi(n, n') = \Phi(n - n'), \tag{2.1.8}$$

i.e., the given function of one variable $\Phi(n)$ fully determines the effective characteristics of elastic bonds in the homogeneous chain. Besides, from (2.1.7), $\Phi(n)$ must be even

$$\Phi(n) = \Phi(-n)\,. \tag{2.1.9}$$

If $\Phi(n)$ differs from zero for $|n| \leqslant N$ only, then this means that every particle interacts with N neighbors to the right and with N neighbors to the left. The case $N = 1$ corresponds to the simplest model of interaction between nearest neighbors. In real mechanical systems, the action at a distance is always restricted, i.e., N is finite. This is also true for models of nonionic crystals. However, in the case of ionic crystals, the atoms interact according to Coulomb's law and $N \to \infty$. As a rule we shall study an arbitary but finite action at a distance.

---

[1] An antisymmetric component would not given any contribution to the energy.

Let $u(n, t) = u_0 =$ const (translation of the chain as a whole). Then distances between the particles do not change and therefore the forces acting on particles are to be equal to zero. According to (2.1.6) and taking (2.1.8) into account, we obtain the corresponding conditions for $\Phi(n)$

$$\sum_{n'} \Phi(n - n')\, u_0 = 0 \quad \text{or} \quad \sum_{n} \Phi(n) = 0\,. \tag{2.1.10}$$

This relation determines the self-action constant

$$\Phi(0) = -\sum_{n}{}' \,\Phi(n)\,, \tag{2.1.11}$$

where $\sum_{n}{}'$ means summation over all $n \neq 0$.

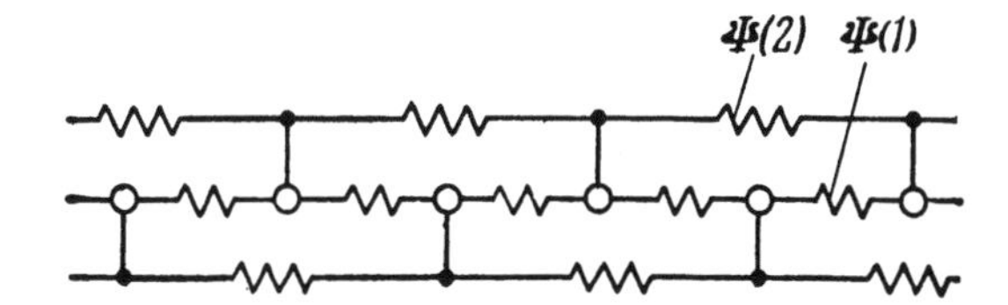

Fig.2.1. The chain with two-neighbor interaction

For example, let the particles be connected by springs with stiffnesses $\Psi(n)$ (see Fig. 2.1, where the interaction of only nearest and next-nearest neighbors is shown for simplicity). Then

$$\begin{aligned} \Phi(n) &= -\,\Psi(n), \quad n \neq 0\,, \\ \Phi(0) &= \sum_{n} \Psi(n)\,, \end{aligned} \tag{2.1.12}$$

and the equations of motion (2.1.6) can also be written in the form of finite differences

$$\begin{aligned} &m\ddot{u}(n, t) + \sum_{n'} \Phi(n - n')\, u(n', t) \\ &\quad = m\ddot{u}(n, t) - \sum_{n'} \Psi(n')\,[u(n - n', t) - u(n, t)] = q(n, t)\,. \end{aligned} \tag{2.1.13}$$

Observe that for real mechanical systems all $\Psi(n) > 0$ and hence $\Phi(0) > 0$, $\Phi(n) < 0$ $(n \neq 0)$. However, a system may be constructed, in principle, for which some bonds would have negative stiffnesses. We shall return to this question later.

The expression (2.1.2) for the elastic energy of the chain, if one takes into account (2.1.8) and (2.1.12), may be rewritten now in the form

$$\Phi = \frac{1}{2} \sum_{nn'} \Phi(n - n')\, u(n)\, u(n') = \frac{1}{4} \sum_{nn'} \Psi(n - n')\,[u(n) - u(n')]^2\,. \tag{2.1.14}$$

The quantity

$$\varphi(n) = \frac{1}{4} \sum_{n'} \Psi(n - n') [u(n) - u(n')]^2$$
$$= \frac{1}{4} \sum_{n'} \Psi(n') [u(n - n') - u(n)]^2 \qquad (2.1.15)$$

may be interpreted as an average energy corresponding to a particle at a point n (the energy of a bond connecting particles n and n' is divided equally between them). Evidently, $\varphi(n)$ is invariant with respect to translations and may be considered as an analog of the energy density in the local theory.

The energy density $\varphi(n)$ is in some cases more convenient to deal with than the total energy Φ. Indeed, $\varphi(n)$ is defined for all $u(n)$ (due to the boundedness of the action at a distance) while Φ has sense only for $u(n)$ which decrease sufficiently rapidly as $n \to \infty$.

It is known that the Lagrangian density in the local theory is defined non-uniquely, i.e., up to total derivatives (divergences) which do not influence the equations of motion.

An analogous situation takes place for a discrete medium. A component of the Lagrangian density $\varphi_1(n)$ connected with the elastic energy Φ may differ the $\varphi(n)$ introduced above by any quadratic function of displacement $f(n)$ such that $\sum_n f(n) = 0$. For example, according to (2.1.14), one may assume

$$\varphi_1(n) = \frac{1}{2} \sum_{n'} \Phi(n - n') \, u(n) \, u(n') . \qquad (2.1.16)$$

However, $\varphi_1(n)$ here could not be interpreted as the energy density since it is not invariant with respect to translations.

The requirement of translational invariance in the local theory distinguishes uniquely among all possible Lagrangian desnities the one which has the meaning of the energy density. In the model under consideration, the energy density is not already a uniquely determined physical quantity. As we shall see later, such nonuniqueness of the definition of a number of important quantities appears to be a characteristic feature of nonlocal theories and, in particular, is also related to the energy flux, stress, and, more generaly, to the energy-momentum tensor.

There are two reasons for this nonuniqueness. The first is connected with the possibility of different ways of distributing energy between the particles or, equivalently, with ways of imaginary cuttings of the bonds (of course, this is due to the nonlocality). Such nonuniqueness may be eliminated, in principle, if we agree to make cuts always in some special way, e.g., to divide the energy of bonds equally as was done above. In such a case the arbitrariness is realted to the agreement only about the way of cutting of bonds.

The second reason is more significant. The equations of motion of a system and its energy are completely determined by the given force constants $\Phi(n)$, which from this point of view carry the complete information about the system. But $\Phi(n)$ themselves are only the effective characteristics of the real bonds and by no means determine the last ones in a unique way. In fact, given force constants can, firstly, correspond to the real bonds inhomogeneous in length and, secondly, the bonds could have complex joints of the type shown in Fig. 2.2 (so-called unpaired bonds). Although they can be substituted by equivalent homogeneous paired[2] bonds, the latter would have the meaning of some effective quantities only.

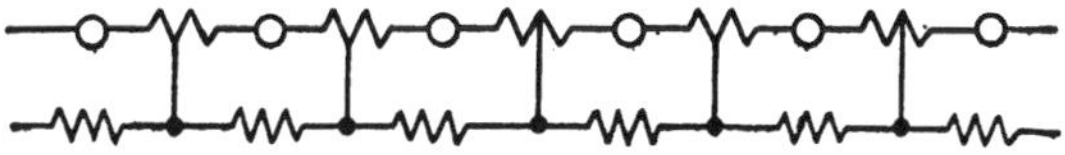

Fig.2.2. The chain with unpaired bonds

This is even more important in crystals where the physical fields of various nature with different characteristics of action at a distance may correspond to the given force constants of harmonic approximation.

In all such cases, for a physically unique determination of energy in some finite region and hence also the energy density, it is necessary to have additional information about the structure of the bonds or about the nature of fields which is not contained in the functional Φ and the equations of motions. Without such information we must consider the physically unique definition of such notions as the density and flow of energy, stress, etc., to be impossible within the scope of the nonlocal theory. The degree of nonuniqueness of these quantities will be investigated in detail below.

We are forced to discuss in detail these, as a matter of fact, methodological questions because experience has shown that they are of certain difficulty for those who are accustomed to the concepts of local field theories.

Before we start to analyze the equations of motions of the chain and analogous more complicated systems, let us sum up the results.

The discrete structure and forces of action at a distance are the characteristic properties of the model under consideration. The discrete structure is always connected with the interaction at a distance, but it is not difficult to imagine a model of a continuous medium with action at a distance. Particularly, it can be obtained as a limiting case of the chain with $N \gg 1$. Other models which can conveniently be described in terms of the nonlocal theory of modium with microstructure are of interest also.

In this connection, it is worthwhile to develop an analytical technique which permits one to consider the discrete and continuous, nonlocal, weakly nonlocal, and local models within the scope of a unified formalism. One of the essential elements of such a formalism is the concept of a quasicontinuum which will be developed in the next two sections. Because little information on this concept

[2]Later we shall see that not paired but rather triple effective bonds are the general case for a three-dimensional lattice.

appears in the literature, it will be considered in somewhat more detail here than would be necessary in a purely applied aspect. One may omit the formalism during the first reading and restrict oneself to the basic relations between functions of discrete arguments and their analytic representations.

2.2 One-Dimensional Quasicontinuum

The basic idea is to set up a one-to-one correspondence between functions of discrete arguments and a class of analytical functions as well as between the operations on them. This will permit us to use a well-developed analytical technique for the description of discrete structures and to give the corresponding theories a form which is analogous to the theory of a continuous medium.

Let the function $u(na)$ be given by its values, generally complex, at the points na (knots) of the x axis. Let us first assume that the distance a between the knots is a constant, independent of n. For convenience of notation we shall write the values of the functions in the knots as $u(n)$.

Let us pose the problem of interpolation of $u(n)$ by a smooth function $u(x)$. Obviously, $u(x)$ is defined to within any $\psi(x)$, which vanishes in all knots, i.e., $u(x)+\psi(x)$ will also be an interpolating function (Fig. 2.3). It is natural to make an attempt to choose from the whole set of interpolating functions the most smooth ones filtering out rapidly oscillating components. Later a rigorous formulation of this condition which provides uniqueness of interpolation will be given. Simultaneously with the function $u(x)$, we shall consider its image under Fourier transform (or, more briefly, Fourier image) which we shall also denote by $u(k)$, but with argument k. On the one hand, such notations are highly convenient when one has to work often with Fourier transforms, and on the other hand, they emphasize that $u(x)$ and $u(k)$ are to be considered as the representations of one and the same function u in x and k spaces. We shall retain these notations, except in specially distinguished cases.

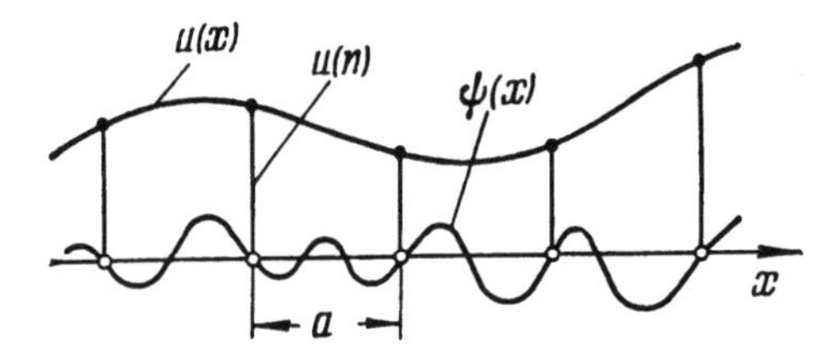

Fig.2.3. Nonuniqueness of interpolation

The correspondence $u(x) \leftrightarrow u(k)$ is given by the formulae[3]

$$u(k) = \int e^{-ikx} u(x)\, dx, \tag{2.2.1}$$

[3]Here the signs of the arguments of the exponents are chosen in accordance with tradition in the physical literature.

$$u(x) = \frac{1}{2\pi} \int e^{ikx} u(k)\, dk\,, \tag{2.2.2}$$

where the integration is carried out over the whole x or k axis.

For the Fourier image to exist in the class of usual functions, it is necessary for $u(x)$ to decrease sufficiently rapidly as $|x| \to \infty$, for example, faster than $|x|^{-1}$. The function $u(n)$ must satisfy the corresponding conditions. Later these restrictions will be eliminated when proceeding to the generalized functions.

In accordance with (2.2.2), $u(x)$ can be represented as a superposition of Fourier harmonics, their frequency of oscillation increasing infinitely as $|k| \to \infty$. The above-mentioned condition of "smoothness" of $u(x)$ is obviously equivalent to restricting $u(k)$ to a minimum neighborhood of the origin in k space. Let us show that the interpolating function $u(x)$ is determined uniquely, if we require its Fourier image $u(k)$ to differ from zero only on the segment $B[-\pi/a \leqslant k \leqslant \pi/a]$ (Fig. 2.4).

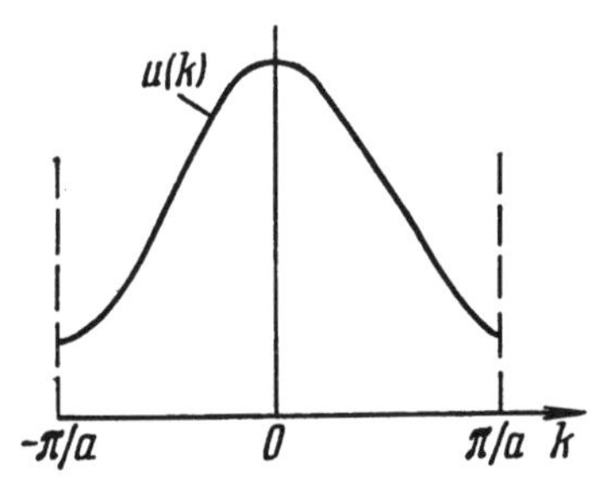

Fig.2.4. Fourier image of an interpolating function

Expanding $u(k)$ in a Fourier series on the segment B and taking into account that $u(k) = 0$ when $k \,\overline{\in}\, B$, we have

$$u(k) = B(k) \sum_n c_n e^{-inak}\,, \tag{2.2.3}$$

where $B(k)$ is a characteristic function of the segment B, i.e., $B(k) = 1$ when $k \in B$ and $B(k) = 0$ when $k \,\overline{\in}\, B$. Assuming that c_n decreases sufficiently rapidly with increasing n, so that termwise integration of the series is possible, and substituting $u(k)$ in (2.2.2), we find

$$u(x) = \sum_n c_n \delta_B(x - na)\,, \tag{2.2.4}$$

where[4]

$$\delta_B(x) \overset{\text{def}}{=} \frac{1}{2\pi} \int_B e^{ikx}\, dk = \frac{\sin(\pi x/a)}{\pi x}\,. \tag{2.2.5}$$

---

[4]Here and later $\overset{\text{def}}{=}$ means: is equal to by definition.

It is easy to see that $\delta_B(x) = \delta_B(-x)$ and

$$\delta_B(0) = a^{-1}, \quad \delta_B(na) = 0, \quad \text{if} \quad n \neq 0\,. \tag{2.2.6}$$

It follows that if we set $c_n = au(n)$, i.e.,

$$u(k) = aB(k) \sum_n u(n)\mathrm{e}^{-inak}\,, \tag{2.2.7}$$

$$u(x) = a \sum_n u(n)\, \delta_B(x - na)\,, \tag{2.2.8}$$

then $u(x)$ is the required interpolating function, taking the given values in the knots and having its Fourier image $u(k)$ concentrated in the segment B. The convergence of the series is ensured by sufficiently rapid decrease of $u(n)$ as $|n| \to \infty$.

It is known [2.1,2] that the Fourier images of finite functions, i.e., those differing from zero only in a finite domain, are entire analytic functions of exponential type. Since, by definition $u(k)$ differs from zero only on the segment B, then $u(x)$ can be continued into the complex region as an entire function of exponential type $\leqslant \pi/a$.

Let us mention also the relations

$$u(n) = \frac{1}{2\pi} \int_B \mathrm{e}^{inak} u(k)\, dk = \int \delta_B(x - na)\, u(x)\, dx\,. \tag{2.2.9}$$

The first is merely an expression for the coefficients of the Fourier series (2.2.7) and the second can be obtained from Parseval's equality

$$\int \overline{q(x)}\, u(x)\, dx = \frac{1}{2\pi} \int \overline{q(k)}\, u(k)\, dk\,, \tag{2.2.10}$$

if we set $q(x) = \delta_B\,(x - na)$. Indeed, taking into account (2.2.5) and (2.1.5), we have $\overline{q(k)} = \exp(inak)\, B(k)$.

Thus, under the conditions pointed out, for any function $u(n)$, the functions $u(x)$ and $u(k)$ are in one-to-one correspondence. Let $N \ni u(n)$, $X \ni u(x)$, and $K \ni u(k)$ be the corresponding linear functional spaces (their exact definition will be given below). Then (2.2.7–9) establish a one-to-one correspondence $N \leftrightarrow X \leftrightarrow K$, which preserves the linear structure, i.e., these spaces are linearly isomorphic. To each operation (not necessarily linear) applied to the functions of a discrete argument there corresponds uniquely an operation applied to their images which are functions of the continuous argument. It is essential that as a rule, this correspondence occurs to be in a natural fashion.

Let us consider some of these operations. Let $q(n)$ be a fixed function decreasing sufficiently rapidly as $|n| \to \infty$. Then the expression

$$\langle q\,|\,u\rangle \stackrel{\text{def}}{=} \sum \overline{q(n)}\, u(n) \tag{2.2.11}$$

is a complex number depending linearly on the function $u(n)$, i.e., $q(n)$ determines the linear functional q over the space N.

One can justify the following relations

$$\langle q\,|\,u\rangle = \int \overline{q(x)}\, u(x)\, dx = \frac{1}{2\pi} \int \overline{q(k)}\, u(k)\, dk\,, \tag{2.2.12}$$

where

$$\begin{aligned} q(x) &= \sum_n q(n)\, \delta_B(x - na)\,, \\ q(k) &= B(k) \sum_n q(n)\, \mathrm{e}^{-inak}\,. \end{aligned} \tag{2.2.13}$$

Functions $q(n)$, $q(x)$, and $q(x)$ will be called the kernels of the functional $\langle q\,|\,u\rangle$ in n, x, and k representations.

Problem 2.2.1: Prove (2.2.12). [Hint: Use (2.2.7-9) and Parseval's equality (2.2.10).]

Note that here the accepted correspondence between $q(n)$ and its images $q(x)$ and $q(k)$ differs from (2.2.7,8) by a factor a. This has been done for convenience of physical interpretation of $q(x)$ and $q(n)$. Later $u(n)$ will usually be the displacement and $q(n)$ the force. Then $q(x)$ may be interpreted as the average force density and $q(k)$ as its Fourier image.

Let us consider as an example the functional δ defined by the function

$$\delta(n) \stackrel{\text{def}}{=} \begin{cases} 1, & \text{if} \quad n = 0\,, \\ 0, & \text{if} \quad n \neq 0\,. \end{cases} \tag{2.2.14}$$

Obviously,

$$\langle \delta\,|\,u\rangle = u(n)\,|_{n=0} = u(x)\,|_{x=0} = \int_B u(k)\, dk\,. \tag{2.2.15}$$

Thus the correspondence

$$\delta(n) \leftrightarrow \delta_B(x) \leftrightarrow 1(k) \tag{2.2.16}$$

is made.[5]

It follows from (2.2.15) that $\delta_B(x)$ plays, in the functional space X, the role of the usual δ-function and in particular

$$\int \delta_B(x)\, dx = 1\,. \tag{2.2.17}$$

[5]Here we deviate from the notation above because $\delta(k)$ stands for the δ function in k space.

However, as distinct from the usual δ function, $\delta_B(x)$ is not singular. When $a \to 0$, the value $\delta_B(0) = a^{-1}$ goes to infinity and $\delta_B(x)$ turns into Dirac's δ function. Since in the following it will always be clear from the context which δ function is referred to, we shall omit the index B and denote kernels of the functional δ in the n, x, and k representations $\delta(n)$, $\delta(x)$, and $1(k)$, respectively.

The functional determined by the functions

$$q(n) = a \leftrightarrow q(x) = 1 \leftrightarrow q(k) = 2\pi\delta(k) \tag{2.2.18}$$

is another example.

We have the identities

$$a\sum_n u(n) = \int u(x)dx = u(k)|_{k=0} , \tag{2.2.19}$$

i.e., the summation in N corresponds exactly to the integration in X (in distinction from the approximate equality $a\sum\limits_n \approx \int \cdots dx$ for arbitrary functions).

Note that sometimes it is conventient to interpret the expression $\langle q|u\rangle$ also as a scalar product of q and u invariant with respect to n, x, and k representations, and satisfying the usual condition

$$\langle q|u\rangle = \overline{\langle u|q\rangle} . \tag{2.2.20}$$

Let us consider now functions of two variables $\Phi(n, n')$ defined on a square lattice with the step a. Taking the scalar product of the function $\Phi(n, n')$ with $u(n)$ from the left and with $v(n')$ on the right, we obtain the form $\langle u|\Phi|v\rangle$, which is linear in the second argument and antilinear in the first.[6] The invariance of the form $\langle u|\Phi|v\rangle$ with respect to n, x, and k representations follows from the invariance of the scalar product:

$$\begin{aligned}\langle u|\Phi|v\rangle &\stackrel{\text{def}}{=} \sum_{nn'} \overline{u(n)}\, \Phi(n, n')\, v(n') = \iint \overline{u(x)}\, \Phi(x, x')\, v(x')\, dx\, dx' \\ &= \frac{1}{2\pi}\iint \overline{u(k)}\, \Phi(k,k')\, v(k')\, dk\, dk' ,\end{aligned} \tag{2.2.21}$$

where $u(x) \leftrightarrow u(n)$, $v(x) \leftrightarrow v(n)$, according to (2.2.8), and

$$\begin{aligned}\Phi(k, k') &= \frac{1}{2\pi}\sum_{nn'} \Phi(n, n')\, e^{-ia(kn-k'n')} \\ &= \frac{1}{2\pi}\iint \Phi(x, x')\, e^{-i(kx-k'x')}\, dx\, dx' .\end{aligned} \tag{2.2.22}$$

Thus, $\Phi(k, k')$ is the Fourier image of $\Phi(x, x')$ in the first argument and the inverse Fourier image in the second.

[6]If $F(\alpha u + \beta v) = \bar{\alpha}F(u) + \bar{\beta}F(v)$, where α,β are complex constants, then the functional $F(u)$ is called antilinear.

The functions $\Phi(n, n')$, $\Phi(x, x')$ and $\Phi(k, k')$ can be considered as the corresponding representations of the kernel of the form $\langle u|\Phi|v\rangle$. In general the latter is not defined for all pairs u, v. A natural domain of definition of the form depends on the properties of its kernel.

The relation

$$\langle u|\Phi|v\rangle = \overline{\langle v|\Phi^+|u\rangle} \tag{2.2.23}$$

defines the form Φ^+, which is Hermitian conjugate to Φ. For the kernel of the form Φ^+, we have

$$\begin{aligned} &\Phi^+(n, n') = \overline{\Phi(n, n')}, \quad \Phi^+(x, x') = \overline{\Phi(x', x)}\,, \\ &\Phi^+(k, k') = \overline{\Phi(k', k)}\,. \end{aligned} \tag{2.2.24}$$

In cases for which $\Phi(n, n')$ is finite, the form Φ is defined over all u, v. Its kernel $\Phi(k, k')$ can be continued analytically in the entire complex plane, and for complex k's the last of the relations (2.2.24) is to be replaced by

$$\Phi^+(k, k') = \Phi(-k', -k)\,. \tag{2.2.24a}$$

A form for which $\Phi^+ = \Phi$ is called Hermitian. Obviously the corresponding quadratic form $\Phi[u]$ always has real values. Indeed,

$$\Phi[u] \overset{\text{def}}{=} \langle u|\Phi|u\rangle = \overline{\langle u|\Phi^+|u\rangle} = \overline{\langle u|\Phi|u\rangle}\,. \tag{2.2.25}$$

A Hermitian form is called positive (positive definite) if $\Phi[u] \geqslant 0$ ($\Phi[u] > 0$) when $u \neq 0$.

We often encounter forms which have kernels of the type

$$\Phi(n - n'),\ \Phi(x - x'),\ \Phi(k)\,\delta(k - k')\,,$$

where

$$\Phi(k) = a^{-1} \sum_n \Phi(n)\, \mathrm{e}^{-\mathrm{i}nak} = \int \Phi(x)\, \mathrm{e}^{-\mathrm{i}nak}\, dx\,. \tag{2.2.26}$$

In such cases, we shall use the terms "difference-type form" or "difference-type kernel". More rigorously, these might be called "translationally invariant forms" or the kernels thereof. For the case of operators, this concept reduces to the more familiar convolution-type operator.

The last equality in (2.2.21) now takes the form

$$\langle u|\Phi|v\rangle = \frac{1}{2\pi} \int \overline{u(k)}\, \Phi(k)\, v(k)\, dk\,. \tag{2.2.27}$$

Admitting some liberty, the functions $\Phi(n)$, $\Phi(x)$, and $\Phi(k)$ are also called the (one-dimensional) kernel of the difference type form.

The scalar product introduced above is the simplest example of a positively definite difference type form. Its kernel has, obviously, the representations $\delta(n-n')$, $\delta(x-x')$, $\delta(k-k')$.

One can associate with each form Φ the linear operator

$$\Phi\colon u \rightarrow q = \Phi u\,, \tag{2.2.28}$$

defining it by the relation (for an arbitary function v)

$$\langle\overline{q|v}\rangle = \langle v|\Phi|u\rangle\,. \tag{2.2.29}$$

This implies the following representations:

$$\begin{aligned} q(n) &= \sum_{n'} \Phi(n, n')\, u(n')\,, \\ q(x) &= \int \Phi(x, x')\, u(x')\, dx'\,, \\ q(k) &= \int \Phi(k, k')\, u(k')\, dk'\,, \end{aligned} \tag{2.2.30}$$

i.e., the kernel of the form is also the kernel of the operator determined by the form.

The Hermitian conjugate operators Φ and Φ^+ correspond to the Hermitian conjugate forms Φ and Φ^+. The Hermitian positive and positive definite operators are defined analogously.

An operator having the difference kernel $\Phi(n-n')$ is associated with a discrete convolution in n representation:

$$q(n) = \Phi(n) * u(n) \stackrel{\text{def}}{=} \sum_{n'} \Phi(n-n')\, u(n')\,, \tag{2.2.31}$$

an integral convolution and a product in x- and k- representations, respectively:

$$q(x) = \Phi(x) * u(x) \stackrel{\text{def}}{=} \int \Phi(x-x')\, u(x')\, dx',\; q(k) = \Phi(k)\, u(k)\,. \tag{2.2.32}$$

As above, we shall use the terms difference-type operator or kernel in such cases.

Obviously, the functional δ, introduced above, is the (one-dimensional) kernel of the identity operator and in particular in the x representation

$$\delta(x) * u(x) = u(x)\,. \tag{2.2.33}$$

In the next section, which can be omitted in the first reading, the properties of the quasicontinuum are investigated in more deatil.

2.3 One-Dimensional Quasicontinuum (Continued)

Let us attempt to generalize and at the same time clarify the scheme introduced above. It will permit one firstly to extend the class of admissible functions and secondly to characterize more completely the properties of the quasi-continuum and its distinctions from the usual Euclidean space.

First of all, it is desirable to weaken the restrictions on the behavior of the functions $u(n)$ as $|n| \to \infty$ because later we shall deal with periodic and increasing functions. In order to do this let us consider the connection between the n and k representations of a function u. It follows from (2.2.7) that $au(n)$ can be interpreted as the Fourier coefficients of the function $u(k)$ on the segment B. If $u(n)$ does not decrease as $|n| \to \infty$ then the Fourier series diverges in the usual sense, but can converge in the sense of generalized functions. This shows the necessity of introducing into the consideration a suitable space of generalized functions $u(k)$.

Now and later on we shall use B to denote the segment $-\pi/a \leqslant k \leqslant \pi/a$ with identified point π/a and $-\pi/a$, i.e., we shall consider k as belonging to the circle B having radius a^{-1}. Let $K(B)$ be a space of infinitely differentiable test functions $\varphi(k)$, $k \in B$, whose Fourier coefficients are denoted by $a\varphi(n)$,

$$\varphi(k) = a \sum_n \varphi(n)\, e^{-inak}, k \in B\,, \tag{2.3.1}$$

$$\varphi(n) = \frac{1}{a} \int_B \varphi(k)\, e^{inak}\, dk\,. \tag{2.3.2}$$

For the $\varphi(k)$ to be infinitely differentiable, it is obviously necessary and sufficient that $\varphi(n) \to 0$ faster than any degree of $|n|$ as $|n| \to \infty$. We denote by $N(B)$ the corresponding space of functions $\varphi(n)$ determined over the x-axis at points na. The formulae (2.3.1,2) establish the isomorphism $N(B) \leftrightarrow K(B)$.

The space $K'(B)$ of generalized functions $u(k)$, i.e., the space of linear continuous functionals (u, φ) on $K(B)$, can be represented as the space of the formal Fourier series

$$u(k) = a \sum_n u(n)\, e^{-inak}, \quad k \in B\,, \tag{2.3.3}$$

whose coefficients $u(n) \in N'(B)$ increase no faster than a power of n, i.e., $|u(n)| < C|n|^m$, where $C > 0$ is a constant and m is an integer. For example,

$$\delta(k) = \frac{a}{2\pi} \sum_n e^{-inak}, \quad k \in B\,. \tag{2.3.4}$$

If $u(k)$ is a usual function, for which Fourier series (2.3.3) converges, then it determines a functional $u \in K'(B)$

$$(u, \varphi) \overset{\text{def}}{=} \frac{1}{2\pi} \int \overline{u(k)}\, \varphi(k)\, dk = a \sum_n \overline{u(n)}\, \varphi(n)\,, \tag{2.3.5}$$

where the second equality follows from the identity

$$(e^{inak}, e^{imak}) = \frac{1}{a}\,\delta(n-m)\,. \tag{2.3.6}$$

The action of any functional $u \in K'(B)$, by definition, is also given by (2.3.5) where the first equality is already purely symbolic.

We can now interpret $N'(B)$ as the space of functionals on $N(B)$ defining their action by the equality

$$(u, \varphi) \stackrel{\text{def}}{=} a \sum_n \overline{u(n)}\,\varphi(n)\,. \tag{2.3.7}$$

Then the functional (u, φ) will be invariant with respect to n and k representations, this fact giving one more interpretation to (2.3.3), establishing the isomorphism $N'(B) \leftrightarrow K'(B)$.

Finally, let us introduce the inverse Fourier images $X(B)$ and $X'(B)$ of spaces $K(B)$ and $K'(B)$. The space $X(B)$ obviously consists of analytic functions $\varphi(x)$, represented in the form of series[7]

$$\varphi(x) = a \sum_n \varphi(n)\,\delta(x - na) \tag{2.3.8}$$

with rapidly decreasing coefficients $\varphi(n) \in N(B)$.

The generalized function $u(x) \in X'(B)$, i.e., an inverse Fourier image of $u(k) \in K'(B)$, is defined by Parseval's equality

$$\int \overline{u(x)}\,\varphi(x)\,dx = \frac{1}{2\pi} \int \overline{u(k)}\,\varphi(k)\,dk\,. \tag{2.3.9}$$

It can be shown (see e.q. [2.1.]) that the generalized functions $u(x) \in X'(B)$ are regular, i.e., they are generated by usual functions, the last ones increasing as $|x| \to \infty$ no faster than a power of $|x|$ and are analytically continued in the complex plane as entire functions of exponential type $\leqslant \pi/a$.

Substitution of the segment B by the circle permits one to exclude from the $X'(B)$ functions which vanish at all the knots. It is easy to show that all these functions can be represented in the form of finite linear combinations of terms of the form $x^m \sin(\pi x/a)$. For the Fourier images of the last ones, we have

$$x^m \sin\frac{\pi x}{a} \to \pi i \left(i\frac{d}{dk}\right)^m \left[\delta\left(k+\frac{\pi}{a}\right) - \delta\left(k-\frac{\pi}{a}\right)\right], \tag{2.3.10}$$

where $\delta(k)$ is the δ function in $K'(B)$. But when the points π/a and $-\pi/a$ are identified, the term in square brackets is identically equal to zero and hence the corresponding generalized functions in $X'(B)$ are equal to zero too. In other words, the tranaformation of B into the circle permits us to factorize, in a natural way, $X'(B)$ with respect to zero functionals.

---

[7]Remember that $\delta(x) = \delta_B(x)$ here and later.

Now let us show that $u(x)$ has the given values $u(n)$ at the points $x = na$. We have

$$u(x)|_{x=na} = \int \delta(na - x)\, u(x)\, dx = \frac{1}{2\pi} \int e^{inak}\, u(k)\, dk\,. \tag{2.3.11}$$

But according to (2.3.3, 5, 6) this expression is equal to $u(n)$, as desired.

Thus, we have established that the functions $u(n) \in N'(B)$ are interpolated by analytic functions $u(x) \in X'(B)$. It remains to verify that the radius of the circle B cannot be less than a^{-1}. To show this, let us consider the function $u(n) = (-1)^n$. Its x and k representations are

$$u(x) = \cos\frac{\pi x}{a}\,, \tag{2.3.12}$$

$$u(k) = \pi\left[\delta\left(k + \frac{\pi}{a}\right) + \delta\left(k - \frac{\pi}{a}\right)\right] = 2\pi\delta\left(k \pm \frac{\pi}{a}\right), \tag{2.3.13}$$

i.e., we have the example of $u(n) \in N'(B)$ with the Fourier image $u(k) \in K'(B)$ concentrated at the point $k = \pm\pi/a$.

Thus we have proved:

Theorem.[8] The formulae (2.3.3, 9) establish the concrete isomorphism of spaces

$$N'(B) \leftrightarrow X'(B) \leftrightarrow K'(B)\,, \tag{2.3.14}$$

under which $u(x) \in X'(B)$ interpolate $u(n) \in N'(B)$.

Note 1 The theorem allows one to regard $u(n)$, $u(x)$, and $u(k)$ as representations of one and the same function u in isomorphic spaces with different functional bases. We shall correspondingly speak about the n, x, and k representations. It is also natural to call the first representation discrete, the second quasicontinuous, and the third the Fourier representation.

Note 2 As it is evident from (2.3.3), $u(k)$ is uniquely determined by its Fourier series. It follows that if the usual function $u(k)$ corresponds to the generalized function $u(k)$, and has a discontinuity of the first kind at a point k_0, then $u(k_0) = 1/2[u(k_0 - 0) + u(k_0 + 0)]$. This allows us to give a meaning to the expressions $u(k)\,\delta(k - k_0)$ by setting it to be equal to $u(k_0)\,\delta(k - k_0)$. In particular, the equality

$$k\delta(k \pm \pi/a) = 0 \tag{2.3.15}$$

is valid and it will be interpreted below in the x representation as well.

---

[8]This can be considered as a generalization of the Kotelnikov-Shannon theorem well known in information theory [2.3]

Let us now establish the correspondence between operations in $N'(B)$, $X'(B)$, and $K'(B)$.

First of all note that all the formulae of the previous section are also true for the generalized functions, in particular, the expressions (2.2.11, 12) for the linear functional q and (2.2.21) for the functional Φ. It is only necessary to impose stronger conditions on the kernels $q(n)$ and $\Phi(n, n')$; for example, it is sufficient to require them to be finite, i.e., to differ from zero only in a finite domain.

Let us consider in more detail a few difference type operators and the corresponding kernels. The displacement operator T_m in $N'(B)$,

$$T_m\colon u(n) \to u(n+m), \tag{2.3.16}$$

is defined on the whole $N'(B)$. For its kernel we have

$$T_m(n) = \delta(n+m),\ T_m(x) = \delta(x+ma),\ T_m(k) = \mathrm{e}^{\mathrm{i}mak} \tag{2.3.17}$$

and hence

$$T_m\colon u(x) \to u(x+ma),\ u(k) \to \mathrm{e}^{\mathrm{i}mak}\, u(k)\,. \tag{2.3.18}$$

The operator T_m is defined on $X'(B)$ only for displacements divisible by an elementary cell; however, it is naturally extended to the displacement operator T_y with arbitrary y,

$$T_y\colon u(x) \to u(x+y),\ u(k) \to \mathrm{e}^{\mathrm{i}yk}\, u(k)\,. \tag{2.3.19}$$

This also defines T_y on $N'(B)$. Its kernal $T_y(n)$ can be found from the correspondence with $T_y(x)$ or $T_y(k)$. We have

$$T_y(n) = \frac{a}{2\pi}\int_B \mathrm{e}^{\mathrm{i}(na+y)k}\, dk = \frac{a}{\pi}\,\frac{(-1)^n \sin(\pi y/a)}{na+y}\,. \tag{2.3.20}$$

The differentiation operator $\partial\colon u(x) \to u'(x)$ is defined on the whole $X'(B)$. For its kernel, we have

$$\begin{aligned} &\partial(x) = \delta'(x)\,, && \partial(k) = \mathrm{i}k\,,\\ &\partial(n) = \frac{(-1)^n}{an} \quad (n \neq 0), && \partial(n)|_{n=0} = 0\,. \end{aligned} \tag{2.3.21}$$

As an example we give (2.3.15) in the n and k representations,

$$\begin{aligned} &-\mathrm{i}\partial(n) * \left[\frac{1}{2\pi}(-1)^n\right] = 0,\\ &-\mathrm{i}\frac{d}{dx}\left(\frac{1}{2\pi}\cos\frac{\pi x}{a}\right) = \frac{\mathrm{i}}{2a}\sin\frac{\pi x}{a} = 0\,. \end{aligned} \tag{2.3.22}$$

[Remember that sin $(\pi x/a)$ in $X'(B)$ is identified with zero].

From (2.3.21) it follows that in $N'(B)$ "differentiation" is an essentially non-local operation. For the "local" in $N'(B)$ finite difference operator

$$\nabla : u(n) \to u(n+1) - u(n) \tag{2.3.23}$$

we have

$$\begin{aligned} &\nabla(n) = \delta(n+1) - \delta(n)\,, \quad \nabla(x) = \delta(x+a) - \delta(x)\,, \\ &\nabla(k) = \mathrm{e}^{\mathrm{i}ak} - 1\,. \end{aligned} \tag{2.3.24}$$

Taking (2.3.21) into account, we obtain the invariant operational identities

$$\nabla = \mathrm{e}^{a\partial} - 1, \quad a\partial = \ln(1+\nabla)\,. \tag{2.3.25}$$

Let us consider the equations

$$\partial u = f, \quad u(x) = 0, \quad \text{as} \quad x \to -\infty\,, \tag{2.3.26}$$

$$\nabla v = g, \quad v(x) = 0 \quad \text{as} \quad x \to -\infty\,. \tag{2.3.27}$$

It is easily seen that, f and g being given, u and u are defined uniquely, i.e., there exist the operators

$$\partial^{-1} : f \to u = \partial^{-1} f : u(x) = \int_{-\infty}^{x} f(x)\, dx\,, \tag{2.3.28}$$

$$\nabla^{-1} : g \to v = \nabla^{-1} g : v(n) = \sum_{m=-\infty}^{n-1} g(m)\,. \tag{2.3.29}$$

Explicit expressions can be given for their kernels. In particular

$$\partial^{-1}(k) = \frac{1}{\mathrm{i}k} + \pi\delta(k), \quad \nabla^{-1}(k) = \frac{1}{\mathrm{e}^{\mathrm{i}ak} - 1} + \frac{\pi}{a}\,\delta(k)\,, \tag{2.3.30}$$

$$\partial^{-1}(x) = \theta(x), \quad \nabla^{-1}(x) = \frac{1}{\pi}\,\beta\left(1 - \frac{x}{a}\right)\sin\frac{\pi x}{a}\,, \tag{2.3.31}$$

where $\theta(x)$ is the Heaviside function

$$\theta(x) = \begin{cases} 1 & x > 0\,, \\ 0 & x < 0\,, \end{cases} \tag{2.3.32}$$

and $\beta(x)$ is expressed through Euler's ψ function or gamma function [2.4]

$$\beta(x) = \sum_{n=0}^{\infty} \frac{(-1)^n}{x+n} = \frac{1}{2}\left[\psi\left(\frac{x+1}{2}\right) - \psi\left(\frac{x}{2}\right)\right], \; \psi(x) = \frac{d}{dx}\ln\Gamma(x)\,. \tag{2.3.33}$$

Problem 2.3.1: Verify the formulae (2.3.30, 31).

Problem 2.3.2: Show that the kernel $\partial^{-1}(x)$ could also be represented in the form

$$\partial^{-1}(x) = \frac{1}{\pi} \operatorname{Si}\left(\frac{\pi x}{a}\right) + \frac{1}{2},$$

where Si x is the integral sine. [Hint: The last expression is the inverse Fourier image of $\partial^{-1}(k)$ and belongs to $X'(B)$ in distinction to $\theta(x)$. The equivalence of both the forms for the kernel $\partial^{-1}(x)$ results from the condition that $f(k) \in K'(B)$].

We have here a far-reaching analogy between the invariant operators of continuous and discrete integration. Let us pass on to the nonlinear operators and consider firstly the product

$$f = u \cdot v\,. \tag{2.3.34}$$

For $u(n)$, $v(n) \in N'(B)$, by definition

$$f(n) = u(n)\, v(n) \tag{2.3.35}$$

and hence $f(n) \in N'(B)$.

Before writing the explicit expression for $f(k)$, let us introduce in B the operation of displacement (along the circle) and for any $u(k)$, $v(k) \in K'(B)$ define the integral convolution

$$u(k) * v(k) \stackrel{\text{def}}{=} \int_B u(k - k')\, v(k')\, dk' = v(k) * u(k)\,. \tag{2.3.36}$$

In particular, it is easy to verify that

$$(\mathrm{e}^{\mathrm{i}nak}) * (\mathrm{e}^{\mathrm{i}mak}) = \frac{2\pi}{a}\, \delta(n - m)\, \mathrm{e}^{\mathrm{i}nam}\,. \tag{2.3.37}$$

Taking into account (2.3.3) and (2.3.37), we find

$$f(k) = \frac{1}{2\pi}\, u(k) * v(k)\,. \tag{2.3.38}$$

The function $f(x)$ can be found now either as the inverse Fourier image of $f(k)$ or from the correspondence $f(n) \leftrightarrow f(x)$. It is necessary to emphasize that, generally speaking, $f(x) \neq u(x)v(x)$, where the multiplication is understood in the usual sense. We shall write the product defined above in the x representation as

$$f(x) = u(x) \cdot v(x) = v(x) \cdot u(x) . \tag{2.3.39}$$

To the usual product $u(x)v(x)$ corresponds the convolution on the axis k in the k representation whose support, generally speaking, goes beyond the limits of the segment $B[-\pi/a \leqslant k \leqslant \pi/a]$, but belongs to the segment $[-2\pi/a \leqslant k \leqslant 2\pi/a]$. In other words $X'(B)$ is not closed with respect to this operation. Conversely, the functions from $X'(B)$ form a commutative ring with respect to the product (2.3.39). If the supports of $u(k)$ and $v(k)$ belong to the segment $-\pi/2a \leqslant k \leqslant \pi/2a$, then their convolution coincides with the usual one and hance $u(x) \cdot v(x) = u(x)v(x)$. This equality is also true if one of the factors is a polynomial.

Let us also point out the obvious relations

$$\begin{aligned} &\delta(x - na) \cdot \delta(x - ma) = a^{-1}\delta(n - m)\, \delta(x - na) , \\ &\delta(x) \cdot \delta(y) = a^{-1}\delta(x), \quad u(x) \cdot \delta(x) = u(0)\, \delta(x) . \end{aligned} \tag{2.3.40}$$

It is easy to verify the validity of the following analog of Leibnitz's formula:

$$\nabla(u \cdot v) = \tilde{u} \cdot \nabla v + \nabla u \cdot v, \tag{2.3.41}$$

$$\tilde{u}(n) = \frac{1}{2}\, [u(n) + u(n + 1)]. \tag{2.3.42}$$

At the same time, it seems that there is no simple analog of Leibnitz's formula for the derivative of $u \cdot v$.

Taking into account that $\overline{u(x)} \leftrightarrow \overline{u(-k)}$, we have

$$\begin{aligned} \int \overline{u(x)} \cdot v(x)\, dx &= \frac{1}{2\pi}\, \overline{[u(-k)} * v(k)]_{k=0} \\ &= \frac{1}{2\pi} \int_B \overline{u(k)}\, v(k)\, dk = \int \overline{u(x)}\, v(x)\, dx = \langle u \,|\, v \rangle . \end{aligned} \tag{2.3.43}$$

It follows that in the scalar product and in expressions of the type (2.3.35), the products can be understood both in the usual sense and in the sense of (2.3.39).

Let a nonlinear function $F[u]$ be defined in $N'(B)$, its values $f(n) = F[u(n)]$ belonging to $N'(B)$. It is convenient to consider F as a nonlinear operator

$$F\colon u(n) \to f(n) = F[u(n)] . \tag{2.3.44}$$

We have in the x and k representations, correspondingly,

$$F\colon u(x) \to f(x) = F[u(x)] , \tag{2.3.45}$$

$$F\colon u(k) \to f(k) = F[u(k)] , \tag{2.3.46}$$

where it is assumed as usual

$$u(n) \leftrightarrow u(x) \leftrightarrow u(k), \quad f(n) \leftrightarrow f(x) \leftrightarrow f(k)\,. \tag{2.3.47}$$

Here, generally speaking,

$$f(x) \neq F(u(x)), \quad f(k) \neq F[u(k)]\,.$$

If $F[u]$ is given as an analytical function, then the expressions for the operators $F[u(x)]$ and $F[u(k)]$ can be obtained, expanding $F[u]$ in a series and interpreting the product in the suitable way. For $F[u] = u^2$, for example,

$$F[u(x)] = u(x)\cdot u(x), \quad F[u(k)] = \frac{1}{2\pi}\, u(k)*u(k)\,.$$

Let us note that, if the argument of $F[u]$ is a polynomial (in x-representation), then $f(x) = F[u(x)]$. In fact, in this case, the support of $u(k)$ is a point and the powers of $u(x)$ coincide with the usual ones.

Let us now generalize the formalism considered to the case of irregularly situated knots. Let $u(x_n)$ be defined at the points $\ldots < x_n < x_{n+1} < \ldots$ of the x axis (n runs through all integers), $|u(x\ )|$ being not greater than $C|n|^p$. Let us introduce a suitable $\xi(x)$ such that $\xi(x_n) = na$ and set

$$u(\xi) = a \sum_n u(x_n)\, \delta(\xi - na)\,. \tag{2.3.48}$$

Then $u(x) = u[\xi(x)]$ will be the desired interpolating function.

If, for example, a set M of knots for which $x_n \neq na$ is bounded, then the condition $\xi(x) \in X'(B)$ defines $\xi(x)$ uniquely.

Problem 2.3.3: Show that in this case

$$\xi(x) = x - \sigma(x)\,\pi(x) \sum_{n\in M} \frac{x_n - na}{\sigma(x)(x - x_n)}\,, \tag{2.3.49}$$

where

$$\sigma(x) = \sum_{n\in M} \delta(x - na), \quad \pi(x) = \prod_{n\in M} (x - x_n)\,. \tag{2.3.50}$$

Up to here, we have restricted ourselves to the consideration of functions which increase on the x axis no faster than a polynomial. In a number of problems, one has to deal with increasing exponentially functions. Let

$$f(x) = e^{\beta x} u(x), \quad \mathrm{Im}\{\beta\} = 0\,. \tag{2.3.51}$$

The Fourier image of $f(x)$ is $u(k + i\beta)$. It follows that if $u(x) \in X'(B)$, then the argument of Fourier images of functions of the type (2.3.51) is the complex variable $k = k' + ik''$, belonging to the cylinder $\mathscr{B}$, spanned on the circle $\boldsymbol{B}$, i.e.,

$$k \in \mathscr{B} \Rightarrow k' \in B, \quad -\infty < k'' < \infty . \tag{2.3.52}$$

The corresponding spaces of functions will be denoted as $N'(\mathscr{B})$, $X'(\mathscr{B})$, and $K'(\mathscr{B})$. The interpolation theorem and the above-developed formalism are obviously extended to these spaces.

Putting off the general definition of the quasicontinuum to Part II, we shall temporarily consider the quasicontinuum as the one-dimensional x space and the given class of admissible functions $X'(\mathscr{B})$.

2.4 Equation of Motion and Elastic Energy Operator

In this section we introduce the general model of the medium with simple structure. The simple chain investigated in Sect. 2.1 will serve as the starting point.

When using the results of Sect. 2.2 and, in particular, the equalities (2.2.11, 12, 21), let us write the Lagrangian of the chain (2.1.5) in the form which is invariant with respect to n, x, and k representations,

$$L = \frac{1}{2} \langle \rho \dot{u}(t) | \dot{u}(t) \rangle - \frac{1}{2} \langle u(t) | \Phi | u(t) \rangle + \langle u(t) | q(t) \rangle . \tag{2.4.1}$$

Here ρ and $q(t)$ have the meaning of mass density and external forces density [compare the correspondence (2.2.13) and the subsequent agreement about functionals]. To the displacement $u(t)$ there correspond $u(n, t)$, $u(x, t)$, and $u(k, t)$ in n, x, and k representations. The external forces density $q(t)$ and the elastic energy operator Φ have a similar invariant meaning. To preserve a one-to-one correspondence with the chain, it is necessary to assume that all the functions of space variable x belong to the space $X'(\mathscr{B})$, i.e., we are dealing with the elastic quasicontinuum.

If this assumption is omitted, then the Lagrangian (2.4.1) will describe the most general model of the linear elastic continuum of simple structure with nonlocal interaction between particles.

In what follows, as far as possible, we shall develop the theory in a form suitable for the description of both the continuous and the discrete media. One of the advantages of such an approach is the convenience of the comparison of the results with their physical interpretations.

In parallel with the functions of time $u(t)$, we shall consider their Fourier images $u(\omega)$ connected by the relations

$$u(\omega) = \int u(t)\, e^{i\omega t}\, dt , \tag{2.4.2}$$

$$u(t) = \frac{1}{2\pi} \int u(\omega)\, e^{-i\omega t}\, d\omega . \tag{2.4.3}$$

Note that the signs in the arguments of exponents in the direct and inverse Fourier transforms are opposite to those in the transforms (2.2.1) and (2.2.2). This allows us to represent $u(x, t)$ as a superposition of plane waves propagating in the positive direction of the x axis when $k, \omega > 0$;

$$u(x, t) = \frac{1}{(2\pi)^2} \iint u(k, \omega)\, e^{i(kx-\omega t)}\, dk\, d\omega\,. \tag{2.4.4}$$

The equation of motion of the medium in the (x, t) and (k, ω) representations takes the forms

$$\rho(x)\, \ddot{u}(x, t) + \int \Phi(x, x')\, u(x', t)\, dx' = q(x, t)\,, \tag{2.4.5}$$

$$-\frac{\omega^2}{2\pi}\, \rho(k) * u(k, \omega) + \int \Phi(k, k')\, u(k', \omega)\, dk' = q(k, \omega)\,. \tag{2.4.6}$$

The first equation is obtained in the usual way from the Lagrangian (2.4.1). The second can be obtained by Fourier transform in (x, t). In the case of the quasicontinuum, these equations can also be obtained from the discrete representation (2.1.6) using (2.2.30). In a number of cases, it is convenient to write the equations of motion in the (x, ω) and (k, t) representations that can be obtained from (2.4.5) in the obvious way. In the following, we shall also use the operator form of the equation

$$\Phi_\omega u = q, \quad \text{where} \quad \Phi_\omega \stackrel{\text{def}}{=} -\rho\omega^2 + \Phi\,, \tag{2.4.7}$$

which is invariant with respect to x, n (for the quasicontinuum), and k representations.

Let us now consider conditions which the elastic energy operator Φ must satisfy, and their consequences.

Hermiticity: The operator Φ is defined by the quadratic form

$$\begin{aligned} 2\Phi &= \langle u|\Phi|u\rangle = \iint u(x)\, \Phi(x, x')\, u(x')\, dx\, dx' \\ &= \frac{1}{2\pi} \iint \overline{u(k)}\, \Phi(k, k')\, u(k')\, dk\, dk' = \sum_{nn'} u(n)\, \Phi(n, n')\, u(n') \end{aligned} \tag{2.4.8}$$

(the last equality is only for the quasicontinuum). It follows that its kernel must satisfy the conditions [compare (2.1.7)]

$$\Phi(x, x') = \Phi(x', x), \quad \Phi(k, k') = \overline{\Phi(k', k)}, \quad \Phi(n, n') = \Phi(n', n)\,. \tag{2.4.9}$$

The kernels $\Phi(x, x')$ and $\Phi(k, k')$ are connected by the relation [compare (2.2.22)]

$$\Phi(k, k') = \frac{1}{2\pi} \iint \Phi(x, x')\, e^{-i(kx-k'x')}\, dx\, dx'\,, \tag{2.4.10}$$

and for the quasicontinuum

$$\Phi(x, x') = \sum_{nn'} \Phi(n, n')\, \delta(x - na)\, \delta(x' - n'a)\,,$$

$$\Phi(k, k') = \frac{1}{2\pi} \sum_{nn'} \Phi(n, n')\, e^{-ia(nk - n'k')}, \quad k \in B, \tag{2.4.11}$$

where $\delta(x) = \delta_B(x)$ is given by the expression (2.2.5).

Since the kernel $\Phi(x, x')$ is real, we have, for real k and k',

$$\Phi(-k, -k') = \overline{\Phi(k, k')}\,. \tag{2.4.12}$$

Note that, generally speaking, the form (2.4.8) is defined only on displacements decreasing sufficiently rapidly as $|x| \to \infty$. However, the operator Φ given by the kernels (2.4.9) can be extended naturally to a wider class of functions. In connection with this, the domain of definition of the equations of motions is wider than that of the quadratic form of energy.

Finiteness of Action at a Distance: We shall assume that the action at a distance is bounded by some characteristic radius of interaction l, i.e., $\Phi(x, x') = 0$ when $|x - x'| > l$ for arbitrary x [for quasicontinuum $\Phi(n, n') = 0$ when $|n - n'| > l$ for arbitrary n].

If the medium is bounded by a characteristic dimension $2L$, then one can also set $\Phi(x, x') = 0$ when $|x| > L$ for any x', while it is natural to accept that $l \ll L$. The domain of the plane x, x' in which certainly $\Phi(x, x') = 0$ is obviously complementary to the hexagonal shown in Fig. 2.5.

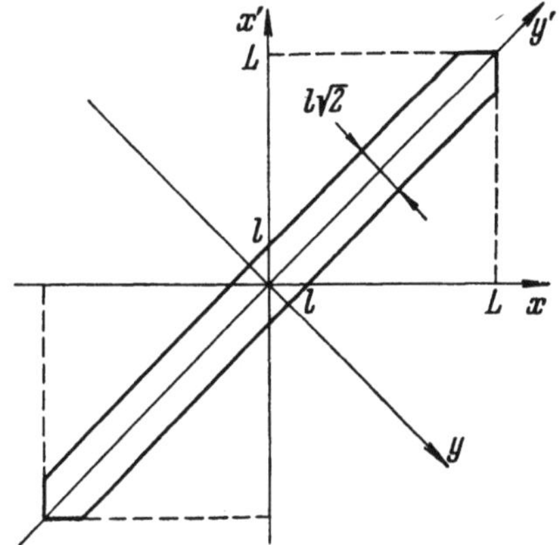

Fig.2.5. Support of the function $\Phi(x,x')$

Let us introduce new variables in the planes x, x' and k, k', corresponding to the rotation of the former axes through angle $\pi/4$

$$y = \frac{1}{\sqrt{2}}(x - x'), \quad y' = \frac{1}{\sqrt{2}}(x + x')\,,$$

$$\kappa = \frac{1}{\sqrt{2}}(k + k'), \quad \kappa' = \frac{1}{\sqrt{2}}(k' - k) \tag{2.4.13}$$

and set $\Phi[x(y, y'), x'(y, y')] = \tilde{\Phi}(y, y')$. Then according to (2.4.10) the function $\Phi(x, x')$ will correspond to

$$\begin{aligned}\tilde{\Phi}(\kappa, \kappa') &= \Phi(k(\kappa, \kappa'), k'(\kappa, \kappa')) \\ &= \frac{1}{2\pi} \iint \tilde{\Phi}(y, y')\, e^{-i(\kappa y - \kappa' y')} dy\, dy' .\end{aligned} \tag{2.4.14}$$

From (2.4.9) and (2.4.13) it follows that $\tilde{\Phi}(y, y')$ is an even function of y and $\tilde{\Phi}(y, y') = 0$ when $|y| < l/\sqrt{2}$. But then $\tilde{\Phi}(\kappa, \kappa')$ is an entire function in κ of exponential type $\leqslant l/\sqrt{2}$. For the bounded medium, $\tilde{\Phi}(\kappa, \kappa')$ is also an entire function of κ', but it is of exponential type $\leqslant L/\sqrt{2}$ and hence $\Phi(k, k')$ is an entire function of point (k, k'). It is essential that, for an unbounded medium $\tilde{\Phi}(\kappa, \kappa')$, as a rule, cannot be analytical in κ'[compare below the case of the homogeneous medium (2.4.32)].

Invariance with Respect to Translations: Let us accept by definition that only the interaction between the particles of the medium contributes to the elastic energy Φ given by (2.4.8). It follows that the elastic energy Φ is to be invariant under the transformation $u(x) \to u(x) + u_0$, where $u_0 =$ constant, since this transformation does not change distances between the particles of the medium.

The direct application of this condition for the unbounded medium meets with some difficulties, because the energy Φ is not defined a priori on displacements, not vanishing as $|x| \to \infty$. Therefore, let us first consider the case of a bounded medium.

The condition of invariance of (2.4.8) with respect to the above transformation yields

$$2\langle u_0|\Phi|u\rangle + \langle u_0|\Phi|u_0\rangle = 0 \tag{2.4.15}$$

or taking into account the arbitrariness of $u(x)$ and u_0

$$\int \Phi(x, x')\, dx' = 0 . \tag{2.4.16}$$

From here follows the possibility of representing of $\Phi(x, x')$ in the form

$$\Phi(x, x') = \psi(x)\, \delta(x - x') - \Psi(x, x'), \tag{2.4.17}$$

where

$$\Psi(x, x') = \Psi(x', x), \quad \psi(x) \overset{\text{def}}{=} \int \Psi(x, x')\, dx' , \tag{2.4.18}$$

and in other respects, $\Psi(x, x')$ is an arbitrary function.

In the case of the quasicontinuum

$$\Phi(n, n') = \psi(n)\, \delta(n - n') - \Psi(n, n')\,, \tag{2.4.19}$$

$$\Psi(n, n') = \Psi(n', n), \quad \psi(n) \stackrel{\text{def}}{=} \sum_{n'} \Psi(u, n')\,. \tag{2.4.20}$$

The substitution of these expressions in the equation of motion shows that $\Psi(n, n')$ and $\Psi(x, x')$ can be interpreted as the stiffness and the stiffness density of the elastic bond, connecting the points, n, n' or x, x', respectively, and the first term on the right-hand sides of (2.4.17) and (2.4.19) as self-action. It follows from the boundedness of the action at a distance, that $\Psi(x, x') = 0$ when $|x - x'| > l$. The corresponding operator Ψ will be called the elastic bond operator.

We note the relations

$$\Phi(k, k') = \frac{1}{2\pi} \iint \left[\cos\frac{(k' - k)(x - x')}{2} - \cos\frac{(k + k')(x - x')}{2}\right] e^{\frac{i}{2}(k'-k)(x+x')}\, \Psi(x, x')\, dx\, dx' \tag{2.4.21}$$

or, considering (2.4.13, 14),

$$\begin{aligned}\tilde{\Phi}(\kappa, \kappa') &= \frac{1}{2\pi} \iint (\cos \kappa' y - \cos \kappa y)\, e^{i\kappa' y'} \tilde{\Psi}(y, y')\, dy\, dy' \\ &= \tilde{\Psi}(\kappa', \kappa') - \tilde{\Psi}(\kappa, \kappa')\,.\end{aligned} \tag{2.4.22}$$

The representation

$$\Phi(k, k') = kk'c\,(k, k') \tag{2.4.23}$$

where $c(k, k')$ is an analytic function of the same type as $\Phi(k, k')$, and $c(k, k')$ is uniquely defined by this relation is also valid. In the x representation

$$\Phi(x, x') = \partial\partial' c\,(x, x')\,. \tag{2.4.24}$$

The function $c(x, x')$ can be expressed with the help of $\Psi(x, x')$,

$$c(x, x') = \iint \theta(x - z)\, [\theta(x' - z) - \theta(x' - z')]\, \Psi(z, z')\, dz\, dz'\,. \tag{2.4.25}$$

Here $\theta(x)$ is the Heaviside function (2.3.32).

It will be clear from the following that $c(x, x')$ and $c(k, k')$ are the kernels of the operator of elastic moduli.

Now, the equation of motion (2.4.5) has the form

$$\rho(x)\, \ddot{u}(x, t) - \partial \int c(x, x')\, \partial' u(x', t)\, dx' = q\,(x, t)\,. \tag{2.4.26}$$

Thus for a bounded medium, the condition of invariance of energy with respect to translations is equivalent to any of the relations (2.4.16, 17, 23, 24). By definition, these relations are considered to be true also for the limiting cases of the unbounded medium. The second term in (2.4.15), vanishing for an arbitrary bounded medium, must also be considered to be zero for the unbounded medium, i.e., the energy of pure translation is zero.

Local Approximation: For physical reasons it is evident that for the displacements $u(x, t)$ varying slowly over distances of the order l the nonlocal theory, in the zeroth approximation in l, must transform into the usual local theory of elasticity. To carry out this transition, it is suffcient to expand $u(x', t)$ in a series in the neighborhood of point $x' = x$ and keep the first two terms

$$u(x', t) = u(x, t) + (x' - x)\,\partial u(x, t) + \cdots \tag{2.4.27}$$

A substitution in (2.4.26) yields

$$\rho(x)\,\ddot{u}(x, t) - \partial[c_0(x)\,\partial u(x, t)] = q(x, t)\,, \tag{2.4.28}$$

where

$$c_0(x) = \int_{-l}^{l} c(x, x')\,dx' \tag{2.4.29}$$

is the elastic modulus of the local theory. One can obtain this also by expanding $c(x, x')$ in a series in multipoles along the x' axis and keeping the first term only.[9]

If we assume that $\Psi(x, x')$ [and hence $\Phi(x, x')$ and $c(x, x')$] changes slowly in distances of the order of l along the y' axis (Fig. 2.5), then it is easy to show that the transition to the local theory is equivalent to the expansion of these functions in multipoles along the y axis. In this way, more symmetrical expressions for the coefficients and the connections between them are obtained. In particular

$$c(x, x') = c_0(x)\,\delta(x - x')\,,$$

$$c_0(x) = \frac{1}{2}\iint (x' - x'')^2\,\Psi(x' - x'')\,\delta\left(x - \frac{x' + x''}{2}\right)dx'\,dx''\,. \tag{2.4.30}$$

Stability by definition, is equivalent to the requirement for (2.4.8) to be positive definite for any admissible displacement, different from the translation.

Substitution of (2.4.17) in (2.4.8) yields

$$\Phi = \frac{1}{4}\iint \Psi(x, x')\,[u(x) - u(x')]^2\,dx\,dx'\,. \tag{2.4.31}$$

If follows that nonnegativity of $\Psi(x, x')$, i.e., stability of all bonds, is sufficient for stability. However, it is not difficult to show that this condition is not

---

[9]See, for example, [2.5] about the expansion into series in multipoles.

necessary. If $\Psi(x, x') \geqslant 0$, then the medium will be called absolutely stable. Other criteria of stability will be considered below.

Homogeneity: In the case of a homogeneous medium, the operator Φ must be invariant with respect to translation and hence,

$$\Phi(x, x') = \Phi(x - x'), \quad \Phi(k, k') = \Phi(k)\,\delta(k - k'), \tag{2.4.32}$$

where, according to (2.4.9) and (2.4.10),

$$\Phi(x) = \Phi(-x), \quad \Phi(k) = \overline{\Phi(k)} = \Phi(-k), \tag{2.4.33}$$

$$\Phi(k) = \int \Phi(x)\, \mathrm{e}^{-ikx}\, dx\,. \tag{2.4.34}$$

Besides, for the quasicontinuum, (2.1.8), (2.1.9) and

$$\Phi(k) = \frac{1}{a} \sum_n \Phi(n)\, \mathrm{e}^{-inak}, \quad k \in B \tag{2.4.35}$$

are fulfilled.

From (2.4.16, 17, 23) we have

$$\int \Phi(x)\, dx = 0, \quad \Phi(k)\,|_{k=0} = 0\,, \tag{2.4.36}$$

$$\Phi(x) = \psi_0 \delta(x) - \Psi(x), \quad \psi_0 = \int \Psi(x)\, dx\,, \tag{2.4.37}$$

$$\Phi(k) = k^2 c(k) = \psi_0 - \Psi(k) = \int \Psi(x)(1 - \cos kx)\, dx\,. \tag{2.4.38}$$

The equations (2.4.5, 6) take now the form ($\rho = \text{const}$)

$$\rho \ddot{u}(x, t) + \int \Phi(x - x')\, u(x', t)\, dx' = q(x, t)\,, \tag{2.4.39}$$

$$[-\rho\omega^2 + \Phi(k)]\, u(k, \omega) = q(k, \omega)\,. \tag{2.4.40}$$

When passing on to local theory, from (2.4.30) we find for the elastic modulus of the homogeneous medium

$$c_0 = \int_0^l x^2 \Psi(x)\, dx\,. \tag{2.4.41}$$

In the case of a homogeneous chain with interaction of N neighbors

$$c_0 = a \sum_{n=1}^{N} n^2 \Psi(n)\,. \tag{2.4.42}$$

2.5 Strain and Stress, Energy Density, and Energy Flux

The representation of the elastic energy operator Φ in the form (2.4.24) permits us to shape the nonlocal model into the form which is analogous to the model of classical one-dimensional elastic continuum.

As usual, let us define the strain ε by the relation[10]

$$\varepsilon(x) \stackrel{\text{def}}{=} \partial u(x) \quad \text{or} \quad \varepsilon(k) = iku\,(k) \tag{2.5.1}$$

and introduce the quantity

$$\sigma(x) \stackrel{\text{def}}{=} \int c(x, x')\, \varepsilon(x')\, dx' \quad \text{or} \quad \sigma(k) = \int c(k, k')\, \varepsilon(k')\, dk' \,. \tag{2.5.2}$$

In the operator form

$$\sigma = c\varepsilon \,. \tag{2.5.3}$$

The equation of motion (2.4.5) can now be written in the form

$$\rho\ddot{u} - \partial\sigma = q \,. \tag{2.5.4}$$

It follows from here that σ can be interpreted as stress and (2.5.3) as an operator Hooke's law.

Let us show that one can introduce the elastic energy density, which is expressed through stress and strain in the usual way. For this purpose, taking into account (2.4.24), let us transform the expression for Φ

$$2\Phi = \langle u|\,\Phi\,|u\rangle = \langle u|\,\partial\partial' c\,|u\rangle = \langle \partial u|\,c\,|\partial u\rangle = \langle \varepsilon|\sigma\rangle = \int \varepsilon(x)\, \sigma(x)\, dx \,. \tag{2.5.5}$$

It follows from here that the quantity

$$\varphi(x) \stackrel{\text{def}}{=} \frac{1}{2}\, \sigma(x)\, \varepsilon(x) \tag{2.5.6}$$

can be interpreted as the elastic energy density.

Note that σ and φ are invariant with respect to translations, as they should be, due to their physical meaning.

Earlier, when considering the chain, we saw that the elastic energy density might be introduced in different ways and none of them could be given any preference from the physical point of view. This conclusion remains true for the energy density also in the general case of nonlocal theory.

Let us introduce two point quantities

$$\begin{aligned} &\varepsilon(x, x') \stackrel{\text{def}}{=} u(x) - u(x') \,, \\ &\sigma(x, x') \stackrel{\text{def}}{=} \Psi(x, x')\, \varepsilon(x, x') = -\,\Phi(x, x')\, \varepsilon(x, x') \,, \\ &\varphi(x, x') \stackrel{\text{def}}{=} \frac{1}{2}\, \sigma(x, x')\, \varepsilon(x, x') = \frac{1}{2}\, \Psi(x, x')\, [u(x) - u(x')]^2 \,, \end{aligned} \tag{2.5.7}$$

having the obvious meaning of relative displacement of points x and x', stress

[10] Here and later on, the dependence of field quantities on t or ω is not explicity shown.

and elastic energy of bonds connecting these points (more exactly $\sigma(x, x')$ is the force with which the point x acts on the point x').

Let $\theta(x)$ be the Heaviside function (2.3.32). Then the total stress at the point x due to the bonds

$$\sigma(x) \overset{\text{def}}{=} - \iint \theta(x - x')\, \sigma(x', x'')\, dx'\, dx'' \tag{2.5.8}$$

and the average energy per particle at point x

$$\varphi(x) \overset{\text{def}}{=} \frac{1}{2} \int \varphi(x, x')\, dx' = \frac{1}{4} \int \Psi(x, x')\, \varepsilon^2(x, x')\, dx' \tag{2.5.9}$$

may be similarly interpreted as the stress and the elastic energy density. Obviously, they differ from the corresponding quantities (2.5.2) and (2.5.6). Furthermore, if due to the fact that the model is one-limensional the stresses differ only by a constant (moreover, if the boundary conditions are taken into account, they, coincide), then the energy densities are more significantly different—by a quantity, the integral of which over the axis is equal to zero. This ambiguity has the same nature as that in the chain (see Sect. 1). At the same time, under given structure of bonds, the quantities (2.5.7) are defined uniquely in the linear nonlocal theory and have a clear physical meaning. They could be named the two point strain, stress, and elastic energy density. In the anharmonic theory, the corresponding multipoint functions would have an analogous meaning.

Energy flux is a more difficult matter. Let us separate out from the medium an arbitrary domain V and write the energy balance equation for it. In this connection, in view of the presence of action at a distance, one must consider all the elastic bonds crossing the boundary of the domain V, to be cut. We have

$$\dot{T} + \dot{\Phi} + S_+ + S_- = N\,. \tag{2.5.10}$$

Here $\dot{T}$ and $\dot{\Phi}$ are rates of change of kinetic and elastic energies, localized in V, N is the power of external forces, S_+ and S_- are fluxes of energy through the boundaries of the domain V in the positive and negative directions of the x axis. The quantities

$$\dot{T} = \frac{1}{2} \frac{\partial}{\partial t} \int_V \rho(x)\, \dot{u}^2(x)\, dx, \quad N = \int_V q(x)\, \dot{u}(x)\, dx \tag{2.5.11}$$

are uniquely determined functions of the domain V (and time t) and are independent of a way in which bonds are cut. Hence the sum $\dot{\Phi} + S_+ + S_-$ possesses this property, though each term of the sum depends on the way the bonds have been cut and in that sense is not a physically uniquely determined quantity.

Contracting the domain V to the point x, we pass to the differential form of the energy balance equation

$$\dot{\tau}(x,t) + \dot{\varphi}(x,t) + \frac{\partial S(x,t)}{\partial x} = \nu(x,t)\,, \tag{2.5.12}$$

where τ, ρ, ν are the corresponding densities of energy and power, $S(x, t)$ is the energy flux through the elastic bonds crossing the point x. As above, only the sum $\dot{\varphi} + \partial S/\partial x$ has an invariant meaning.

Let us investigate in detail the degree of ambiguity of the flux $S(x, t)$ arising here. For this let us consider firstly, the flux of energy through an arbitrary elastic bond, which connects the points x' and x''. Let ξ $(0 \leqslant \xi \leqslant 1)$ be the relative distance of a point belonging to the bond from the point x'. Let us cut the bond at the point ξ and replace the part of the bond $\xi < \xi' \leqslant x''$ by a force $f(x', x'', \xi)$ which acts on the point ξ. Then, it is clear that the flux of energy in the direction from x' to x'' is equal to (the dependence on time is not shown explicitly here and later on)

$$S(x', x'', \xi) = -f(x'\, x'', \xi)\, \dot{u}(x', x'', \xi)\,, \tag{2.5.13}$$

where $\dot{u}(x', x'', \xi)$ is the velocity of the point ξ.

Since the elastic bonds are considered to be noninertial, then the stress $\sigma(x', x'')$ is constant along the bond and velocities are distributed along the bond linearly, i.e., $f(x', x'', \xi) = -\sigma(x', x'')$

$$\dot{u}(x', x'', \xi) = \dot{u}(x') + \xi[\dot{u}(x'') - \dot{u}(x')]\,. \tag{2.5.14}$$

The total flux of energy $S(x, \xi)$ through a point x is found by summation of the fluxes over all the bonds, to which the point x belongs:

$$S(x, \xi) = \iint_{x'<x<x''} S(x'\, x'', \xi)\, dx'\, dx''\,. \tag{2.5.15}$$

Let us introduce functions which we shall deal with repeatedly later:

$$\begin{aligned} &\theta(x', x'') \overset{\text{def}}{=} \theta(-x')\,\theta(x''), \quad \theta_+(x', x'') \overset{\text{def}}{=} \theta(x', x'') + \theta(x'', x')\,, \\ &\theta_-(x'\, x'') \overset{\text{def}}{=} \theta(x', x'') - \theta(x'', x') = \theta(x'') - \theta(x')\,. \end{aligned} \tag{2.5.16}$$

It is easy to see that they take the values shown in Fig. 2.6.

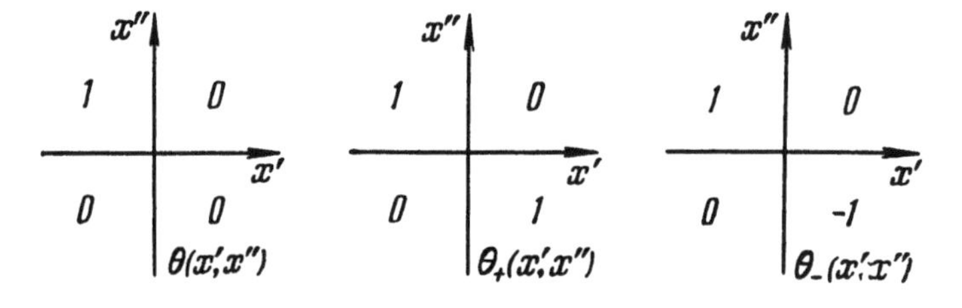

Fig.2.6. Functions θ, θ_+, and θ_-

The expression for $S(x, \xi)$ can be represented now in the form

$$S(x, \xi) = \iint \theta(x' - x, x'' - x)\, S(x', x'', \xi)\, dx'\, dx'' \\ = - \int f(x, x', \xi)\, \dot{u}(x')\, dx' , \qquad (2.5.17)$$

where, taking (2.5.13) and (2.5.14) into account,

$$f(x, x', \xi) = \int [\theta(x' - x, x'' - x) - \xi\theta_+(x' - x, x'' - x)]\, \sigma(x'', x')\, dx'' . \qquad (2.5.18)$$

Consider now the equation of motion (2.5.5) and multiply it on both sides by $\dot{u}(x)$. We have

$$\dot{\tau}(x) + \dot{u}(x) \int \Phi(x, x')\, u(x')\, dx' = \nu(x) . \qquad (2.5.19)$$

Comparison with (5.12) shows that the relation

$$\dot{\varphi}(x, \xi) + \frac{\partial S(x, \xi)}{\partial x} = \dot{u}(x) \int \Phi(x,x')\, u(x')\, dx' , \qquad (2.5.20)$$

must be fulfilled, where $\varphi(x, \xi)$ is the elastic energy density, the form of which depends on the manner of bond cutting.

Problem 2.5.1. Using the identity

$$\frac{\partial}{\partial x} \theta(x' - x, x'' - x) = [\delta(x' - x) - \delta(x'' - x)]\, \theta(x'' - x') \qquad (2.5.21)$$

and the relation (2.5.20), show that

$$\varphi(x, \xi) = \int [\xi\theta(x' - x) + (1 - \xi)\, \theta(x - x')]\, \varphi(x, x')\, dx' . \qquad (2.5.22)$$

Problem 2.5.2. Verify that, if $\xi = 1/2$, then

$$f(x, x') = \frac{1}{2} \int \theta_-(x' - x, x'' - x)\, \Phi(x', x'')\, \varepsilon(x', x'')\, dx'' , \qquad (2.5.23)$$

and the energy density (2.5.9) corresponds to the energy flux

$$S(x) = - \int f(x, x')\, \dot{u}(x')\, dx' . \qquad (2.5.24)$$

Problem 2.5.3. Show that the energy flux

$$S(x) = - \sigma(x)\, \dot{u}(x) \\ - \frac{1}{2} \iint \theta_-(x' - x, x'' - x)\, c(x', x'')\, \partial u(x'')\, \partial \dot{u}(x')\, dx'\, dx'' , \qquad (2.5.25)$$

corresponds to the energy density (2.5.6), the second term vanishing in the local approximation (2.4.30).

Problem 2.5.4. Write down an expression for the energy density and flux in the n representation and verify that the choice of elementary cells of the form shown symbolically in Fig. 2.7a, b, corresponds to the cases $\xi = 1/2$ and $\xi = 0$.

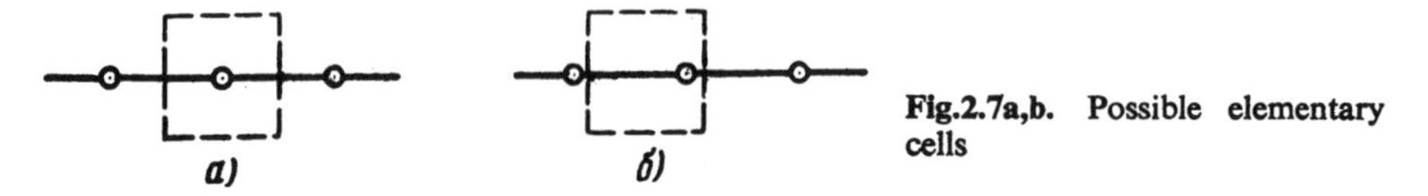

Fig.2.7a,b. Possible elementary cells

Thus, we see that to a change of a way of cutting bonds there corresponds a redistribution of the contributions of bond energy $\varphi(x', x'')$ in the density $\varphi(x)$ and in the discrete case it means a different choice of elementary cells.

Processes which are periodic in time are of fundamental interest. Let

$$u(x, t) = u(x)\,e^{-i\omega t}, \qquad (2.5.26)$$

where $u(x)$ is a complex amplitude. For the periodic processes, let us find the average value of the energy flux.

If $s(t) = f(t)g(t)$, where $f(t)$ and $g(t)$ vary in time in accordance with the law (2.5.26), then, as it is well known, the average value of $s(t)$ is equal to

$$\langle s \rangle_t = \frac{1}{2}\,\mathrm{Re}\,\{f\bar{g}\}. \qquad (2.5.27)$$

First of all, let us consider the average value of the energy flux through an elastic bond. It is easy to see that its value should not depend on the choice of a section. Indeed, the average value $\dot{\varphi}(x)$ in any section of the bond for periodic processes is equal to zero and hence the mean energy flux must be constant along the bond.

Problem 2.5.5. Taking into account the identity

$$\langle u(x')\,\dot{u}(x'')\rangle_t = -\,\langle \dot{u}(x')\,u(x'')\rangle_t = -\frac{\omega}{2}\,\mathrm{Im}\,\{u(x')\,\overline{u(x'')}\} \qquad (2.5.28)$$

verify that the average energy flux along the bond is independent of ξ and is equal to

$$\langle S(x', x'')\rangle_t = \frac{\omega}{2}\,\Phi(x', x'')\,\mathrm{Im}\,u(x')\,\overline{u(x'')}\}. \qquad (2.5.29)$$

Problem 2.5.6. The velocity v of the propagation of energy along the bond is equal by definition to the ratio of the energy flux to the energy density. Show that

$$\langle v \rangle_t = -2\omega \frac{|x' - x''| \operatorname{Im}\{u(x')\,\overline{u(x'')}\}}{|\varepsilon(x', x'')|^2} \tag{2.5.30}$$

According to (2.4.17), the total mean flow is equal to

$$\langle S(x) \rangle_t = \iint \theta(x' - x', x'' - x) \langle S(x', x'') \rangle_t \, dx' \, dx'' \tag{2.5.31}$$

and hence, is independent of the way of cutting bonds. This is also seen from (2.5.20), because $\langle \dot{\varphi}(x, \xi) \rangle_t = 0$ Taking into account (2.5.29), we find

$$\langle S(x) \rangle_t = \frac{\omega}{4} \iint \theta_-(x' - x', x'' - x)\, \Phi(x', x'') \operatorname{Im}\{u(x')\, u(x'')\}\, dx' \, dx'' \tag{2.5.32}$$

or

$$\langle S(x) \rangle_t = \frac{\omega}{2} \int \operatorname{Im}\{\overline{f(x, x')}\, u(x')\}\, dx' , \tag{2.5.33}$$

where (compare (2.5.23))

$$f(x, x') = \frac{1}{2} \int \theta_-(x' - x, x'' - x)\, \Phi(x', x'')\, u(x'')\, dx'' . \tag{2.5.34}$$

Thus, in the nonlocal theory, $\langle S(x) \rangle_t$, as distinct from $S(x, t)$, is a uniquely determined quantity, which does not depend on the way of cutting bonds.

Problem 2.5.7. Using (2.5.20), verify that $\langle S(x) \rangle_t =$ constant in the domain where external forces are absent.

2.6 Boundary Problems

Let us consider the junction of two media with different characteristics (the characteristics of the second medium will be denoted by an asterisk). The junction of chains with nearest-neighbor interactions and the junction in the general case are shown in Fig. 2.8.

In each medium one can distinguish boundary layers S and S^* with widths of the order of l and l^* respectively, where parameters are perturbed by interaction with the neighboring medium. Let us denote by D and D^*, the regions of unperturbed parameters and set $V = D + S$, $V^* = D^* + S^*$.

Let us write the equations of motion (2.4.7) in the boundary regions S and S^*:

$$-\omega^2 \rho_S u_S + S\Phi u = q_S, \quad -\omega^2 \rho_{S^*} u_{S^*} + S^* \Phi u = q_{S^*} . \tag{2.6.1}$$

Here, S, S^* are the operators of multiplication by the characteristic functions of

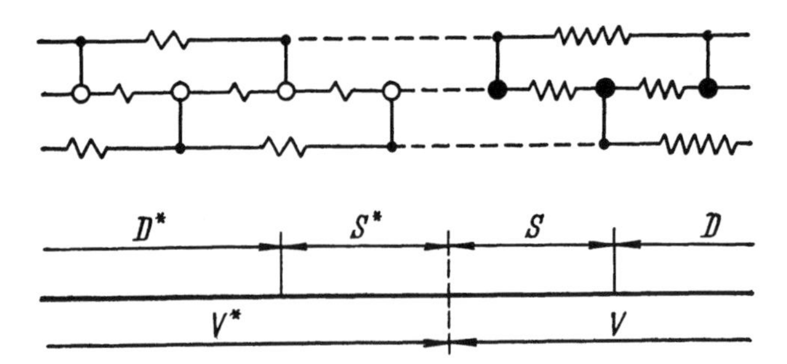

Fig. 2.8a,b. Junction of **(a)** discrete and **(b)** continuous media

the boundary regions, $u_S = Su$, etc. Due to the presence of action at a distance, the terms $S\Phi u$ and $S^*\Phi u$ bring about the connection of these equations between themselves and with the equations for the unperturbed regions. Equations (2.6.1) are the analog of the junction conditions in the local theory of a continuous medium. In particular, when the perturbed bonds are broken, these equations imply the force boundary conditions.

Proceeding to the formulation of basic boundary problems, let us decompose the elastic bond operator Ψ into a sum of operators

$$\Psi = \Psi_V + \Psi_{V^*} + \Psi_{VV^*} ,$$

$$\Psi_V \stackrel{\text{def}}{=} V\Psi V, \quad \Psi_{V^*} \stackrel{\text{def}}{=} V^*\Psi V^*, \quad \Psi_{VV^*} \stackrel{\text{def}}{=} V\Psi V^* + V^*\Psi V . \tag{2.6.2}$$

Here Ψ_V, Ψ_{V^*} characterize the interaction of points of the corresponding media, and Ψ_{VV^*} characterizes the interaction between the media (in Fig. 2.8, to each operator there corresponds its own type of bonds). Substituting (2.6.2) in (4.17), we find the corresponding decomposition of the operator Φ

$$\Phi = \Phi_V + \Phi_{V^*} + \Phi_{VV^*} , \tag{2.6.3}$$

where, for example, (I is the identity operator)

$$\Phi_V = \psi_V I - \Psi_V, \quad \Psi_V(x) = \int \Psi_V(x, x')\, dx' . \tag{2.6.4}$$

If the media do not interact, then $\Psi_{VV^*} = 0$ and hence, $\Phi_{VV^*} = 0$. In this case, the equations (2.4.7) split up into independent equations for both media. For the region V, we have

$$-\omega^2 \rho u_V + \Phi_V u_V = q_V . \tag{2.6.5}$$

It is not difficult to see that Φ_V admits the representation

$$\Phi_V = D\Phi + \Gamma, \quad \Gamma \stackrel{\text{def}}{=} S\Phi_V . \tag{2.6.6}$$

This permits us to write (2.6.5) in the form of an equivalent system[11]

$$D\Phi_{\omega}u \equiv -\omega^2 \rho u_D + D\Phi u = q_D\,, \tag{2.6.7}$$

$$\Gamma_{\omega}u = -\omega^2 \rho u_S + \Gamma u = q_S\,. \tag{2.6.8}$$

The first equation connects the displacement with the forces in the region D. As $l \to 0$, it transforms into the equation of motion of the usual (local) theory of elasticity. The second equation connects the displacements with the forces in the boundary region S and can be obtained from the junction conditions (2.6.1). As $l \to 0$ it transforms into the usual force boundary conditions. This analogy allows us to call the formulated above boundary value problem (2.6.7), (2.6.8), the first basic problem of the non-local theory of elasticity. The integral equation (2.6.5) is equivalent to this problem.

In the above we considered a semibounded region. Generalization for a bounded region V is obvious.

Let us point out that static problem for the bounded region can be reduced to a Fredholm integral equation of the second kind with symmetric positive definite kernel, provided all elastic bonds are stable, i.e. if $\Psi(x, x') \geqslant 0$. This equation is obtained from (2.6.5) by the standard substitution of variables and has the form

$$v(x) - \int_V K(x, x')\, v(x')\, dx' = q^0(x), \quad x \in V\,, \tag{2.6.9}$$

where

$$v(x) = \psi_V^{\frac{1}{2}}(x)\, u(x), \quad q^0(x) = \psi_V^{-\frac{1}{2}}(x)\, q(x)\,,$$

$$K(x, x') = [\psi_V(x)\, \psi_V(x')]^{-\frac{1}{2}}\, \Psi(x, x')\,.$$

The distinctive property of the equation (2.6.9) is that $K(x, x') = 0$ when $|x - x'| > l$, the parameter l being as a rule, small in comparison with dimensions of the region V.

Let us now proceed to the second basic problem, when displacements rather than forces are given in the boundary region i.e.

$$D\Phi_{\omega}u = q_D, \quad u_S = h\,, \tag{2.6.10}$$

where the function $h(x)$ is defined on S. To this problem there corresponds the equivalent integral equation

$$-\omega^2 \rho u_D + D\Phi u_D = f_D \quad (f_D = q_D - D\Phi h)\,. \tag{2.6.11}$$

---

[11] It is convenient to omit the index V in denoting the displacement in the region V wherever the corresponding cutting is contained in the operators.

In the static case, this equation, as in the first problem, can be reduced to a Fredholm equation of the second kind with a symmetric kernel.

Let us especially note an important case of homogeneous medium, when

$$\Psi(x, x') = \Psi(x - x') \quad \text{and} \quad \psi(x) = \psi_0, \quad \rho(x) = \rho_0$$

are constants. Equation (2.6.11) becomes a Fredholm equation with a difference-type kernel

$$(-\omega^2\rho_0 + \psi_0)\, u(x) - \int_D \Psi(x - x')\, u(x')\, dx' = f(x) \quad (x \in D)\,. \qquad (2.6.12)$$

The essential difference between the first and the second basic problems is to be emphasized. The first problem for the homogeneous medium cannot be reduced to a integral equation with difference-type kernel because of the intrinsic inhomogeneity of the elastic bonds in the boundary layer. As will be seen later, in the nonlocal theory of elasticity this leads to the fact that the solution of the first problem for the homogeneous medium is much more difficult than is the second one.

Developing further the analogy with the usual theory of elasticity, let us obtain the Green's formula in the nonlocal theory.

It follows from (6.4) that $\Phi_V = \Phi_V^+$ i.e., Φ_V is a self-adjoint operator. Taking into account (2.6.6), we obtain the operator identity

$$D\Phi + \Gamma = \Phi D + \Gamma^+\,.$$

Let us apply both sides of the equality to u and add an inertial term in the form

$$\omega^2\rho u_V = \omega^2\rho u_V + \omega^2\rho u_S\,.$$

Taking into account (2.4.7), (2.6.7) and (2.6.8) we find

$$\Phi_\omega u_D = q_D + \Gamma_\omega u - \Gamma_\omega^+ u_S\,. \qquad (2.6.13)$$

Here $\Gamma_\omega u = q_S$ is the analog of a simple layer, $\Gamma_\omega^+ u_S$ is the analog of a double layer with density u_S. As $l \to 0$ they transform into the usual layers.

Let $G_\omega(x, x')$ be a fundamental solution of the equation (2.4.7), i.e. an arbitrary function satisfying the equation

$$\int \Phi_\omega(x, x'')\, G_\omega(x'', x')\, dx'' = \delta(x - x')\,. \qquad (2.6.14)$$

Obviously, all such functions differ from each other by solutions of the corresponding homogeneous equation. Applying the operator G_ω generated by the function $G_\omega(x, x')$ to (2.6.13), we obtain the Green's formula

$$u_D = G_\omega q_D + G_\omega q_S - G_\omega \Gamma_\omega^+ u\,, \qquad (2.6.15)$$

i.e. the representation of a solution in the form of the sum of a volume potential and potentials of simple and double layers.

A fundamental solution obeying additional boundary conditions is called the Green's function of the corresponding boundary problem. In particular, for a Green's function of the first basic problem, the following condition is fulfilled

$$\int \Gamma_\omega(x, x'') G_\omega(x'', x') = \delta(x - x') \quad (x, x' \in S), \tag{2.6.16}$$

and the solution has the form

$$u_V(x) = \int_D G_\omega(x, x') q_D(x') dx' + \int_S G_\omega(x, x') q_S(x') dx', \quad x \in V. \tag{2.6.17}$$

For the second fundamental problem

$$u_D(x) = \int_D G_\omega(x, x') q_D(x') dx' - \int_S dx' u_S(x') \int_D dx'' G_\omega(x, x'') \Gamma_\omega^+(x'', x'), \tag{2.6.18}$$

where $G_\omega(x, x') = 0$, if x or x' belongs to S.

Note that the Green's functions are resolvents of the corresponding integral equations (2.6.5) and (2.6.11).

Using the Green's formula (2.6.15), we can also formulate the analog of the mixed boundary value problem of theory of elasticity.

In conclusion, we emphasize that the results obtained are correct also for the discrete medium, the integral equation for the bounded region being equivalent to a system of algebraic equations.

2.7 The Dispersion Equation

Beginning with this section and continuing to the end of this chapter we consider the homogeneous medium.

First of all we shall be interested in free vibrations of the unbounded medium. Firstly, they are of interest from the physical point of view, especially the propagating nondecaying waves. Secondly, as will subsequently be seen, the solutions of the problems, which concern forced vibrations, scattering of waves on local inhomogeneities and also the boundary problems can be constructed from free vibrations. However, for this it is necessary to have a complete set of nondecaying as well as exponentially decaying (increasing) free vibrations. Therefore, let us first of all find all the solutions of the equation of motion (2.4.39) or (2.4.40) with the right hand-side equal to zero, i.e. the equation

$$\rho \ddot{u}(x, t) + \int \Phi(x - x') u(x', t) dx' = 0 \tag{2.7.1}$$

or in the (k, ω) representation

$$[-\rho\omega^2 + \Phi(k)]\, u(k, \omega) = 0\,. \tag{2.7.2}$$

The operator under consideration is invariant with respect to displacements along x and t. Hence, it is natural to construct elementary solutions with the help of eigenfunctions of the displacement operators, i.e. of exponential functions. Thus, we search for an elementary solution is the form of a wave

$$u(x, t) = A\mathrm{e}^{\mathrm{i}\,(kx-\omega t)}\,, \tag{2.7.3}$$

where A, ω, k are arbitrary, in general complex constants. Here, the standard notations for frequency and wave number are used. They are not to be confused with the arguments of the Fourier-image of the function $u(x, t)$, for which we have in this case

$$u(k', \omega') = 4\pi^2 A\delta(k' - k)\,\delta(\omega' - \omega)\,, \tag{2.7.4}$$

Substitution of (7.3) in (7.1) leads to the dispersion equation

$$\Phi(k, \omega) = -\rho\omega^2 + \Phi(k) = 0\,, \tag{2.7.5}$$

which connects the quantities ω and k, for which (2.7.3) is the solution. This equation is the analog of the characteristic equation for a differential operator, but as distinct from the latter can have an infinite number of solutions for a fixed value of ω.

The dispersion equation is often written also as

$$\omega^2 = \omega^2(k), \quad \omega^2(k) \overset{\mathrm{def}}{=} \frac{1}{\rho}\,\Phi(k)\,. \tag{2.7.6}$$

Problem 2.7.1. Using (2.4.35) and (2.1.2), show that for a chain with two interacting neighbors

$$\omega^2(k) = \frac{4}{m}\,\Psi(1)\sin^2\frac{ak}{2} + \Psi(2)\sin^2 ak\Big], \quad |k| \leqslant \frac{\pi}{a}\,. \tag{2.7.7}$$

Obviously, the dispersion equation may be also obtained by substituting (2.7.4) in (2.7.2) and equating to zero the coefficients of δ-functions.

Problem 2.7.2. Show that all the solutions of (2.7.1) can be represented in the form of a superposition of elementary solutions (2.7.3), satisfying (2.7.6), i.e. the general solution has the form

$$u(x, t) = \int_L A(k)\,\mathrm{e}^{\mathrm{i}[xk - t\omega(k)]}\,dk\,, \tag{2.7.8}$$

where $A(k)$ is an arbitrary function, L is a contour in the complex k plane (for the quasi-continuum—in the cylinder $\mathscr{B}$) for which the integral exists. (Hint:

Use the fact that all the solutions of the equation $(k - a)f(k) = 0$ have the form $f(k) = A\,\delta(k - a)$).

Two quantities having the dimension of velocity

$$v_f \overset{\text{def}}{=} \frac{\omega}{k}, \quad v_g \overset{\text{def}}{=} \frac{d\omega}{dk}. \tag{2.7.9}$$

are connected with the elementary wave (2.7.3).

Both velocities have an explicit physical meaning for real ω and k. The phase velocity v_f is a purely kinematic quantity, defining the velocity of propagation of the phase of the wave $kx - \omega t = \text{const}$. Now ω and k can be independent (if the wave is caused by external force). The group velocity v_g is defined only in the presence of a functional dependence $\omega = \omega(k)$ and characterizes the velocity of propagation of wave packet (in the case of modulated wave—the velocity of modulation). For complex ω and k, the group velocity has no clear physical meaning.

If the velocities v_f and v_g depend on the wave number k, then it is said that spatial dispersion takes place,[12] the medium is called dispersive and equation (2.7.6) the dispersion law. In the case of the ordinary elastic medium the dispersion law is linear, the two velocities (2.7.9) are identical and are constant for arbitrary wave numbers, i.e. the spatial dispersion is absent. Models with a non-linear law $\omega(k)$ are called models with spatial dispersion. The most typical examples of media with spatial dispersion are crystal lattices and plasmas [2.6].

In this chapter we confine ourselves to time-harmonic vibrations; for this case the displacement $u(x, t)$ has the representation (2.5.26) with a real frequency ω. Complex frequencies are useful only for a description of energy dissipation.

For an analytical $\Phi(k)$ we shall consider both reel and complex wave numbers. To a reel k, $k > 0$ $(k < 0)$, there corresponds in (2.7.3) an undamed wave propagating in the positive (negative) direction of the x axis. To a complex wave number with $\text{Im}\{k\} > 0$ there corresponds a wave decreasing exponentially in the positive direction of the x axis and increasing in the negative direction.

The condition of reality of the frequency or, equivalently, the condition $\text{Im}\{\Phi(k)\} = 0$ leeds to the equations to be satisfied by k in order that the wave (2.7.3) be a solution. Setting $k = \alpha + i\beta$, with reel α and β, we can write this equation in the form

$$\text{Im}\{\Phi(k)\} = \text{Im}\{\int \Phi(x)\,e^{-ikx}\}\,dx = -\int \Phi(x) \sin \alpha x \sinh \beta x\,dx = 0\,. \tag{2.7.10}$$

One of the solutions of this equation is $\beta = 0$, i.e. to real values of k correspond real frequencies. The structure of other solutions will be clarified below.

Let us construct an expression for the energy flux, carried by the elementary

---

[12]Time (frequency) dispersion will be discussed later.

wave (2.7.3). For any free harmonic vibration, the time average energy flux $\langle S\rangle_t$ must be obviously constant in each section of x because external forces are absent. In particular, it follows from here that if $u(x) \to 0$ as $x \to \infty$, then $\langle S\rangle_t = 0$. Thus, the wave (2.7.3) with $\operatorname{Im}\{k\} \neq 0$ does not transfer energy (in average). However, for the wave obtained by a superposition of two waves with complex values of k, but increasing in different directions, generally speaking $\langle S\rangle_t \neq 0$.

Using (2.5.29), we find

$$\begin{aligned}\langle S(x', x'')\rangle_t &= \frac{\omega|A|^2}{2}\Phi(x'-x'')\operatorname{Im}\{e^{i(kx'-\bar{k}x'')}\}\\ &= \frac{\omega|A|^2}{2}\Phi(x'-x'')\,e^{-\beta(x'+x'')}\}\sin\alpha(x'-x'')\,.\end{aligned} \tag{2.7.11}$$

Substitution in (2.5.31), after obvious change of variables and an integration in the region $x' \leqslant x \leqslant x''$ yields

$$\langle S\rangle_t = -\frac{\omega|A|^2}{4\beta}e^{-2\beta x}\int\Phi(x')\sin(\alpha x')\sinh\beta x'\,dx'\,. \tag{2.7.12}$$

In accordance with (2.7.10) when $\beta \neq 0$ we have $\langle S\rangle_t = 0$, as it should be. For a real $k = \alpha$, we have

$$\langle S\rangle_t = -\frac{\omega|A|^2}{4}\int\Phi(x)\sin(kx)\,dx = \frac{\omega|A|^2}{4}\frac{d}{dk}\Phi(k)\,. \tag{2.7.13}$$

Taking into account (2.7.6), we obtain finally

$$\langle S\rangle_t = \frac{1}{2}|A|^2\rho\omega^2(k)\frac{d\omega(k)}{dk}\,. \tag{2.7.14}$$

As expected, the flux does not depend on the point x.

It is easily seen that the coefficient of $\omega'(k)$ is equal to the average energy density. Indeed, for the kinetic energy we have

$$\left\langle\frac{1}{2}\rho\dot{u}^2\right\rangle_t = \frac{1}{4}\rho\omega^2(k)\operatorname{Re}\{u\bar{u}\} = \frac{1}{4}\rho|A|^2\omega^2(k)\,. \tag{2.7.15}$$

The average value of the elastic energy density $\langle\varphi\rangle_t$ must also be equal to this.

Problem 2.7.3. Show this by averaging (2.5.9) in time.

However, the velocity v of propagation of energy is by definition the ratio of the energy flux to the energy density and hence, $v = v_g$, i.e. for real values of ω and k the energy propagates with the group velocity. Obviously, this is not true for complex wave numbers.

Let us dwell briefly upon the case when $\Phi(k)$ is determined only for real k (for example, as a result of measuring $v(k)$) and cannot be continued in the complex plane. Let us assume that $\omega(k)$ is bounded but has discontinuities of the first kind at a finite number of points $k_\iota > 0$ (and also at symmetric points—k_ι). Let $[\Phi]_\iota$ be the discontinuities of the function $\Phi(k)$. Then $\Phi(k)$ can be written in the form

$$\Phi(k) = \Phi_0(k) + \sum_\iota [\Phi]_\iota [\theta(k - k_\iota) + \theta(-k - k_\iota)], \tag{2.7.16}$$

where $\Phi_0(k)$ is a continuous function. Similarly to (2.4.38) we have

$$\Phi_0(k) = \int \Psi_0(x) (1 - \cos kx) \, dx, \tag{2.7.17}$$

where $\Psi_0(x)$ is the characteristic of elastic bonds corresponding to $\Phi_0(k)$.

If the sum in the right hand side of (2.7.16) is represented in a similar form, then we obtain for the elastic bonds, corresponding to $\Phi(k)$

$$\Psi(x) = \Psi_0(x) + \frac{1}{\pi x} \sum_\iota [\Phi]_\iota \sin k_\iota x. \tag{2.7.18}$$

In the n representation for the quasicontinuum

$$\Psi(n) = \Psi_0(n) + \frac{1}{\pi n} \sum_\iota [\Phi]_\iota \sin k_\iota na. \tag{2.7.19}$$

An important conclusion follows from here: to discontinuities of the first kind, of the dispersion curve there corresponds an infinite action at a distance, decreasing as $|x|^{-1}$, when $|x| \to \infty$. This result will be used in Sect. 3.7 when interpreting the medium with complex structure in terms of an equivalent medium of simple structure.

In conclusion, let us consider two spectral characteristics of vibrations, which are widely used in the theory of crystal lattice.

Let us firstly assume that we deal with a chain of a finite but very large number of atoms N. Then wave numbers have discrete values, intervals between neighboring ones being of the order of $(Na)^{-1}$. The discrete dispersion "curve" which defines N eigenfrequencies $\omega(k)$ of normal vibrations of the chain corresponds to them. The quantity $\nu(\omega)\Delta\omega$, which characterizes the relative number of eigenfrequencies in the interval $(\omega, \omega + \Delta\omega)$ is obviously proportional to the corresponding interval of wave numbers Δk.

As $N \to \infty$ the spectrum of vibrations becomes continuous and we can define the frequency distribution function $\nu(\omega)$ by the relation

$$\nu(\omega) \stackrel{\text{def}}{=} \frac{a}{\pi} \left| \frac{dk}{d\omega} \right|, \tag{2.7.20}$$

where the normalizing factor is due to the condition that the integral of $\nu(\omega)$ is equal to unity.

Let us define analogously the quantity $g(\omega^2)\, d\omega^2$ as the limit of the relative number of the squares of frequencies lying in the interval $(\omega^2, \omega^2 + d\omega^2)$. The spectral characteristic $g(\omega^2)$ is called the density of vibrations. Obviously, the frequency distribution function $\nu(\omega)$ and the density of frequencies $g(\omega^2)$ are connected by the relation

$$\nu(\omega) = 2\omega g(\omega^2) . \tag{2.7.21}$$

Problem 2.7.4. Verify that for the chain with interaction of nearest neighbors, $\nu(\omega) = g(\omega) = 0$ where $\omega > \omega_{max}$ and when $0 \leqslant \omega < \omega_{max}$,

$$\nu(\omega) = \frac{2}{\pi\sqrt{\omega_{max}^2 - \omega^2}}, \quad g(\omega^2) = \frac{1}{\pi\omega\sqrt{\omega_{max}^2 - \omega^2}} \tag{2.7.22}$$

We see from this example that the spectral characteristics have singularities at the end of the interval of continuous spectrum.

If the dispersion law (2.7.6) defines a nonmonotonic dependence of ω on k, singularities also appear within the interval for the frequencies corresponding to zero values of the group velocity.

Problem 2.7.5. Show that the density of vibrations can be represented in the form

$$g(\omega^2) = \frac{a}{2\pi} \int \delta[\omega^2 - \omega^2(k)]\, dk . \tag{2.7.23}$$

The representation (2.7.23) is the most convenient one for various generalizations particularly, when passing to three dimensional models.

Thus, the spectral characteristics can be found if the dispersion law $\omega(k)$ is known; however, generally speaking, the latter contains more information about the vibrations. It is significant that for the calculations of some crystal properties, for example, for the construction of thermodynamics, it is sufficient to know only the law of distribution of vibrations in frequencies, i.e. the spectral characteristics [2.7, 8].

2.8 Kernel of Operator Φ_ω in the Complex Region

Let us proceed to a more detailed study of the solutions of the dispersion relation (2.7.5). For this let us consider the analytical properties of the kernel $\Phi(k, \omega)$ of the operator Φ_ω

Assuming the action at a distance to be finite and taking into account (2.4.38), we have

$$\Phi(k) = 2 \int_0^l \Psi(x)(1 - \cos kx)\, dx . \tag{2.8.1}$$

Correspondingly for the chain with N interacting neighbors

$$\Phi(k) = \frac{2}{a} \sum_{n=1}^{N} \Psi(n)(1 - \cos kna) = \frac{4}{a} \sum_{n=1}^{N} \Psi(n) \sin^2 \frac{kna}{2} \quad (k \in B) . \quad (2.8.2)$$

As was pointed out above a function of the form (2.8.1) when l is finite can be continued to the complex plane as an entire function of exponential type $\leqslant l$. In the case of the quasicontinuum (of a chain), the region of definition of the entire function $\Phi(k)$ is the complex cylinder $\mathscr{B}$ obtained by the identification of the straight lines $\mathrm{Re}\{k\} = \pm\pi/a$. In either of these cases, the behavior of $\Phi(k)$ in the complex region is implicitly determined by that of $\Phi(k)$ on the real axis, i.e. by the dispersion curve $\omega = \omega(k)$. The latter evidently determines also the structure of the elastic bonds. For example, in the case of the chain, to obtain $\Psi(n)$, it is sufficient to expand $\Phi(k)$ into a Fourier series. Let us also note that with finite action at a distance, the dispersion curve and its tangent cannot have discontinuities, since $\omega^2(k)$ is an analytic function.

As was shown in Sect. 2.4, when the condition of invariance of energy with respect to translation is fulfilled, the expansion of $\Phi(k)$ into series in k starts with the term k^2 [compare Eqs. (2.8.1) and (2.8.2)]. Hence, for long wavelengths we have $\omega(k) \sim k$ which correspond to the usual law of dispersion for longitudinal vibrations of a bar or transverse vibrations of a string. However, there are also possible models, in which elastic bonds occur not only between particles of the medium but also with the external rigid foundation. In the longwave approximation for such models, $\omega(k) \approx \omega_0(1 + l^2k^2)$, where l is a characteristic scaling parameter and the frequency ω_0 is determined by the stiffness of the elastic bonds with the foundation. In the following it will be assumed that bonds with the foundation are absent, i.e. the conditions of translational invariance are fulfilled. In Chap. 4 will we consider some consequences of the breaking of translational invariance, within the framework of the energy method.

From (2.8.2), it is easy to see that for the chain with interaction of finite number of neighbors the function $\Phi(k)$ is always bounded for real values of k and hence there exists a maximum frequency $\omega_{\max}$ for nondecaying waves, which can be reached for the limiting values of $k = \pm\pi/a$ as well as for the internal points. In Fig. 2.9 the dispersion curves for chains with interaction of one (1) and two (2) neighbors are shown as an example.

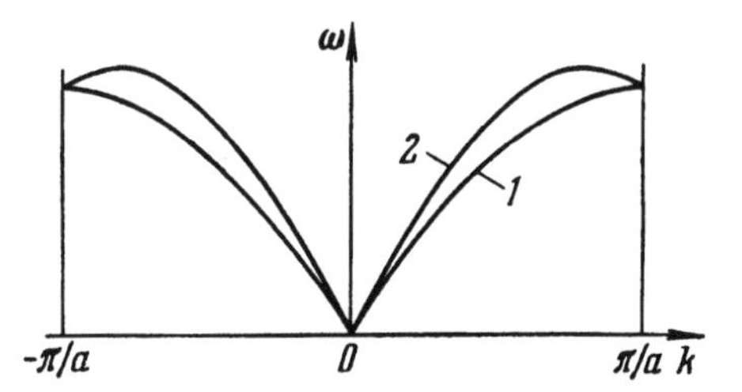

Fig.2.9. Typical dispersion curves

Thus the chain plays the role of a filter of low frequencies: signals with frequencies $\omega < \omega_{max}$ will propagate and signals with higher frequencies will decay exponentially.

This phenomenon is usually connected with discrete structure; however, it follows from (2.8.1), that ω_{max} appears also for the nonlocal model of the continuous medium under physically reasonable restrictions on $\Psi(x)$.

The presence of the maximum (one or more) in the curve $\omega = \omega(k)$ for real values of k, has as a consequence the existence of zones of frequencies in which to each frequency there correspond several nondecaying waves which propagate in one and the same direction. However, for example, in accordance with (2.7.6) and (2.8.2), we have for the chain with interaction of nearest neighbors.

$$\omega^2(k) = \omega_{max}^2 \sin^2 \frac{ka}{2}, \quad \omega_{max}^2 = \frac{3\Psi}{a\rho}, \tag{2.8.3}$$

and a maximum is obtained at boundary values of $k = \pm\pi/a$. A unique wave propagating in the given direction corresponds to each frequency $\omega < \omega_{max}$.

This also takes place in the local theory but for another reason: $\omega(k)$ depends on k linearly and hence ω_{max} does not exist.

Besides ω_{max}, there exists another characteristic frequency $\omega(\infty)$. Comparing (2.4.38), (2.7.6) and (2.8.1) and taking into account that $\Psi(k) \to 0$ as $k \to \infty$ we find

$$\omega^2(\infty) = \frac{2}{\rho}\int_0^l \Psi(x)\, dx = \frac{\psi_0}{\rho}. \tag{2.8.4}$$

For stable bonds $\Psi(x) \geqslant 0$ and $\omega^2(\infty) > 0$.

The boundary frequency $\omega(\pm\pi/a)$ is an analog of $\omega(\infty)$ in the case of a chain. The frequencies $\omega(\pm\pi/a)$ and ω_{max} may, in some cases, coincide as is seen in the example of a chain with nearest neighbor interaction.

It is easy to see from (2.8.1) that, for stable bonds $\Phi(k)$ and $\omega^2(k)$ are positive for all real values of $k \neq 0$. In view of (2.3.27) and (2.7.6), for all admissible displacements different from translation, we find that

$$\Phi = \frac{1}{2}\langle u|\Phi|u\rangle = \frac{\rho}{4\pi}\int \overline{u(k)}\, \omega^2(k)\, u(k)\, dk > 0. \tag{2.8.5}$$

However, the condition $\omega^2(k) > 0$ for $k \neq 0$, $\mathrm{Im}\{k\} = 0$, which ensures stability of the medium could be fulfilled even in the case in which some bonds are unstable. This should not cause any surprise inasmuch as the characteristics of individual bonds are only effective quantities, which depend upon the basic model notions. It is $\Phi(k)$ (or, what is the same, $\omega^2(k)$) that is the real invariant quantity. Examples of real physical models for which some effective bonds are unstable, will be demonstrated later.

Problem 2.8.1. Using (2.7.7) verify that for the stability of the medium in the case of the chain with interaction of two neighbors, it is necessary and sufficient to require that

$$\Psi(1) > 0, \quad \beta \overset{\text{def}}{=} \Psi(2)/\Psi(1) > -\frac{1}{4}\,. \tag{2.8.6}$$

In the theory of entire functions, it is shown that functions of exponential type are determined by the distribution of their roots to within a constant multiplier. Let us study in detail the roots of the functions $\Phi(k)$ and $\Phi(k, \omega)$ in the complex plane.

It can be shown that $\Phi(k)$ has a countable number of roots $k_m (m = 0, 1, \ldots)$ and a finite number of roots is contained in any bounded region, i.e. $k = \infty$ is the only cluster point. As $m \to \infty$ we have $|k_m| \to \infty$ but the imaginary part of k_m increases slower than the real one: $\operatorname{Im}\{k_m\} \sim \ln|k_m|$.

If it is assumed that $\Psi(x) \geqslant 0$ (or $\Psi(n) \geqslant 0$), then it follows from (2.8.1) or (2.8.2) that $\Phi(k)$ cannot have roots on the real or imaginary axes except the two-fold root $k_0 = k_1 = 0$. Other roots are distributed in quadruples at the vertices of rectangles. In fact, from the reality (when $\operatorname{Im}\{k\} = 0$) and evenness of $\Phi(k)$, it follows that, if λ is a root, then $\bar{\lambda}$,-λ,-$\bar{\lambda}$ will also be roots. However, as pointed out above, the condition $\Psi(x) \geqslant 0$ is not a necessary one and, therefore, in the general case $\Phi(k)$ can have purely imaginary roots.

Problem 2.8.2. For the chain with interaction of two neighbors when $-1/4 < \beta < 0$, verify that two roots are situated on the imaginary axis.

Let us agree to enumerate the roots in the nondecreasing order of their moduli, so that

$$\operatorname{Im}\{k_{2m}\} > 0, \quad k_{2m+1} = -k_{2m} \quad (m \neq 0)\,, \tag{2.8.7}$$

i.e. roots with even (odd) numbers are situated in the upper (lower) halfplane. Then for $\Phi(k)$ there is the expansion in roots [2.9]

$$\Phi(k) = c_0 k^2 \prod_{m=1}^{\infty} \left(1 - \frac{k^2}{k_{2m}^2}\right), \tag{2.8.8}$$

where, as easily seen, the constant c_0 is defined by the expression

$$c_0 = \frac{1}{2}\frac{d^2}{dk^2}\Phi(k)\big|_{k=0} = \int_0^l x^2 \Psi(x)\, dx\,. \tag{2.8.9}$$

Comparison with (2.4.41) shows that c_0 coincides with the elastic modulus of the corresponding local approximation.

In the case of the chain, it is evident from (2.8.2), that $\Phi(k)$ is a polynomial of degree N with respect to $\cos ka$, or that is the same, $\sin^2(ka/2)$. But $\operatorname{Re}\{k\} \leqslant \pi/a$

and hence, $\Phi(k)$ has exactly $2N$ roots, i.e. the following representation is valid:

$$\Phi(k) = -2p(N)\frac{\Psi(N)}{a}(1 - \cos ka)\prod_{m=1}^{N-1}(\cos ka - \cos k_{2m}a), \tag{2.8.10}$$

where $p(n)$ is the coefficient of $\cos^N ka$ in the expansion of $\cos Nka$.

There is another form, analogous to (2.8.8), namely

$$\Phi(k) = \frac{4c_0}{a^2}\sin^2\frac{ka}{2}\prod_{m=1}^{N-1}\left(1 - \frac{\sin^2\frac{ka}{2}}{\sin^2\frac{k_{2m}a}{2}}\right). \tag{2.8.11}$$

The constant c_0 coincides with (2.4.42) and is the elastic modulus of the corresponding local approximation.

Problem 2.8.3. Verify that c_0 and the roots of $\Phi(k)$ in the case of the chain, are connected by the relation

$$c_0 = 2^{N-1}p(N)\,a\Psi(N)\prod_{m=1}^{N-1}\sin^2\frac{k_{2m}a}{2}. \tag{2.8.12}$$

Above it was assumed tacitly that all the roots are simple ones. Generalization to the case of multiple roots is obvious.

For boundary problems, the functions $\Phi_\pm(k)$ obtained by factorization of $\Phi(k)$ will be also shown to be useful. The problem of factorization of functions of the type (2.8.1) was considered in the work [2.10]. It follows from the results obtained there, that if $\Psi(x) \in L_1(-l, l)$ then

$$\Phi(k) = \Phi_+(k)\,\Phi_-(k), \tag{2.8.13}$$

where $\Phi_\pm(k)$ are bounded on the real axis, do not have roots in the upper (lower) halfplane and are connected with $\Phi(k)$ by the relations

$$\Phi_\pm(k) = \exp\left[\frac{1}{2\pi i}\int\frac{\ln\Phi(k')}{\pm k - k'}dk'\right], \quad \mathrm{Im}\{k\} \gtrless 0. \tag{2.8.14}$$

Factorization (2.8.13) may be given a more clear meaning, by splitting the product (2.8.8) into two factors, so that to each of them contribute roots located either in the upper or the lower half planes.

The following representations are also valid

$$\Phi_+(k) = \int_0^l \Psi_+(x)(1 - e^{-ikx})\,dx,$$

$$\Phi_-(k) = \int_{-l}^0 \Psi_-(x)(1 - e^{-ikx})\,dx = \Phi_+(-k), \tag{2.8.15}$$

where $\Psi_\pm(x)$ are connected with $\Phi_\pm(x)$, the inverse Fourier images of $\Phi_\pm(k)$, in the obvious way, and can eventually be expressed in terms of $\Psi(x)$. It is significant that the functions $\Phi_\pm(x)$ are concentrated in the intervals $(0, l)$ and $(-l, 0)$, respectively (cf. Appendix 3, where the proof for the discrete case is given).

Analogous formulae for the expansion in terms of the roots occur for $\Phi(k, \omega)$. When $\omega \neq 0$, instead of (2.8.8), it is necessary to write

$$\Phi(k, \omega) = -\rho\omega^2 + \Phi(k) = -\rho\omega^2 \prod_{m=0}^{\infty} \left(1 - \frac{k^2}{k_{2m}^2(\omega)}\right), \tag{2.8.16}$$

and instead of (2.8.10) and (2.8.11),

$$\begin{aligned}\Phi(k, \omega) &= 2p(N)\frac{\Psi(N)}{a}\prod_{m=0}^{N-1}[\cos ak - \cos ak_{2m}(\omega)] \\ &= -p\omega^2 \prod_{m=0}^{N-1}\left(1 - \frac{\sin^2(ak/2)}{\sin^2(ak_{2m}(\omega)/2)}\right),\end{aligned} \tag{2.8.17}$$

where $k_m(\omega)$ are the roots of the function $\Phi(k, \omega)$ or, equivalently, the solutions of the dispersion equation (2.7.5). Here we preserve the agreement (2.8.7) concerning the numeration of the essentially complex roots and for the real roots we set $k_{2m}(\omega) > 0$ and $k_{2m+1}(\omega) = -k_{2m}(\omega)$.

As will be shown in the next section, the structure of the general solution of the equations of motion is completely determined by the roots of the function $\Phi(k, \omega)$. For this reason, the study of trajectories of the motion of the roots $k_m(\omega)$ as a function of the frequency ω is of interest. A graphic representation of the behavior of the trajectories can be obtained from the example of the chain for which the trajectories belong to the complex cylinder $\mathscr{B}$.

According to (2.8.3), when $N = 1$ and $\omega < \omega_{\max}$ there are two real roots which separate from point $k = 0$ and coincide again at $k = (\pm\pi/a)$, when ω changes from 0 to $\omega(\pm\pi/a) = \omega_{\max}$. When $\omega > \omega_{\max}$, the roots are essentially complex and they separate along the generator of the cylinder $\mathrm{Re}\{k\} = \pm\pi/a$, with increasing ω (Fig. 2.10).

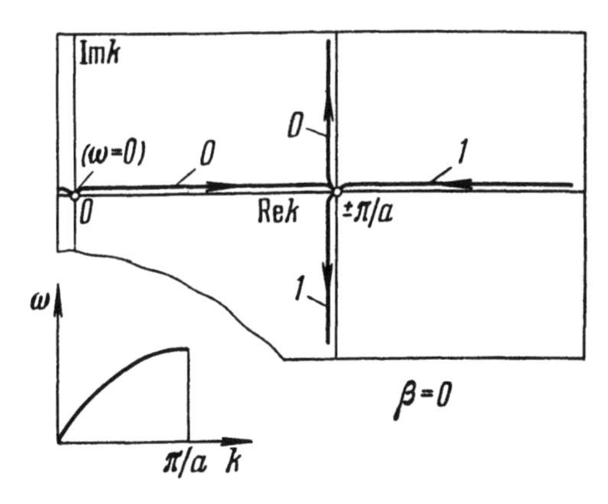

Fig.2.10. Trajectories of roots

When $N = 2$ in accordance with (2.7.7), there are four roots and their trajectories depend on the dimensionless parameter β, defined by the expression (2.8.6).

The three topologically different diagrams of the trajectories on the evolvent of the cylinder are shown in Fig. 2.11 (the dispersion curves are shown there too).

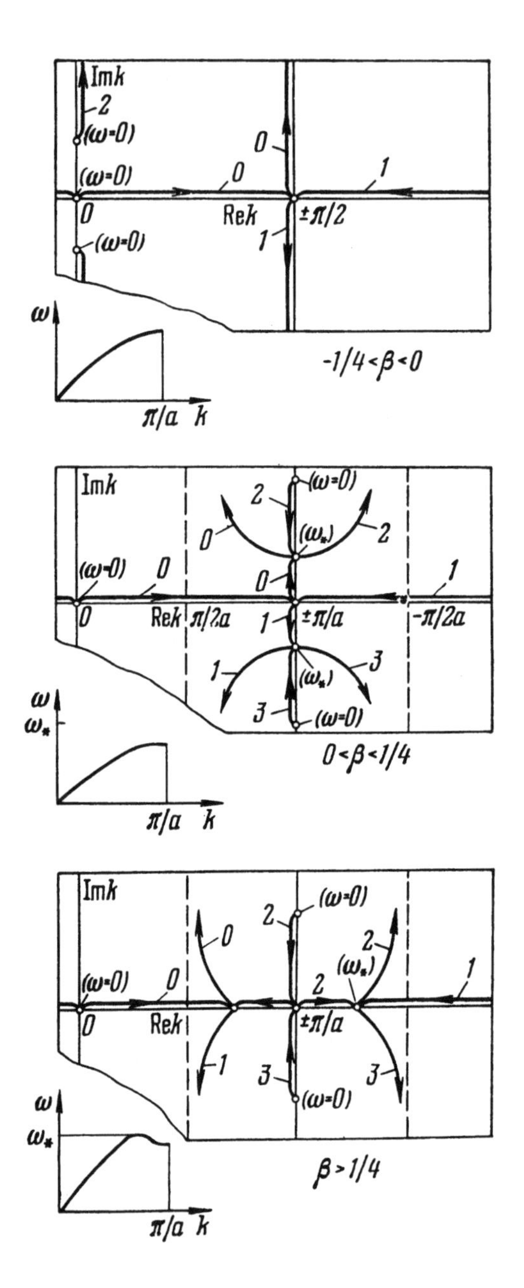

Fig.2.11. Phase diagrams

Such diagrams characterize qualitatively different chains, and we will call them the phase diagrams.

Thus, the region $\beta > -1/4$, corresponding to stable chains (Problem 2.8.1) splits into three subregions with the same phase portraits. The last ones are determined by the set of singular points of the trajectories and their nature.

When $\omega = 0$, there are the singular points $k_{0.1}(0) = 0$ and conjugate points $k_2(0)$ and $k_3(0)$, located either on the imaginary axis ($\beta < 0$) or on the line $\mathrm{Re}\{k\} = \pi/a$ ($\beta > 0$). For $\beta \to \pm 0$ (interaction of the nearest neighbors), the points $k_2(0)$ and $k_3(0)$ go to infinity.

The motion of the roots as the frequency increases is easily traced in the figures. For convenience, the straight line $\mathrm{Re}\{k\} = \pm\pi/a$, the real and imaginary axes will be called directors. Firstly, the roots can move along the directors. For $\beta > 0$, starting with some frequency ω^* the roots leave the directors and asymptotically approach the straight lines $\mathrm{Re}\{k\} = \pm 2\pi/a$.

Singular points of the "saddle" type at which the trajectories intersect, forming multiple roots correspond to the frequency ω^*. Such points are of considerable significance for the phase portraits. Particularly, if these points are situated on the real axis, then they correspond to the intermediate extrema of the dispersion curve.[13]

Phase portraits for the interaction of any number of neighbors can be constructed in an analogous way. They give a complete qualitative information about the system and permit to build a detailed classification. At the same time, the absence of quantitative information about the connection between ω and k is their defect; in particular, dispersion curves have to be added to the phase portraits. In this connection we will give another representation of functional dependence $\omega(k)$ which unites the dispersion curves for real values of k with the most important information about their behavior in the region of complex values of k.

Let us introduce dimensionless quantities

$$\Omega = \frac{a\rho}{4\Psi(1)}\omega^2, \quad \zeta = \sin^2\frac{ak}{2}, \quad \beta_n = \frac{\Psi(n)}{\Psi(1)}. \tag{2.8.18}$$

Then in view of (2.8.2), the dispersion equation (2.7.5), takes the form

$$\Omega - \sum_{n=1}^{N} \beta_n \sin^2\frac{ank}{2} = 0, \tag{2.8.19}$$

or, in the variables Ω, ζ,

$$\Omega - \sum_{n=1}^{N} \gamma_n \zeta^n = 0, \tag{2.8.20}$$

[13]Besides these important singular points, obviously, there are also singular points of the "saddle" type at the intersections of generators with the real axis.

where the parameters γ_n are connected with the parameters β_n in the obvious way.

Note that to real k's, the imaginary k-axis and the generator $\mathrm{Re}\{k\} = \pi/a$, there correspond the segment $0 \leqslant \zeta \leqslant 1$, the interval $-\infty < \zeta < 0$ and the interval $1 < \zeta < \infty$, respectively. Thus the most interesting region of variation of k along the directors is mapped on the real axis ζ and the curve (2.8.20) in the plane Ω, ζ gives essential information about the solution of the dispersion equation.

Problem 2.8.4. Show that the extrema of the curve $\Omega(\zeta)$ when $\Omega \geqslant 0$ correspond to the important singular points of the "saddle" type on the phase portrait, the trajectories leaving the directors for maxima and approaching the directors for the minima.

The curves $\Omega(\zeta)$ corresponding to the diagrams 2.10 and 2.11 are shown in Fig. 2.12. The parts of the curves on the segment $0 \leqslant \zeta \leqslant 1$ represent the dispersion curves in another scale. In the given case, the stability condition is obviously equivalent to the condition $\Omega'(0) = 0$. The displacement of the maximum of the parabola (saddle point), as β changes, also becomes visible. Imaginary frequencies correspond to negative values of Ω.

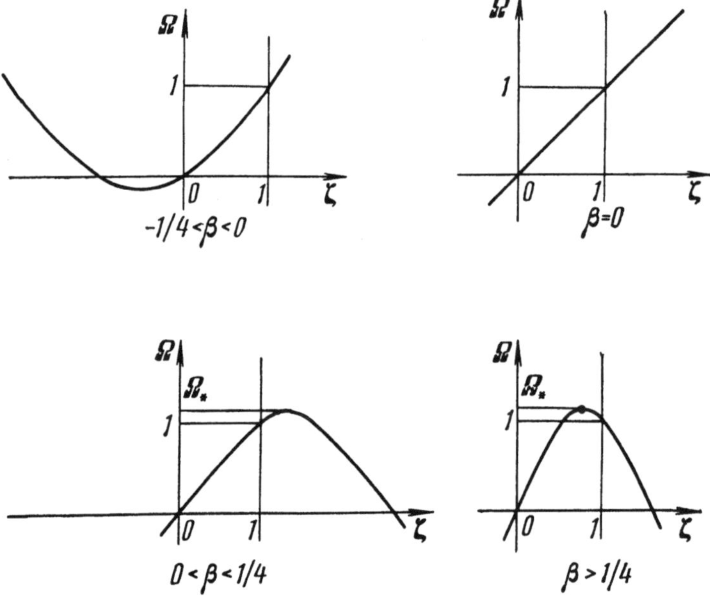

Fig.2.12. Curves $\Omega(\zeta)$

Later we shall see that such dispersion curves are most useful for the analysis of more complex systems.

2.9 Green's Function and Structure of the General Solution

Let us construct a general solution of the equations of motion for the homogeneous medium. Since we restrict ourselves to harmonic processes in time, let us write the equations of motion (2.4.39) in the (x, ω) representation

$$-\rho\omega^2 u(x, \omega) + \int \Phi(x - x')\, u(x', \omega)\, dx' = q(x, \omega)\,. \tag{2.9.1}$$

As usual, let us represent the solution of this equation as a sum of a particular solution and the general solution of the corresponding homogeneous equation.

In order to construct the particular solution, let us introduce the Green's function $G(x, \omega)$, which satisfies the equation

$$-\rho\omega^2 G(x, \omega) + \int \Phi(x - x')\, G(x', \omega)\, dx' = \delta(x) \tag{2.9.2}$$

or in the operator form

$$\Phi_\omega G_\omega = I\,, \tag{2.9.3}$$

where I is the identity operator and G_ω is the Green's operator.

Thus the Green's function $G(x, \omega)$ is the kernel of the operator G_ω, inverse to the operator Φ_ω. In order to determine $G(x, \omega)$ uniquely, it is necessary to specify the boundary conditions, as $|x| \to \infty$.

On the basis of the physical meaning of the function $G(x, \omega)$, which describes vibrations of the medium, induced by a force at the origin of the coordinate system (sources at infinity are absent), let us accept the natural conditions of radiation as the boundary conditions. In the problem under consideration they are specified in the following way:

a) Nondecaying waves go to infinity. For example, if the wave has the form $\exp(ikx)$ as $x \to \infty$ with real value of k, then we must have $k > 0$ (compare the representation of a wave in the form (2.7.3)).

b) As $|x| \to \infty$ the waves with essentially complex value of k must decay. The direction of their 'propagation' does not play any role since they do not carry energy. For example, for the wave $\exp(ikx)$, as $x \to \infty$, we require that $\operatorname{Im}\{k\} > 0$, but the sign of $\operatorname{Re}\{k\}$ can be arbitrary.

c) In the static case ($\omega = 0$), the asymptotic Green's function $G(x)$ must coincide with the static Green's function

$$G_0(x) = -\frac{|x|}{2c_0}\,, \tag{2.9.4}$$

which corresponds to the local model.

Let us write (2.9.2) in the k representation

$$[-\rho\omega^2 + \Phi(k)]\, G(k, \omega) = 1\,. \tag{2.9.5}$$

Hence, the formal solution has the form

$$G(k,\omega) = \Phi^{-1}(k,\omega) = \frac{1}{\rho[\omega^2(k) - \omega^2]}\,. \tag{2.9.6}$$

but it is necessary to complete (9.6) by introducing the boundary conditions on $G(x, \omega)$.

We have

$$G(x,\omega) = \frac{1}{2\pi}\int \frac{e^{ixk}}{\Phi(k,\omega)}\,dk = \frac{1}{2\pi\rho}\int \frac{e^{ixk}}{\omega^2(k)-\omega^2}\,dk\,, \tag{2.9.7}$$

where any curve in the complex plane which does not pass through the poles and is properly closed at infinity may be chosen as the contour of integration, without the boundary conditions. Changing the paths around the poles k_m adds to $G(x, \omega)$ solutions of the homogeneous equation of the type $\exp[ik_m(\omega)x]$. Since to every root $k_{2m}(\omega)$ there corresponds a root $k_{2m+1}(\omega) = -k_{2m}(\omega)$, it is always possible to choose paths around the poles in different ways for $x > 0$ and $x < 0$, such that the poles would give contributions of the form $\exp[ik_{2m}(\omega)|x|]$ when $k_{2m}(\omega) > 0$ for real roots and when $\operatorname{Im}\{k_{2m}(\omega)\} > 0$ for essentially complex roots. The boundary conditions for $G(x, \omega)$ are satisfied because of this.

Bearing in mind the agreement about the enumeration of the roots $k_m(\omega)$ (Sect. 2.8), we conclude that only roots of the form $k_{2m}(\omega)$, which lie either in the upper half plane or on the right halfaxis (in the case of statics, only one of the roots $k_0 = 0$), explicitly contribute to the expression for $G(x, \omega)$.

Let us now proceed to the explicit construction of $G(x, \omega)$ and start with the case of statics. Firstly, instead of (2.8.8) for $\Phi(k)$, let us consider the polynomial

$$\Phi_N(k) \stackrel{\text{def}}{=} c_0 k^2 \prod_{m=1}^{2N} \left(1 - \frac{k^2}{k_{2m}^2}\right). \tag{2.9.8}$$

The expansion of the rational function $\Phi_N^{-1}(k)$ into simple fractions, as easily verified, has the form

$$\Phi_N^{-1}(k) = \frac{1}{c_0 k^2} + 2\sum_{m=1}^{2N} \frac{k_{2m}}{\Phi_N'(k_{2m})}\frac{1}{k^2 - k_{2m}^2}\,. \tag{2.9.9}$$

If we now formally pass to the limit $N \to \infty$, then we obtain the expansion of the static Green's function $G(k)$ into a series of simple fractions

$$G(k) = \Phi^{-1}(k) = \frac{1}{c_0 k^2} + 2\sum_{m=1}^{\infty} \frac{k_{2m}}{\Phi'(k_{2m})}\frac{1}{k^2 - k_{2m}^2}\,. \tag{2.9.10}$$

Using the factorization (2.8.12), the series can be shown to converge in any closed bounded region, containing no poles, and hence it represents a meromorphic function $G(k)$; for large m's there is an estimate

$$\frac{1}{\Phi'(k_{2m})} = \frac{1}{\mathrm{i}\omega_*^2 l} + o\left(\frac{1}{k_{2m}}\right), \tag{2.9.11}$$

where ω_* is defined by (2.8.4).

Substituting (2.9.10) into (2.9.7) with $\omega = 0$ and taking into account the boundary conditions, we find

$$G(x) = -\frac{|x|}{2c_0} + \sum_{m=1}^{\infty} \frac{\mathrm{i}e^{\mathrm{i}k_{2m}|x|}}{\Phi'(k_{2m})}. \tag{2.9.12}$$

Problem 2.9.1. Obtain this expression by integration in the complex plane.

The first term in (2.9.12) coincides with the static Green's function (2.9.4) of the corresponding local model and makes the main contribution into the asymptotics of $G(x)$ as $|x| \to \infty$, as it should be. The exponential terms are due to nonlocality and decay at distances of the order of the radius of the action at a distance l. Obviously, the main contribution to the asymptotics of $G(k)$ is made by the shallowest exponents which correspond to the first roots of the function $\Phi(k)$.

As $x \to 0$, the Green's function $G(x)$ has the δ-functional singularity. To explain its origin, it is sufficient to notice that $\Phi(k) \to \rho\omega_*^2 < \infty$ (as $k \to \infty$, $\mathrm{Im}\{k\} = 0$) in accordance with (2.8.1) and (2.8.4). Hence $G(k) \to \rho^{-1}\,\omega_*^{-2} > 0$ i.e.

$$G(x) \sim \rho^{-1}\omega_*^{-2}\delta(x), \quad \text{as } x \to \infty. \tag{2.9.13}$$

An informal explanation of the divergence of $G(x)$ can be obtained, if we recall the physical meaning of the considered model of continuum with action at a distance. As shown above, one of the possible interpretations of the model is the case $l = Na \gg a$, when the discrete structure of the chain can be neglected. In the k representation, this means the replacement of the complex cylinder by a plane and in the x representation this means the replacement of the regular (discrete) δ function by the singular $\delta(x)$. Thus, the divergence of $G(x)$ is a direct consequence of the extrapolation of the approximate model, valid only for distances of the order of l, to the region of distances much smaller than l, or equivalently, to the region $k \gg l^{-1}$.

In connection with this it becomes obvious that in the expansions in terms of roots (2.8.8) and (2.9.10) only a finite number of terms can have real physical meaning and in the expression (2.9.12) for $G(x)$, it is necessary to consider only a finite number of exponents.

Later we will be faced with similar situations, when the formal extention of the actual region of applicability of the approximate model leads to some effects of nonphysical nature.

Problem 2.9.2. Show that for the chain with interaction of N neighbors, the expansion of $G(k)$ in simple fractions can be represented in the form

$$G(k) = \frac{a^2}{2c_0(1 - \cos ka)} - \sum_{m=1}^{N-1} \frac{\sin k_{2m}a}{\Phi'(k_{2m})(\cos ka - \cos k_{2m}a)}, \tag{2.9.14}$$

and the expression for $G(x)$ coincides with (2.9.12) but contains $N - 1$ exponents.

Problem 2.9.3. Show that for the chain with interaction of two neighbors

$$G(n) = \frac{a}{2c_0}\left(-|n| + \frac{2\beta}{\sqrt{1+4\beta}}\, \mathrm{e}^{\mathrm{i}k_2a|n|}\right), \tag{2.9.15}$$

where β is defined by the expression (2.8.6) and

$$k_2 = \pm\frac{\pi}{a} + \frac{\mathrm{i}}{a}\ln\frac{4\beta}{(\sqrt{1+4\beta} - 1)^2}. \tag{2.9.16}$$

The generalization of the given formulae to the case of the dynamic Green's function is obvious. It is the expansion in roots $k_{2m}(\omega)$, that is the analog of (2.9.10):

$$G(k, \omega) = \Phi^{-1}(k, \omega) = 2\sum_{m=0}^{\infty} \frac{k_{2m}(\omega)}{\Phi'(k_{2m}(\omega))}\, \frac{1}{k^2 - k_{2m}^2(\omega)}. \tag{2.9.17}$$

Substitution in (2.9.7) yields

$$G(x, \omega) = \sum_{m=0}^{\infty} \frac{\exp[\mathrm{i}k_{2m}(\omega)|x|]}{\Phi'[k_{2m}(\omega)]}. \tag{2.9.18}$$

Let us compare this expression with the Green's function

$$G_0(x, \omega) = \frac{\mathrm{i}}{2\omega\sqrt{\rho c_0}} \exp\left(\mathrm{i}\omega\sqrt{\frac{\rho}{c_0}}\,|x|\right) \tag{2.9.19}$$

of the corresponding local approximation.

For sufficiently low frequencies (in comparison with $\omega_{\max}$) we find from (2.7.5) and (2.8.8) that

$$k_0(\omega) \approx \sqrt{\frac{\rho}{c_0}}\,\omega, \quad \Phi'(k_0(\omega)) \approx 2\sqrt{\rho c_0}\,\omega \tag{2.9.20}$$

and hence, the first term of the series (2.9.18) approximately coincides with $G_0(x, \omega)$. But in the neighborhood of the point $\omega = 0$, all the roots $k_{2m}(\omega)$ except $k_0(\omega)$ are essentially complex and thus the main term of the asymptotics of $G(x, \omega)$, when $\omega \ll \omega_{\max}$ coincides with $G_0(x, \omega)$, as might be expected.

Thus, having constructed the Green's function $G(x, \omega)$ we can express in terms of it the solution $u(x, \omega)$ of the equation (2.7.1) with an arbitrary right-hand side $q(x, \omega)$, decreasing sufficiently rapidly as $|x| \to \infty$. Moreover, $u(x, \omega)$ will naturally satisfy the same conditions of radiation as $G(x, \omega)$ does.

We have

$$u(x, \omega) = \int G(x - x', \omega)\, q(x', \omega)\, dx' . \tag{2.9.21}$$

The fact that this expression satisfies (2.9.1) can be verified either by direct substitution or by writing it in the (k, ω) representation

$$u(k, \omega) = G(k, \omega)\, q(k, \omega) = \Phi^{-1}(k, \omega)\, q(k, \omega) . \tag{2.9.22}$$

Later, we shall often use also the operator form of expressions of the type (2.9.21) and (2.9.22), namely

$$u = G_{\omega} q . \tag{2.9.23}$$

To obtain the general solution of (2.9.1) it is necessary to add to (2.9.21) the general solution of the homogeneous equation. The latter, as shown above, is the superposition of the solutions corresponding to each of the roots $k_m(\omega)$, i.e.

$$u(x, \omega) = \sum_{m=0}^{\infty} (\alpha_m e^{ik_m(\omega)x} + \beta_m e^{-ik_m(\omega)x}) , \tag{2.9.24}$$

where α_m, β_m are arbitrary constants.

2.10 Approximate Models

In the preceding sections we pointed out special distinctions of nonlocal theory from a local one, which appear already in the one-dimensional model of the medium of simple structure. Let us enumerate few of them:

a) a scaling parameter exists;
b) a boundary is substituted by a boundary region;
c) new decaying and nondecaying waves appear, and in the static case there appear exponential solutions of the equations of equilibrium;
d) the velocity of propagation of the waves depends on their length (spatial dispersion);
e) a limiting frequency for nondecaying waves exists;
f) divergence of fields and energy is eliminated for concentrated forces and defects (in the quasicontinuum).

Looking ahead, let us also point out some three-dimensional nonlocal phenomena in the medium of simple structure: new types of surface waves, rotation of the plane of polarization of the waves (gyrotropicity), etc.

It is natural to try to construct approximate models which qualitatively or quantitatively correctly account for some of these phenomena. From this point of view, let us consider various approximate models and the ranges of their application.

Simplest models can be obtained if one replaces $\Phi(k)$ approximately by a polynomial, and one sets

$$\Phi(k) \approx \Phi_s(k), \quad \Phi_s(k) \stackrel{\text{def}}{=} c_0 k^2 P_s(k^2)\,, \tag{2.10.1}$$

where $P_s(k^2)$ is an appropriate polynomial in k^2 of degree s.

In the x representation this corresponds to the substitution of the integral operator Φ by the differential operator Φ_s. The equation of motion takes the form

$$\rho\omega^2 u(x,\omega) + c_0 \partial^2 P_s(-\partial^2)\, u(x,\omega) = -\, q(x,\omega)\,, \tag{2.10.2}$$

which is typical for phenomenological theories with higher derivatives like couple-stress, multipole and similar theories, which will be considered below.

In the zeroth approximation, obviously, it is necessary to set

$$\Phi_0(k) = c_0 k^2, \quad \Phi_0(\partial) = -\, c_0 \partial^2\,, \tag{2.10.3}$$

and we have the usual one-dimensional elastic model, in which no information about nonlocal effects is contained. Its field of applicability: elastic fields for which variations on the distances of the order l can be neglected, for example, extremely long waves; in other words: the region $|k| \ll l^{-1}$ in which the dispersion curve may be approximated by a straight line.

When proceeding to higher approximations there are two possible approaches, upon the choice of which the physical meaning and area of applicability of approximate polynomial models ultimately depend.

For instance, as an approximating polynomial $\Phi_s(k)$, one can take the first term of the series expansion of $\Phi(k)$ in a neighborhood of the point $k = 0$. In particular, the first approximation may be obtained by setting

$$P_1(k^2) = 1 + l^2 k^2\,, \tag{2.10.4}$$

where the scale parameter l depends on the curvature of the dispersion curve at small values of k.

This approximation must describe well the dispersion of sufficiently long nondecaying waves, therefore it may be spoken about as a (first) long wavelength approximation. In the phenomenological approach, when the given model is considered as the initial one, we will call it a model of a medium with weak spatial dispersion. The scale parameter l is to be considered small and the model is, strictly speaking, not applicable at distances less than, or of the order of, l.

It is significant that in the admissible long wavelength region, i.e. for small values of k and ω, the approximate dispersion equation

$$\Phi_1(k,\omega) \stackrel{\text{def}}{=} -\, \rho\omega^2 + \Phi_1(k) = 0 \tag{2.10.5}$$

has as many roots (two), as the equation in the zeroth approximation does, and these roots differ little from the roots of the zeroth approximation.

If one extrapolates $\Phi_1(k)$ to the region of large values of k and ω (short waves), then the roots of $\Phi_1(k, \omega)$ will strongly differ from the roots of $\Phi_0(k, \omega)$ and moreover two additional roots appear. However, generally speaking, these roots will have nothing in common with the first roots of the initial operator $\Phi(k, \omega)$. In particular, the condition of symmetry of the roots in the complex plane indicated in Sect. 2.8 will not be observed. Therefore, this approximation is inadequate in those cases in which at least a qualitively correct consideration of additional solutions caused by nonlocality is necessary, for example, in boundary problems.

Another way to approximate $\Phi(k)$ consists in constructing an approximating polynomial $\Phi_s(k)$ according to the first roots. In this approximation from the exact solutions of the type (2.9.12) or (2.9.18), the rapidly oscillating and decaying components are filtered out and there is left only the information about components which vary most smoothly over distances of the order of l.

Obviously, it is the next approximation (after the zeroth one) which is of the main interest. It is determined by a nonzero root of $\Phi_2(k)$ with the smallest modulus, i.e. by k_2. Let us designate it by λ and for definiteness assume that $\mathrm{Im}\{\lambda\} > 0$ and $\mathrm{Re}\{\lambda\} \geqslant 0$.

One has to distinguish two cases: $\mathrm{Re}\{\lambda\} > 0$ and $\mathrm{Re}\{\lambda\} = 0$. In the first case there are four roots: λ, $-\lambda$, $\bar{\lambda}$ and $-\bar{\lambda}$ with the same moduli. The corresponding approximation $\Phi_1(k)$ for $\Phi(k)$ has the form

$$\Phi_1(k) = c_0 k^2 \left(1 - \frac{k^2}{\lambda^2}\right)\left(1 - \frac{k^2}{\bar{\lambda}^2}\right). \tag{2.10.6}$$

In the x representation a differential operator of sixth order corresponds to (2.10.6):

$$\begin{aligned}\Phi_1(\partial) &= -\, c_0 \partial^2 \left(1 + \frac{\partial^2}{\lambda^2}\right)\left(1 + \frac{\partial^2}{\bar{\lambda}^2}\right) \\ &= -\, c_0 \partial^2 (1 + 2\cos 2a \cdot l^2 \partial^2 + l^4 \partial^4)\,,\end{aligned} \tag{2.10.7}$$

where $a = \arg\lambda$ and $l = |\lambda|^{-1}$ is a scaling parameter.

In the second case there are two imaginary roots; λ and $-\lambda$. The differential operator $\Phi_1(\partial)$ is now of fourth order

$$\Phi_1(\partial) = -\, c_0 \partial^2 (1 - l^2 \partial^2)\,. \tag{2.10.8}$$

Strictly speaking, this case is to be considered as extremely improbable. Moreover, as we know, a necessary condition for the presence of imaginary roots consists in the instability of some of the bonds. Therefore the operator (2.10.7) is to be considered as a more correct approximation to the exact one.

On the other hand, the operator of fourth degree (2.10.8) correctly reflects the same phenomena qualitatively (excluding the oscillating character of the solution) as the operator of sixth degree (2.10.7) does. This permits one to use it in those cases for which the simplicity of the approximate model is the main feature. Later we shall meet with this when solving boundary problems.

As an illustration, let us present the expressions for the static Green's functions of the operators (2.10.7) and (2.10.8)

$$G_1(x) = -\frac{1}{c_0}\left[\frac{|x|}{2} + \operatorname{Im}\left\{\frac{\bar{\lambda}^2}{\lambda(\lambda^2-\bar{\lambda}^2)}\,e^{i\lambda|x|}\right\}\right], \quad \operatorname{Re}\{\lambda\} > 0\,, \tag{2.10.9}$$

$$G_1(x) = -\frac{1}{c_0}\left[\frac{|x|}{2} + \operatorname{Im}\left\{\frac{1}{\lambda}\,e^{i\lambda|x|}\right\}\right], \quad \operatorname{Re}\{\lambda\} = 0\,. \tag{2.10.10}$$

In Fig. 2.13, they are shown by solid curves constructed for $\lambda = \pi(1/3 + i)$ and $\lambda = i\pi/2$ (in dimensionless coordinates). The dotted curve corresponds to the Green's function (2.9.15) for the chain with the interaction of two neighbors when $\beta = 1$; the straight line is the Green's function $G_0(x)$ of the zeroth approximation.

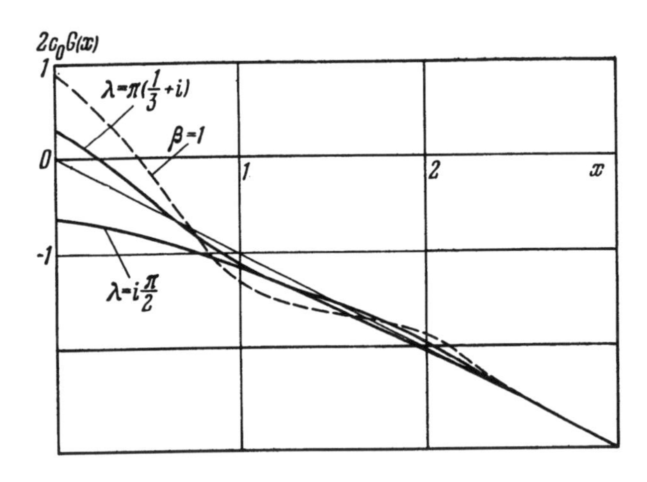

Fig.2.13. Green's curves

The model corresponding to the operator

$$\Phi_1(k,\omega) \overset{\text{def}}{=} -\rho\omega^2 + c_0k^2\left(1-\frac{k^2}{\lambda^2}\right)\left(1-\frac{k^2}{\bar{\lambda}^2}\right), \tag{2.10.11}$$

will be called the approximation by first roots. The advantage of this approximation in comparison with the long wavelength one is in the qualitively correct description of the phenomena for which waves with length of order l, conservation of the main terms of the solutions asymptotics and the possibility of correct formulation of boundary problems play an essential role. But, of course, the long wavelength approximation is usually adequate for the nondecaying waves in the region of small values of k.

Subsequent development of the field of applicability of the approximation by first roots can be obtained by setting [compare Eq. (2.8.16)]

$$\Phi_1(k, \omega) = -\rho\omega^2 \left(1 - \frac{k^2}{\mu^2(\omega)}\right)\left(1 - \frac{k^2}{\lambda^2(\omega)}\right)\left(1 - \frac{k^2}{\bar{\lambda}^2(\omega)}\right), \tag{2.10.12}$$

where $\mu(\omega)$ and $\lambda(\omega)$ approximate the roots $k_0(\omega)$ and $k_2(\omega)$ of the operator $\Phi(k, \omega)$ respectively, while $\operatorname{Im}\{\mu(\omega)\} = 0$. Obviously, for low frequencies ω,

$$\mu(\omega) \approx \sqrt{\frac{\rho}{c_0}}\,\omega, \quad \lambda(\omega) \approx \lambda\,. \tag{2.10.13}$$

With an appropriate choice of the functions $\mu(\omega)$ and $\lambda(\omega)$, a modified approximation by first roots in the form (2.10.12) will correspond well to the exact model in a wide range of frequencies and wave numbers.

Other approximate models, not connected with polynomial approximation are also possible. An approximate prescription of the dispersion curve over a wide range of wavelengths with the help of an appropriate function of wave number k (for example, in interpolating experimental data) can be of interest in a number of cases. This model is good for the description of nondecaying waves but, generally speaking, it is incorrect to continue it in the complex region and to use it for boundary problems.

Models based on approximations of the type

$$\Psi(x) = \sqrt{\frac{2}{\pi}}\, c_0 l^{-3} \exp\left(-\frac{x^2}{2l^2}\right) \tag{2.10.14}$$

or

$$\Psi(x) = \begin{cases} 3c_0 l^{-3} & \text{as} \quad |x| < l\,, \\ 0 & \text{as} \quad |x| > l\,. \end{cases} \tag{2.10.15}$$

can have a wider range of applicability.

To them there correspond

$$\Phi(k) = 2c_0 l^{-2}\left[1 - \exp\left(-\frac{1}{2}\, l^2k^2\right)\right], \tag{2.10.16}$$

$$\Phi(k) = 6c_0 l^{-2}\left(1 - \frac{\sin lk}{lk}\right). \tag{2.10.17}$$

Such models describe qualitatively correctly all the action at a distance phenomena indicated above.

2.11 Solution of Basic Boundary Problems

We consider here a method of solving the basic boundary problems of nonlocal theory and for simplicity we restrict ourselves to the static case.

Let us start with a more simple second basic problem (Sect. 2.6). Let the segment $-l \leqslant x \leqslant 0$ and the semiaxis $0 < x < \infty$ correspond to the boundary layer S and to the region D, respectively (here l is a parameter of action at a distance). Then, setting $\omega = 0$ in (2.6.10), we present the boundary problem in the form

$$\int_{-l}^{\infty} \Phi(x - x')\, u(x')\, dx' = q(x), \quad x \in D\,, \tag{2.11.1}$$

$$u_S(x) = h(x), \quad x \in S\,. \tag{2.11.2}$$

Obviously, it can be reduced to the integral equation

$$\int_{0}^{\infty} \Phi(x - x')\, u(x')\, dx' = q(x) - \int_{-l}^{0} \Phi(x - x')\, h(x')\, dx', \quad x \in D\,. \tag{2.11.3}$$

Equations of such a type were considered in [2.10] with the additional condition $\Phi(k) \neq 0$, which is not fulfilled in the present case since $\Phi(k)$ has a double root $k_1 = k_2 = 0$. Apparently, the method of the paper [2.10] can be appropriately generalized; however, with the aim of greater clarity and mainly in order to obtain a useful approximation, the solution by eigenfunctions of the operator Φ will be constructed in the following.

Let us find the solution of the problem (2.11.1) and (2.11.2) bounded as $x \to \infty$, in the form of a sum

$$u(x) = u^0(x) + v(x)\,, \quad x \in D \cup S\,, \tag{2.11.4}$$

where $u^0(x)$ is the particular solution of (2.11.1), which vanishes as $x \to \infty$ and $v(x)$ is the solution of the corresponding homogeneous equation satisfying the boundary condition

$$v_S(x) = h(x) - u_S^0(x)\,, \quad x \in S\,. \tag{2.11.5}$$

To derive $u^0(x)$ let us introduce the fundamental solution $G^0(x)$ of (2.11.1) connected with the Green's function (2.9.12) by the relation

$$G^0(x) = G(x) - \frac{x}{2c_0}, \quad x \in D \cup S\,. \tag{2.11.6}$$

Then, one obtains

$$u^0(x) = \int_D G^0(x - x')\, q(x')\, dx'\,, \quad x \in D \cup S\,. \tag{2.11.7}$$

Let us represent the solution of the homogeneous equation $\Phi v = 0$ in the form

$$v(x) = \sum_{n=0}^{\infty} v^n e_n(x), \quad x \in D \cup S. \tag{2.11.8}$$

Here as the functional basis $\{e_n\}$ we take the complete set of nonincreasing, as $x \to \infty$, exponential solutions of the homogeneous equation, i.e.

$$e_0(x) = 1, \quad e_n(x) = \mathrm{e}^{\mathrm{i}k_n x}, \quad \mathrm{Im}\{k_n\} > 0, \tag{2.11.9}$$

where $\{k_n\}$ is a set of all roots of $\Phi(k)$, which lie in the upper halfplane, completed by the value $k_0 = 0$.

In order to find the coefficients v^n, it is necessary to expand the values $v_S(x)$ given in S with respect to $e_n(x)$ (more accurately, with respect to restrictions of $e_n(x)$ to S). Of course, it is necessary to impose some conditions of smoothness on $v_s(x)$, which provide the completeness of the basis $\{e_n(x)\}$ (see for example [2.9]).

Since the basis $\{e_n(x)\}$ is not orthonormal, let us construct a system of functions $\{e^m(x)\}$ with support in S, which form a reciprocal basis with respect to $\{e_n(x)\}$:

$$(e^m, e_n) \stackrel{\text{def}}{=} \int e^m(x)\, \mathrm{e}^{\mathrm{i}k_n x}\, dx = e^m(k_n) = \delta_n^m. \tag{2.11.10}$$

Here $e^m(k)$ is the Fourier transform of the function $e^m(x)$, continued by zero outside of S. The system of functions $\{e^m(x), e_n(x)\}$ is known as biorthogonal basis.

Problem 2.11.1. Verify that

$$e^m(k) = \frac{\Phi_-(k)}{\Phi'_-(k_m)\,(k - k_m)}, \tag{2.11.11}$$

where $\Phi_-(k)$ is defined by the expression (8.14).

Problem 2.11.2. Taking into account the properties of $\Phi(k)$, show that $e^m(x)$ defined by the expressions (2.11.10) has its support in S and may be represented in the form

$$e^m(x) = \frac{\mathrm{i}}{\Phi'_-(k_m)} \int_0^{x+1} \Phi_-(x - x')\, \mathrm{e}^{-\mathrm{i}k_m x'}\, dx'. \tag{2.11.12}$$

From (2.11.8) and (2.11.10), we obtain

$$v^n = (e^n, v_S), \quad n = 0, 1, \ldots \tag{2.11.13}$$

Finally combining (2.11.17) and (2.11.8), we find

$$u(x) = \int_D G^0(x - x')\, q(x')\, dx' + \sum_{n=0}^{\infty} v^n e_n(x). \tag{2.11.14}$$

Actually we have constructed the Green's function $G(x, x')$ of the problem (see Sect. 2.6). We suggest that the reader verifies that

$$G(x, x') = G^0(x - x') - \int_S E(x, y)\, G^0(y - x')\, dy\,,$$

$$E(x, y) \overset{\text{def}}{=} \sum_{n=0}^{\infty} e_n(x)\, e^n(y)\,, \tag{2.11.15}$$

where $G(x, x') = G(x', x)$ and $G(x, x') = 0$, if x or x' belongs to S. With the help of the Green's operator, the solution of the problem is written in the form [compare Eq. (2.6.18)]

$$u = Gq - G\Phi h\,. \tag{2.11.16}$$

The expansion (2.11.8) is very convenient for obtaining an approximate solution. In fact, in this case it is sufficient to restrict ourselves to the first terms of the series. In particular, in the approximation by the first roots, the operator Φ is replaced by an operator Φ, given by the expression (2.10.7). This is equivalent to retaining the first three terms in the basis

$$e_0(x) = 1, \quad e_1(x) = \mathrm{e}^{\mathrm{i}\lambda x}, \quad e_2(x) = \mathrm{e}^{-\mathrm{i}\bar{\lambda} x} \tag{2.11.17}$$

with minimum frequency of oscillations and minimum rate of decreasing.

Taking into account that the function $\Phi_-(k)$ in (2.11.11) is determined to within an arbitrary multiplier, we have

$$\Phi_-(k) = k(k - \lambda)\,(k + \bar{\lambda})\,, \tag{2.11.18}$$

wherefrom we find the reciprocal basis, in accordance with (2.11.11),

$$e^0(k) = -\frac{(k - \lambda)\,(k + \lambda)}{|\lambda|^2}\,, \quad e^1(k) = \frac{k(k + \bar{\lambda})}{\lambda(\lambda + \bar{\lambda})}\,, \quad e^2(k) = \frac{k(k - \lambda)}{\bar{\lambda}(\lambda + \bar{\lambda})}\,. \tag{2.11.19}$$

In the x representation, to these functions there correspond the operators of differentiation at the point $x = 0$

$$e^0(\partial) = \frac{(\mathrm{i}\partial + \lambda)\,(-\,\mathrm{i}\partial + \lambda)}{|\lambda|^2}\,, \quad e^1(\partial) = \frac{\mathrm{i}\partial(\mathrm{i}\partial - \bar{\lambda})}{\lambda(\lambda + \bar{\lambda})}\,, \quad e^2(\partial) = \frac{\mathrm{i}\partial(\mathrm{i}\partial + \lambda)}{\bar{\lambda}(\lambda + \bar{\lambda})}\,. \tag{2.11.20}$$

Problem 2.11.3. Verify the relations (2.11.11).

Thus the coefficients of the expansion of v^n are now found is terms of the first derivatives of the functions $v_s(x)$ at $x = 0$.

It is also easy to find an explicit expression for the Green's function $G(x, x')$ of the second basic problem in the approximation by the first roots. To this end, taking into account (2.10.9), we construct $G^0(x)$ and then find $G(x, x')$ by means of (2.11.15).

Problem 2.11.4. Verify that $G(x, x')$ in the approximation by the first roots satisfies the conditions

$$G(x, 0) = \frac{\partial}{\partial x'} G(x, 0) = \frac{\partial^2}{\partial x'^2} G(x, 0) = 0 \,. \tag{2.11.21}$$

In the preceding section we showed the possibility of proceeding to a simpler approximating model (2.10.8), if we assume that the root λ is purely imaginary: $\lambda = \mathrm{i}l^{-1}$. In this case, as easily seen,

$$\begin{aligned} &e_0(x) = 1, \quad e_1(x) = \mathrm{e}^{-x/l}, \\ &e^0(\partial) = 1 + l\partial, \quad e^1(\partial) = -\, l\partial \,. \end{aligned} \tag{2.11.22}$$

The conditions (2.11.21) are replaced by weaker ones

$$G(x, 0) = \frac{\partial}{\partial x'} G(x, 0) = 0 \,. \tag{2.11.23}$$

For comparison, let us give the explicit expressions of the Green's function in the approximation by the first roots for complex λ

$$\begin{aligned} c_0 G(x, x') = {} & \frac{x + x' - |x - x'|}{2} \\ & + \mathrm{Im} \left\{ \frac{\bar{\lambda}}{\lambda^2 - \bar{\lambda}^2} \left[\frac{\bar{\lambda}}{\lambda} (\mathrm{e}^{\mathrm{i}\lambda(x+x')} - \mathrm{e}^{\mathrm{i}\lambda|x-x'|} \right. \right. \\ & - 2 \frac{\bar{\lambda}^2}{\lambda(\lambda + \bar{\lambda})} (1 - \mathrm{e}^{\mathrm{i}\lambda x}) (1 - \mathrm{e}^{\mathrm{i}\lambda x'}) \\ & \left. \left. + 2 \frac{\lambda}{\lambda + \bar{\lambda}} (1 - \mathrm{e}^{\mathrm{i}\lambda x}) (1 - \mathrm{e}^{-\mathrm{i}\bar{\lambda} x'}) \right] \right\} \end{aligned} \tag{2.11.24}$$

and for pure imaginary $\lambda = \mathrm{i}l^{-1}$

$$\begin{aligned} c_0 G\,(x, x') = {} & \frac{x + x' - |x - x'|}{2} \\ & + \frac{1}{2} [\mathrm{e}^{-(x+x')/l} - \mathrm{e}^{-|x-x'|/l}] + (1 - \mathrm{e}^{-x/l}) (1 - \mathrm{e}^{-x'/l}) \,. \end{aligned} \tag{2.11.25}$$

The fulfillment of the conditions (2.11.21) and (2.11.23) and also of the symmetry with respect to the arguments x, x' are directly verified.

The solution of the second basic problem for the chain with interaction of N neighbors is constructed in an analogous way. In this case, the basis $\{e_n(x)\}$ consists of N elements and the displacement $u_s(n)$ is given at N points of the boundary region S. The general problem of constructing the reciprocal basis and the expansion of functions of the discrete argument is considered in Appendix 3.

Let us proceed to the first basic problem

$$\int_{-l}^{\infty} \Phi(x-x')\, u(x')\, dx' = q_D(x), \quad x \in D\,, \tag{2.11.26}$$

$$\int_{-l}^{l} \Gamma(x,x')\, u(x')\, dx' = q_S(x), \quad x \in S\,. \tag{2.11.27}$$

First of all let us note two distinctions of the given problem from that considered above. Firstly, the boundary conditions receive a contribution of displacements in the region $\tilde{S}(-l \leqslant x \leqslant l)$ i.e. in the double boundary region S. Secondly, the boundary operator Γ defined on these displacements has a non-difference kernel $\Gamma(x, x')$. The consequence of the first distinction is the necessity of introducing a new biorthogonal basis for functions with support in $\tilde{S}$. The second, more significant distinction, leads to the fact that in the general case the first basic problem is equivalent to the solution of an infinite system of linear algebraic equations. Therefore a solution in the closed form can be obtained only in the approximation by the first roots.

Let us search for a solution in the form (2.11.4), as we did in the second basic problem. However, the solution $v(x)$ of the homogeneous equation, generally speaking, grows linearly now, as $x \to \infty$, and is determined to within an additive constant. In connection with this it is necessary to replace the first term in (2.11.9) by x/l and widen the basis to the region $\tilde{S}$. Thus we take the new basis $\{e_m\}$ in the form

$$\tilde{e}_0(x) = \frac{x}{l}, \quad \tilde{e}_m(x) = \mathrm{e}^{ikx}, \quad x \in \tilde{S}\,. \tag{2.11.28}$$

To find the reciprocal basis $\{\tilde{e}^m\}$, we can use the expressions (2.11.11), replacing $\Phi_-(k)$ by $\Phi(k)$.

This is connected with the fact that the support of $\Phi(x)$ (as distinct from the support of $\dot{\Phi}_-(k)$) is contained in $\tilde{S}$, as supports of the functions $\tilde{e}^m(x)$.

Problem 2.11.5. Show that

$$\tilde{e}^0(x) = -\frac{l}{c_0}\int_x^l \Phi(x')\, dx', \quad \tilde{e}^m(x) = \frac{\mathrm{i}e^{-\mathrm{i}k_m x}}{\Phi'(k_m)}\int_x^l \mathrm{e}^{\mathrm{i}k_m x'}\, \Phi(x')\, dx', \quad x \in \tilde{S}\,. \tag{2.11.29}$$

Correspondingly, when constructing the particular solution $u^0(x)$ of (2.11.26) in the form (2.11.7) we shall use the Green's function $G(x)$ for the unbounded medium which is given by (2.9.12).

Substituting $u(x)$ in the boundary condition (2.11.27), we obtain

$$\int_S \Gamma(x, x')\, v(x')\, dx' = q(x), \quad x \in S\,, \tag{2.11.30}$$

$$q(x) = q_D(x) - \int_S \Gamma(x - x')\, u^0(x')\, dx', \quad x \in S\,. \tag{2.11.31}$$

But $v(x)$ is the solution of the homogeneous equation $\Phi v = 0$ and hence

$$v(x) = \sum_m v^m \tilde{e}_m(x), \quad x \in S \cup D\,, \tag{2.11.32}$$

where the $\tilde{e}_m(x)$'s are analytical continuations of the elements of the basis $\{\tilde{e}_m\}$ into the region $S \cup D$.

Let us also take into account that $\Gamma(x, x')$ as a function of x and $q(x)$ have their support in S and expand them in terms of the basis $\{e^n(x)\}$ with support in S. As a result, after the substitution of (2.11.32) in (2.11.30), we obtain the infinite system of equations for the coefficients v^m

$$\sum_m \Gamma_{nm} v^m = q_n\,, \tag{2.11.33}$$

where

$$\Gamma_{nm} = \int_{-l}^{0} dx e_n(x) \int_{-l}^{l} \Gamma(x, x')\, \tilde{e}_m(x')\, dx', \quad q_n = \int_{-l}^{0} e_n(x)\, q(x)\, dx\,. \tag{2.11.34}$$

We recall that, according to (2.6.6)

$$\Gamma(x, x') = S(x)\, \Phi_v(x, x')\,, \tag{2.11.35}$$

where the kernel of the operator Φ_v is connected with the kernel of the operator of elastic bonds Ψ by the relations, which follow from (2.6.3)

$$\begin{aligned} \Phi_V(x, x') &= \psi_V(x)\, \delta(x - x') - V(x)\, \Psi(x, x')\, V(x')\,, \\ \psi_V(x) &= V(x) \int_V \Psi(x, x')\, dx'\,, \\ V(x) &= S(x) + D(x), \end{aligned} \tag{2.11.36}$$

and hence it satisfies the condition

$$\int_V \Phi_V(x, x')\, dx' = 0\,. \tag{2.11.37}$$

In the general case, it is necessary to assume that a distortion of elastic bonds in the boundary region S takes place, i.e. $\Psi(x, x')$ with $x, x' \in S$ does not coincide with the elastic bonds $\Psi(x - x')$ in the homogeneous medium. For example, in a crystal, the perturbation of the force constants is caused by changing the interaction potential and the distances between atoms in the boundary region. Thus, we set

$$\Psi(x, x') = \Psi(x - x') + \Psi^1(x, x'), \tag{2.11.38}$$

where $\Psi'(x, x')$ is the perturbation of the bonds, which differs from zero only when $x, x' \in S$. Accordingly, the boundary operator Γ is divided into two components

$$\Gamma(x, x') = \Gamma^0(x, x') + \Gamma^1(x, x'). \tag{2.11.39}$$

We can now calculate the matrix Γ_{nm}. Taking into account that $\Psi(x, x') = 0$ when $|x - x'| > l$ we find, after simple transformation,

$$\Gamma_{nm} = \Gamma^0_{nm} + \Gamma^1_{nm}, \tag{2.11.40}$$

where

$$\Gamma^0_{nm} = \int_{-l}^{0} dx e_n(x) \int_{-l}^{l+x} \Psi(x') [\tilde{e}_m(x) - \tilde{e}_m(x - x')] \, dx', \tag{2.11.41}$$

$$\Gamma^1_{nm} = \int_{-l}^{0} \int_{-l}^{0} \Psi(x, x') \, e_n(x) \, \tilde{e}_m(x') \, dx \, dx'. \tag{2.11.42}$$

For $e_0(x) = 1$ and $\tilde{e}_0(x) = x/l$, we have

$$\Gamma^0_{00} = \frac{1}{l} \int_{l}^{0} dx \int_{-l}^{l+x} x' \Psi(x') \, dx' = -\frac{c_0}{l}, \tag{2.11.43}$$

where

$$c_0 = \int_0^l x^2 \Psi(x) \, dx, \tag{2.11.44}$$

This expression, in accordance with (2.4.41), coincides with the elastic modulus in the zeroth approximation for the homogeneous medium. Thus, in the general case,

$$\Gamma_{00} = -\frac{c_0 + c_1}{l}, \tag{2.11.45}$$

where the quantity

$$c_1 = -\int_0^l \int_0^l x' \Psi^1(x, x') \, dx \, dx' \tag{2.11.46}$$

constituting the correction to the elastic modulus is due to the perturbation of the elastic bonds.

Let us represent the expression for Γ_{0m} with $m \neq 0$, in the form

$$\Gamma_{0m} = \int_S dx \int_V \Phi_V(x, x')\, \tilde{e}_m(x')\, dx' , \tag{2.11.47}$$

where the function $\tilde{e}_m(x)$ is analytically continued in the region V and (by definition) satisfies the equation

$$\int_V \Phi(x - x')\, \tilde{e}_m(x')\, dx' = 0, \quad x \in D . \tag{2.11.48}$$

This equation can be written in the form

$$\int_V \Phi_V(x, x')\, \tilde{e}_m(x')\, dx' = 0, \quad x \in D , \tag{2.11.49}$$

since, if $x \in D$, the operators Φ and Φ_V coincide. This can be easily verified by comparing (2.11.36) and (2.4.37) and taking into account that

$$\Psi(x, x') = \Psi(x - x'), \quad x \in D . \tag{2.11.50}$$

Integrating (2.11.49) over the region D, adding the result to (2.11.47) and changing the order of integration, we obtain

$$\Gamma_{0m} = \int_V dx'\, \tilde{e}_m(x) \int_V \Phi_V(x, x')\, dx', \quad m \neq 0 . \tag{2.11.51}$$

Comparison with (2.11.37), leads to the important result:

$$\Gamma_{0m} = 0, \quad m \neq 0 . \tag{2.11.52}$$

This means that the system (2.11.33) is split into an independent equation for v^0 and a system of equations for the other v^n.

The equation

$$\frac{c_0 + c_1}{l} v^0 = q^0 . \tag{2.11.53}$$

where

$$v^0 = \frac{l}{c_0} \int_{-l}^{l} \Psi(x) \int_0^x v(x')\, dx', \quad q^0 = \int_{-l}^{0} q(x)\, dx , \tag{2.11.54}$$

obviously expresses the equilibrium condition. If $q^0 = 0$, i.e. the system of forces $q(x)$ is balanced, then $v^0 = 0$ and, according to (2.11.32) and (2.11.28), $v(x) \to 0$, as $x \to \infty$.

For the homogeneous strain $v(x) = \varepsilon x$ ($\varepsilon = \text{const}$) we have

$$(c_0 + c_1)\,\varepsilon = q^0\,, \tag{2.11.55}$$

from which it follows that $c_0 + c_1$ may be considered as an effective elastic modulus of the boundary region S in the zeroth approximation.

We shall not investigate the system of equations for v^n when $n > 0$, but we proceed directly to the approximation by the first roots. For simplicity, let us consider a model with a pure imaginary root $\lambda = il^{-1}$. The bases $\{e_n\}$ and $\{e^{n'}\}$ are given by the expressions (2.11.22) and for the bases $\{\tilde{e}_m\}$, $\{\tilde{e}^{m'}\}$ we have

$$\begin{aligned} &\tilde{e}_0(x) = \frac{x}{l}, \quad \tilde{e}_1(x) = \mathrm{e}^{-x/l}, \quad x \in \tilde{S}\,, \\ &\tilde{e}^0(\partial) = l\partial(1 + l\partial), \quad \tilde{e}^1(\partial) = l^2\partial^2\,, \end{aligned} \tag{2.11.56}$$

where, as above, the operator ∂ denotes the differentiation at the point $x = 0$.

According to (2.11.43), (2.11.52) and (2.11.41) we find

$$\begin{aligned} &\Gamma^0_{00} = -\frac{c_0}{l}, \quad \Gamma^0_{01} = 0, \\ &\Gamma^0_{10} = -e\int_0^l (1 - \mathrm{e}^{-x/l})\, x\Psi(x)\, dx, \quad \Gamma^0_{11} = \frac{e^2 l}{2}\int_0^l (1 - \mathrm{e}^{-x/l})^2\, \Psi(x)\, dx\,. \end{aligned} \tag{2.11.57}$$

In the general case, it is necessary to add corrections caused by the perturbation of boundary bonds to these values.

The boundary conditions (2.11.33) take the form

$$\begin{aligned} &l\Gamma_{00}v'(0) + l^2\Gamma_{00}v''(0) = q_0, \\ &l\Gamma_{10}v'(0) + l^2(\Gamma_{10} + \Gamma_{11})\, v''(0) = q_1\,. \end{aligned} \tag{2.11.58}$$

It is significant that all the quantities contained in (2.11.58) are expressed explicitly in terms of the parameters of the exact model. If we consider the given model as the simplest one-dimensional variant of the couple-stress theory, i.e. from a purely phenomenological point of view, then the coefficients Γ_{nm} and the force moments q_n are to be considered as given quantities. But then difficulties concerning their physical interpretation, arise.

Problem 2.11.6. Construct an explicit expression for the Green's function $G(x, x')$ in the approximation under consideration and show that $G(x, x')$ satisfies the boundary conditions

$$\Gamma_{00}\tilde{e}^0(\partial)\, G(x, 0) = 0, \quad [\Gamma_{10}\tilde{e}^0(\partial) + \Gamma_{11}\tilde{e}^1(\partial)]\, G(x, 0) = 0\,, \tag{2.11.59}$$

where the operators act on the second argument of $G(x, x')$.

For the more realistic but also more complicated approximate model with complex root λ it is necessary to write three boundary conditions in which there appear the first three derivatives of the displacements. This corresponds to the sixth order differential operator (2.10.7).

In some cases it may be essential to take into account the oscillating character of the solution which corresponds to complex λ. In particular, this permits us to describe qualitatively correctly the possibility of nonmonotonic variation of the energy of interaction of a defect with the boundary. Such an effect cannot take place when the roots are pure imaginary.

In conclusion, we point out that we considered a one-sided boundary problem, i.e. we assumed the medium to be bounded in a halfplane. It is not difficult to construct a formal generalization to the case of a bounded medium, but this is not required for the physical meaning of the model. In fact, due to a natural assumption about the smallness of the parameter l in comparison with the characteristic length parameter of the medium, we have a typical example of a boundary layer. The exact solution differs noticeably from the solution in zeroth approximation only over distances of the order of l. Therefore the two-sided boundary problem should be solved only in the zeroth approximation and in the approximation by the first roots it is sufficient to restrict oneself by the model of a semibounded medium.

2.12 Notes*

Extensive literature is devoted to one-dimensional chains. The basic results are contained in the excellent monograph by Brillouin and Parodi [B5.3]. The idea of quasi-continuum was introduced independently by Krumhansl [B3.19], Kunin [B3.26] and Rogula [B6.36].

For further references the reader is referred to the works [B3.1, 14, 17, 28, 29, 31, 32].

*The citations refer to the bibliography to be given in [1.12]

3. Medium of Complex Structure

A movement of a particle in a medium of complex structure is described by the displacement of the mass center of the particle and by additional internal degrees of freedom: microrotations and microdeformations. In this chapter starting from simple models, we develop a general theory of media of complex structure. Essential attention is paid to new physical effects, approximate models and a transition to the classical local models.

3.1 Basic Micromodels

The diatomic chain or the corresponding mechanical model with masses and elastic bonds (Fig. 3.1) can serve as a simplest model of medium of complex structure.

It is assumed that the chain is separated into elementary cells, each consisting of two masses, which are connected by elastic bonds with a certain number of neighbors (the interaction of nearest neighbors is shown symbolically in the diagram). Let us first consider a homogenous diatomic chain in which, the structure and characteristics of each cell are the same.

Let n be the number of a cell, m_j ($j = 1{,}2$) be the masses of the particles in the cell, $w(n, j)$ be the displacement of the j — th particle in the n — th cell, $f(n, j)$ be an external force acting on the particle. Then the equations of motion of particles in the n — th cell have the form

$$m_j \ddot{w}(n,j) + \sum_{n'j'} \Phi(n-n', j, j')\, w(n', j') = f(n,j)\,, \tag{3.1.1}$$

where $\Phi(n - n', j, j')$ are force constants, which are determined by the properties of the elastic bonds. Their physical meaning follows directly from the equations: if a displacement of the unit value of the particle (n', j') is the only one which differs from zero, then an external force $\Phi(n - n', j, j')$, which compensates the reaction of elastic bond, must be applied to the particle (n, j). The dependence of the force constants on the difference $n - n'$) is the consequence of the identical structures of the elementary cells. It is assumed that action at a distance is bounded, i.e. $\Phi(n\, j, j') = 0$ when $n > N$.

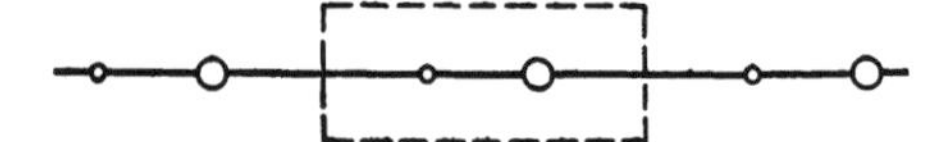

Fig.3.1. The diatomic chain

The equations of motion are the Euler's equations for the Lagrangian

$$L \stackrel{\text{def}}{=} \frac{1}{2} \sum_{nj} m_j \dot{w}^2(n,j)$$

$$- \frac{1}{2} \sum_{nn'jj'} w(n,j)\, \Phi(n-n',j,j')\, w(n',j') + \sum_{nj} f(n,j)\, w(n,j)\,, \tag{3.1.2}$$

wherefrom it follows that the force constants must satisfy the symmetry condition:

$$\Phi(n-n',j,j') = \Phi(n'-n,j',j)\,. \tag{3.1.3}$$

Finally, the force constants must satisfy additional conditions, due to the necessity of invariance of energy with respect to translation, i.e. with respect to transformation

$$w(n,j) \rightarrow w(n,j) + w_0\,, \tag{4.1.4}$$

where w_0 is an arbitrary constant. It can be shown that this is equivalent to the condition of invariance of the equations of motion with respect to translation.

Substituting (3.1.4) in (3.1.1), we find

$$\sum_{n'j'} \Phi(n-n',j,j') = 0\,. \tag{3.1.5}$$

These relations may be also interpreted as the definition of the self-action constants [compare with (2.1.11)]

$$\psi(j) \stackrel{\text{def}}{=} \Phi(0,j,j) = - \sum_{\substack{n\neq 0 \\ j'\neq j}} \Phi(n,j,j')\,. \tag{3.1.6}$$

Problem 3.1.1. Show that (3.1.5) implies the representation

$$\Phi(n-n',j,j') = \psi(j)\, \delta(n-n')\, \delta(j-j') - \Psi(n-n',j,j')\,, \tag{3.1.7}$$

where the $\Psi(n-n',j,j)$ have the meaning of stiffness of elastic bonds and may be given arbitrarily (except the value $\Psi(0,j,j) = 0$).

The physical sense of the model of the medium of complex structure becomes more clear, if one passes to new variables: the displacement $u(n)$ of the center of mass of the n — th cell and the relative displacement $\eta(n)$ of particles in the cell, defined by the relations

$$u(n) \stackrel{\text{def}}{=} \frac{1}{m} [m_1 w(n,1) + m_2 w(n,2)],$$

$$\eta(n) \stackrel{\text{def}}{=} I_*^{-1} [m_1 \xi_1 w(n,1) + m_2 \xi_2 w(n,2)]\,,$$

$$m \stackrel{\text{def}}{=} m_1 + m_2, \quad I_* \stackrel{\text{def}}{=} m_1 \xi_1^2 + m_2 \xi_2^2\,. \tag{3.1.8}$$

Here ξ_j are the coordinates of the particles in the cell with respect to the coordinate of the center of mass, i.e. $m_1\xi_1 + m_2\xi_2 = 0$. The weight factors in the expressions for $u(n)$ and $\eta(n)$ are chosen in such a way that these quantities keep their sense under any ratios of m_1 and m_2 and in particular, in the limiting case $m_2 \ll m_1$ The dimensionless quantity $\eta(n)$ characterizes the average relative deformation of the elementary cell with number n and, as easily verified, is equal to the ratio of the difference of the displacements $w(n, 2) - w(n, 1)$ to the distance $\xi_2 - \xi_1$ between the particles. We shall call $\eta(n)$ the microdeformation in contrast to the macrodeformation, determined by the displacements of the centers of mass.

Problem 3.1.2. Show that generalized forces which correspond to $u(n)$ and $\eta(n)$ are

$$q(n) \overset{\text{def}}{=} f(n, 1) + f(n, 2), \quad \mu(n) \overset{\text{def}}{=} \xi_1 f(n, 1) + \xi_2 f(n, 2) . \tag{3.1.9}$$

In terms of the new variables, the equations of motion take the form

$$\begin{aligned} &m\ddot{u}(n) + \sum_{n'} \Phi^{00}(n - n')\, u(n') + \sum_{n'} \Phi^{01}(n - n')\, \eta(n') + q(n) , \\ &I_* \ddot{\eta}(n) + \sum_{n'} \Phi^{10}(n - n')\, u(n') + \sum_{n'} \Phi^{11}(n - n')\, \eta(n') = \mu(n) , \end{aligned} \tag{3.1.10}$$

where the matrix $\Phi^{ss'}(n)$ $(s, s' = 0, 1)$ is expressed in terms of the force constants $\Phi(n, j, j')$.

Problem 3.1.3. Derive explicit formulae for the transformations $\Phi(n, j, j') \leftrightarrow \dot{\Phi}^{ss'}(n)$. Show that the relations

$$\Phi^{ss'}(n) = \Phi^{ss'}(-n) , \tag{3.1.11}$$

$$\sum_n \Phi^{00}(n) = 0, \quad \sum_n \Phi^{10}(n) = \sum_n \Phi^{01}(n) = 0 . \tag{3.1.12}$$

correspond to the conditions (3.13) and (3.1.5).

It is intuitively obvious that the dynamical variables $u(n)$ and $\eta(n)$ play different roles in the equations of motion. In fact, it will be shown below that, in the region of low frequencies and not too short wavelengths, the main role is played by the displacement $u(n)$ of the centers of mass, i.e. the cell moves in principle as a single unit. Conversely in the region of hiğh frequencies, the relative displacements of particles in a cell are the significant quantities. Ultimately, this is due to a different law of transformation of variables $u(n)$ and $\eta(n)$ with respect to translation: $u(n)$ transforms into $u(n) + u_0$, and microdeformation at the same time is left invariant.

The model under consideration may posses an additional symmetry. Let a deformation be such that the new configuration of the system is similar to the initial one with the coefficient of similarity $1 + \varepsilon_0$ $(\varepsilon_0 \ll 1)$. Such a deformation will be called homogeneous. We leave it to the reader to verify that to the homogenous deformation there corresponds the displacement (to within a translation)

$$w_0(n,j) \stackrel{\text{def}}{=} \varepsilon_0(an + \xi_j)\,, \tag{3.1.13}$$

where a is the distance between the mass centers of neighboring cells. In terms of the variables u, η, taking into acount (3.1.8) we have

$$u_0(n) = \varepsilon_0 an, \quad \eta_0 = \varepsilon_0\,. \tag{3.1.14}$$

If the structure of the system is such that the homogeneous deformation is due to forces applied only on the boundaries of the chain, then (3.1.13) and (3.1.14) must identically satisfy the equations of motion with vanishing right-hand sides. In this case, we say that the equations of motion are invariant with respect to homogeneous deformation.

Substitution of (3.1.13, 14) into (3.1.1, 10) yields the corresponding conditions for force constants (taking into account the translational invariance), i.e.

$$\sum_{nj'} (\xi_{j'} - an)\,\Phi(n,j,j') = 0 \tag{3.1.15}$$

or

$$a\sum_n n\Phi^{10}(n) - \sum_n \Phi^{11}(n) = 0\,, \tag{3.1.16}$$

while the first equation (3.1.10) is satisfied identically.

Obviously, in contrast to the translational invariance, the invariance of the equations with respect to homogeneous deformation does not imply the invariance of the elastic energy.

Several generalizations of the given model are possible. Instead of a diatomic chain, the N-atom chain may be considered. In that case, it is natural to introduce the displacement of mass centers $u(n)$ and the micro-deformations $\eta_p(n)$, $p = 1, \ldots, N-1$ of different orders as kinematic variables. An algorithm for passing to such variables, which is useful for a finite as well as an infinite number of degrees of freedom of the cell, will be constructed below. This will permit one to include in the investigation the case of a continuous distribution of mass in the cell as well.

A departure from the assumption about the periodic cell structure leads to an inhomogenous chain of complex strucutre. However, the description of such a system in terms of the medium of complex structure is expedient only if the parameters of a cell vary sufficiently slowly. Otherwise, the idea of a cell becomes conventional and then there is no reason not to use the model of the medium of simple structure.

Finally, in the case when the radius of action at a distance is much larger than the size of a cell, one can pass from the discrete to the continuum model of the medium of complex structure.

Let us now consider models of another type, for which the internal degrees of freedom are given a priori. The Cosserat model, schematically shown in Fig. 3.2, may serve as the simplest example. Masses m with finite moments of inertia I_* are connected by weightless rods, with a finite number of neighbours (in the diagram-with only nearest neighbours). The transverse displacement $u(n)$ of the masses in one plane and the angle of rotation $\eta(n)$ of the masses in the same plane, are the kinematic variables. The corresponding force variables are the transverse forces $q(n)$ and moments $\mu(n)$.

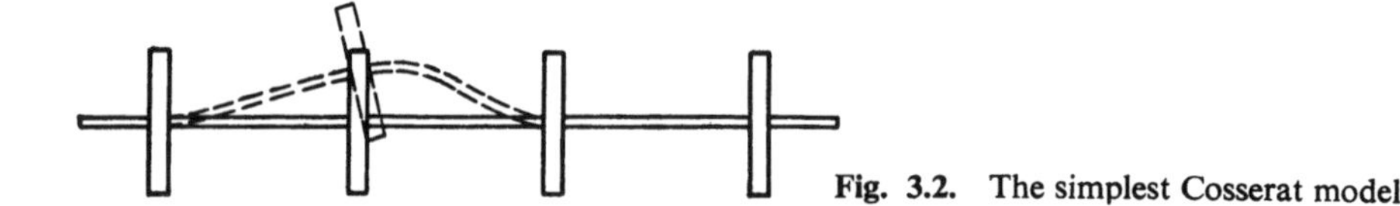

Fig. 3.2. The simplest Cosserat model

In the notations accepted under the assumption that the system is homogenous, the form of the equations of small oscilations of the Cosserat model coincides with that of (3.1.10) for the diatomic chain. The matrix of force constants $\Phi^{ss'}(n-n')$ is determined by the elastic properties of the rods and must satisfy the symmetry conditions (3.1.11) and the conditions (3.1.12) arising from the requirement of translational invariance.

However, unlike the case of a chain, there is an additional transformation property for the new variables $u(n)$ and $\eta(n)$

$$u(n) \to u(n) + \eta_0 an, \quad \eta(n) \to \eta(n) + \eta_0, \tag{3.1.17}$$

which corresponds to an infinitesimal rotation of the system by the angle η_0, relative to which the elastic energy and the equations of motion must be invariant. Comparison with (3.1.14) shows that this transformation is formally identical with the homogenous deformation of the chain and hence, the relations (3.1.16) for the force constants must be fulfilled. However, the additional requirement of invariance with respect to rotation of not only the equations of motion but also of the elastic energy (this condition, of course, need not be fulfilled for homogenous deformation) leads to additional conditions on the force constants. In Sect. 3.9 it will be shown that the complete set of conditions for the Cosserat model has the form

$$\begin{aligned} &a\sum_n n^2\Phi^{00}(n) - 2\sum_n n\Phi^{01}(n) = 0\,, \\ &a\sum_n n\Phi^{10}(n) - \sum_n \Phi^{11}(n) = 0\,, \end{aligned} \tag{3.1.18}$$

i.e. one more condition appears in comparison with (3.1.16).

Let us note that the physical meanings of the conditions (3.1.16) and (3.1.18) are essentially different. The first one corresponds to the choice of the special model of the chain while the second one is universal, i.e. it must be satisfied for any Cosserat model.

In the case of Cosserat model a question naturally arises: what is the analog of the homogenous deformation for the chain? It is not difficult to realize that it is the pure bending, given by the relations

$$u_0(n) = \alpha a^2 n^2, \quad \eta_0 = 2\alpha a n \tag{3.1.19}$$

which plays this role.

The invariance of the euqations of motion (but not energy) with respect to the pure bending means that these expressions must identically satisfy the homogenous equations, but this is possible only if one adds one more condition (besides (3.1.18)) on the force constants

$$a \sum_n n^2 \Phi^{10}(n) - 2 \sum_n n \Phi^{11}(n) = 0 . \tag{3.1.20}$$

We tacitly assumed above that, in the rod system, an initial longitudinal strain is absent. If the latter exists and is sufficiently large, then the picture changes. Under transverse displacements, besides the elastic forces due to the stiffness of the rods, there appear vertical components of strain, which are analogous to those in the case of the string. In order to remain in the frame work of linear approximation, it is necessary to assume that the additional strains are small in comparison with the initial ones. In the transformations corresponding to the rotation, it is necessary now to consider the contribution of longitudinal displacement, and an additional term, proportional to the strain will appear in the relations (3.1.16).

A more general model is shown in Fig. 3.3. Here the kinematic variables consist of the longitudinal and transverse displacements and the angle of rotation. Instead of the system (3.1.10), we have a system of three equations. And the equation for the longitudinal displacements is split off only in the longwave approximation.

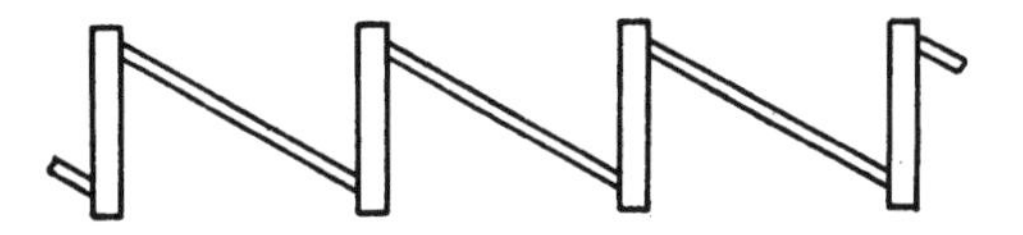

Fig.3.3. Model with longitudinal and transverse displacements

As in the case of the chain, further generalizations may made by increasing the number of internal degrees of freedom (they may also be of nonmechanical nature), due to transition to the continuum, and a rejection of the assumption of homogeneity and invariance with respect to translations and rotations.

Later, when developing the general theory of medium with microstructure, we will illustrate it by examples of concrete models, mentioned above.

3.2 Collective Cell Variables

We saw that models of the type of diatomic chain could be described in two ways: in terms of media of simple and complex strcutures. Let us consider in detail the relation between these different representations of one and the same model.

We assume that the medium is macroscopically homogenous, i.e. it may be divided into elementary cells with identical inertial and elastic properties. Let us introduce a local cell coordinate ξ with the origin at the mass center of the cell and denote the mass density in the cell by $\rho(\xi)$

It is easy to see that the natural generalization of the Lagrangian (3.1.2), to the case of an arbitrary distribution of masses in the cell, has the form

$$
\begin{aligned}
L = {} & \frac{1}{2} \sum_{n} \int \rho(\xi)\, \dot{w}^2(n, \xi)\, d\xi \\
& - \frac{1}{2} \sum_{nn'} \iint w(n, \xi)\, \Phi(n - n', \xi, \xi')\, w(n', \xi')\, d\xi\, d\xi' \\
& + \sum_{n} \int f(n, \xi)\, w(n, \xi)\, d\xi\,.
\end{aligned} \tag{3.2.1}
$$

In order to pass to the discrete distribution, it is sufficient to set

$$
\begin{aligned}
& \rho(\xi) = \sum_{j} m_j \delta(\xi - \xi_j)\,, \\
& \Phi(n, \xi, \xi') = \sum_{jj'} \Phi(n, j, j')\, \delta(\xi - \xi_j)\, \delta(\xi' - \xi_{j'})\,.
\end{aligned} \tag{3.2.2}
$$

In this case the contribution in (3.2.1) will be made by the displacements and the forces at points (n, ξ_j) only, and the Lagrangian takes the form (3.1.2).

Regarding the coordinate ξ as a parameter and using Sect. 2.2, which gives an algorithm for the transition to the quasicontinuum representation, one can replace all the functions of the discrete argument n by their x and k representations. The corresponding expression for the Lagrangian in x representation, is obtained from (3.1.2) by the obvious replacement of sums by integrals.

For what follows, it is convenient to introduce collective cell variables analogously to the case of the diatomic chain. When constructing the algorithm for the transition to the new variables, we will be guided by the following considerations.

There must exist a one-to-one correspondence between the new and the old variables. In particular, the number of degrees of freedom must be retained for the cell with a discrete distritution of masses. The usual expansion of the kinematic variables in power moments and the force variables into multipoles, does not satisfy this condition. It is natural also to demand that the new variables, at least the first moments, possess a clear physical meaning.

In order to satisfy all these requirements, let us construct a special biorthogonal functional basis, depending on the distribution of masses in the cell.

The coefficients of the expansion of kinematic and force variables in the direct and the reciprocal bases, respectively will be the required new variables.

Let us define the moments of inertia of order s $(s = 0, 1, \ldots\ldots)$

$$\rho^{(s)} \overset{\text{def}}{=} \frac{1}{a} \int \rho(\xi)\, \xi^s\, d\xi \quad \text{or} \quad \rho^{(s)} = \frac{1}{a} (\rho, \xi^s), \tag{3.2.3}$$

where a is the demension of the cell and the brackets denote the corresponding scalar product, and let us introduce the diagonal matrices

$$I^{ss'} = \rho^{(2s)} \delta^{ss'}, \quad I^{-1}_{ss'} = \frac{1}{\rho^{(2s)}} \delta_{ss'} . \tag{3.2.4}$$

With the help of the known procedures for orthogonalization, let us construct an orthonormal system of polynominals $e^s(\xi)$ with weight $\rho(\xi)$ and a reciprocal system of functions $e_s(\xi)$, defining them by the relations

$$(\rho e^s, e^{s'}) = a I^{ss'}, \quad (e_s, e^{s'}) = \delta_s^{s'} . \tag{3.2.5}$$

Let us recall the general scheme of step-by-step construction of polynomials $e^s(\xi)$. $e^0(\xi)$ is chosen to be a constant which is normalized by the condition $(\rho e^0, e^0) = a\rho^{(0)}$. The next polynomial $e^1(\xi)$ is sought in the form of a linear combination $\alpha_0 e^0 + \alpha_1 \xi$, and the coefficients α_0, α_1 are found from the conditions of orthogonality of e^1 to e^0 and its normalization. For $e^2(\xi)$ we take the linear combination $\alpha_0 e^0 + \alpha_1 e^1(\xi) + \alpha_2 \xi^2$ etc. It is directly verified that the reciprocal basis $e_s(\xi)$ is expressed in terms of $e^s(\xi)$ by the formulae

$$e_s(\xi) = \frac{1}{a} \rho(\xi)\, I^{-1}_{ss'}\, e^{s'}(\xi) . \tag{3.2.6}$$

Here and later on, the summation over coinciding upper and lower indices is assumed and the sign of summation is omitted (Einstein's summation convention).

Thus, the constructed biorthogonal basis is completely defined by $\rho(\xi)$. In particular when $\rho = \text{const.}$, the polynomials $e^s(\xi)$, coincide with the Legendre polynomials to within the normalization. It is essential that in the case of a finite discrete distribution of masses, the basis turns out to be finite automatically: $s = 0, 1, \ldots, N-1$ where N is the number of degrees of freedom of the cell.

Problem 3.2.1. Show that for arbitrary $\rho(\xi)$ the first elements of the bases, have the form

$$e^0(\xi) = 1, \quad e^1(\xi) = \xi, \quad e_0(\xi) = \frac{\rho(\xi)}{a\rho^{(0)}}, \quad e_1(\xi) = \frac{\xi\rho(\xi)}{a\rho^{(2)}} . \tag{3.2.7}$$

(Hint: take into account that $(\rho, \xi) = 0$ due to the special choice of the origin of the local coordinate system).

Each sufficiently smooth function of the coordinate ξ can be expanded in a series in the bases $e^s(\xi)$ or $e_s(\xi)$. Concerning the conditions of completeness of the basis and the convergence of the series, we refer the readers to special literature on orthogonal polynomials with weights [3.1]. We only point out that these conditions on $\rho(\xi)$ and the restrictions on functions are not restrictive from the physical point of view; we shall consider them to be fulfilled. Then, in the class of admissible functions, the condition of completeness of the biorthogonal basis has the form

$$e^s(\xi)\, e_s(\xi') = \delta(\xi - \xi') . \tag{3.2.8}$$

Let us represent the expansion of all the functions of the coordinate ξ in the elements of the basis, in the form

$$\begin{aligned} &w(n, \xi) = w_s(n)\, e^s(\xi), \quad f(n, \xi) = f^s(n)\, e_s(\xi) , \\ &\Phi(n, \xi, \xi') = \Phi^{ss'}(n)\, e_s(\xi)\, e_{s'}(\xi') . \end{aligned} \tag{3.2.9}$$

The coefficients of the expansion are found by scalar multiplication of each equality by an arbitary element of the basis, reciprocal with respect to one in which the expansion is carried out. In view of the second of the conditions of orthogonality (3.2.5), we have

$$\begin{aligned} &w_s(n) = (e_s, w), \quad f^s(n) = (e^s, f) , \\ &\Phi^{ss'}(n) = \iint \Phi(n, \xi, \xi')\, e^s(\xi)\, e^{s'}(\xi')\, d\xi\, d\xi' . \end{aligned} \tag{3.2.10}$$

The first coefficients of the expansion have a simple physical meaning. Taking into account (3.2.7), we find that w_0 is the displacement of the mass center of the cell, w_1 is its average deformation (microdeformation), f^0 is the average force density, f^1 is the average density of force dipoles (micromoments). Other coefficients of the expansion correspond to microdeformations and micromoments of higher orders.

Problem 3.2.2. Show that all the microdeformations are invariant with respect to translations. [Hint: Use (3.2.8) and take into account that $e^0(\xi) = 1$].

In terms of the new variables, the Lagrangian (3.2.1), in the notations of Sect. 2.2 invariant with respect to n, x and k representations, takes the form

$$2L = \langle \dot{w}_s | I^{ss'} | \dot{w}^{s'} \rangle - \langle w_s | \Phi^{ss'} | w_{s'} \rangle + 2\langle f^s | w_s \rangle . \tag{3.2.11}$$

The corresponding equations of motion in the (x, t) and (k, ω) representations have the form (dependence of w_s and f^s on t or ω is not indicated explicitly)

$$I^{ss'} \ddot{w}_{s'}(x) + \int \Phi^{ss'}(x - x')\, w_{s'}(x')\, dx' = f^s(x) , \tag{3.2.12}$$

$$- \omega^2 I^{ss'} w_{s'}(k) + \Phi^{ss'}(k)\, w_{s'}(k) = f^s(k)\,. \tag{3.2.13}$$

We recall that here $k \in B$, where B is the circle of radius a^{-1}.

The study of these equations will be carried out later, and here let us dwell upon another possibility of describing the initial model in terms of the medium of simple structure: we now take the model to be inhomogenous. With this in mind, let us introduce instead of $w(n, \xi)$ and $f(n,\xi)$ the variables $w(x)$ and $f(x)$ which depend on one spatial coordinate x. In these variables the Lagrangain has the form

$$\begin{aligned} L = \frac{1}{2}\, \rho(x)\, \dot{w}^2(x)\, dx \\ - \frac{1}{2} \int\!\!\int \dot{w}(x)\, \Phi(x, x')\, \dot{w}(x')\, dx\, dx + \int f(x)\, \dot{w}(x)\, dx\,, \end{aligned} \tag{3.2.14}$$

it is to be borne in mind that the conditions of periodicity

$$\rho(x) = \rho(x + a), \quad \Phi(x, x') = \Phi(x + a, x' + a')\,. \tag{3.2.15}$$

must be fulfilled.

From (3.2.14), we find the equations of motion in the (x, t) and (κ, ω) representations (cf. (2.4.6)).

$$\rho(x)\, \ddot{w}(x) + \int \Phi(x, x')\, w(x')\, dx' = f(x)\,, \tag{3.2.16}$$

$$- \frac{\omega^2}{2\pi} \int \rho(\kappa - \kappa')\, w(\kappa')\, d\kappa' + \int \Phi(\kappa, \kappa')\, w(\kappa')\, d\kappa' = f(\kappa)\,. \tag{3.2.17}$$

Here, the argument κ of the Fourier images, belongs to the real axis R_κ (in contrast to $k \in B$).

Thus, equations (3.2.12, 13) and (3.2.16, 17) correspond to different representations of one and the same physical model. It is essential that the medium be homogenous in terms of complex structure and hence the equations of motion, in the (k, ω) representation are reduced to a linear algebraic system. An analysis of the equations is highly simplified, but one has to pay for this by introducing a large (countable in the general case) number of variables. In terms of simple structure, there is one variable, but the medium is inhomogenous, and the equations of motion in the (k, ω) representation remain integral ones. In connection with this, when wave processes are considered and a transition to approximate models is performed, it is the representation of homogenous (periodic) media in terms of the complex structure that proves to be more convenient.

Let us present explicit formulae which connect the representations of simple and complex structures. With this in mind, let the functions of the cell coordinate ξ be continued by zero on the whole x axis. Let $\alpha(x)$ be the charactierstic function of the elementary cell with the number $n = 0$. Then, as it is easily seen,

$$w(n, x) = \alpha(x)\, w(x + na), \quad f(n, x) = \alpha(x) f(x + na),$$
$$\Phi(n - n', x, x') = \alpha(x)\, \alpha(x')\, \Phi(x + na, x' + n'a) \tag{3.2.18}$$

[the dependence upon the difference $n - n'$ follows from (3.2.15)].

Correspondingly, let us continue by zero the functions $e^s(\xi)$ and $e_s(\xi)$ as well.

The Fourier images of all these functions are defined over the axis R_κ(or over the plain $R_\kappa \times R_\kappa$) and may be continued in the complex plane as entire analytical functions.

Taking into account (3.2.10), we have for $w_s(n)$

$$\begin{aligned} w_s(n) &= \int e_s(\xi)\, w(n, \xi)\, d\xi = \int e_s(x)\, w(x + na)\, dx \\ &= \frac{1}{2\pi} \int \overline{e_s(\kappa)}\, \omega(\kappa)\, \mathrm{e}^{\mathrm{i}na\kappa}\, d\kappa\,. \end{aligned} \tag{3.2.19}$$

In accordance with (2.2.7), in the k representation

$$w_s(k) = \frac{a}{2\pi} \int \overline{e_s(\kappa)}\, w(\kappa) \sum_n \mathrm{e}^{-\mathrm{i}na(k-\kappa)}\, d\kappa, \quad k \in B\,. \tag{3.2.20}$$

For an arbitrary $f(k)$, $k \in B$, we denote its periodic continuation on R by $\tilde{f}(\kappa)$. In particular,

$$\tilde{\delta}(\kappa) = \sum_n \delta\left(\kappa - \frac{2\pi n}{a}\right) = \frac{a}{2\pi} \sum_n \mathrm{e}^{-\mathrm{i}na\kappa}\,. \tag{3.2.21}$$

Problem 3.2.3. Prove the formulae ($k, k' \in B$)

$$\begin{aligned} &\int \tilde{\delta}(\kappa - k)\, \tilde{\delta}(k - \kappa')\, dk = \tilde{\delta}(\kappa - \kappa')\,, \\ &\int \tilde{\delta}(k - \kappa)\, \tilde{\delta}(\kappa - k')\, d\kappa = \tilde{\delta}(k - k')\,, \\ &\tilde{f}(\kappa) = \int \tilde{\delta}(\kappa - k) f(k)\, dk, \quad f(k) = \int \tilde{\delta}(k - \kappa)\, \tilde{f}(\kappa)\, d\kappa\,. \end{aligned} \tag{3.2.22}$$

Now (3.2.20) is written in its final form

$$w_s(k) = \int \tilde{\delta}(k - \kappa)\, \overline{e_s(\kappa)}\, w(\kappa)\, d\kappa, \quad k \in B\,. \tag{3.2.23}$$

Let us derive the inverse formula. Taking into account the agreement about the continuation of $w(n, \xi)$ on the x axis, we have

$$w(x) = \sum_n w(n, x - na) \tag{3.2.24}$$

or in the κ representation

$$w(\kappa) = \sum_n w(n, \kappa)\, e^{-\mathrm{i}na\kappa}\,. \tag{3.2.25}$$

But according to (3.2.9)

$$w(n, \kappa) = e^s(\kappa)\, w_s(n)\,, \tag{3.2.26}$$

and taking into account (2.2.7), we finally obtain

$$w(\kappa) = \frac{1}{a}\, e^s(\kappa)\, w_s(\kappa)\,. \tag{3.2.27}$$

Note that the substitutions of (3.2.23) in (3.2.27) and vice-versa lead to the interesting formulae

$$\frac{1}{a}\, \tilde{\delta}(\kappa - \kappa')\, e^s(\kappa)\, \overline{e_s(\kappa')} = \tilde{\delta}(\kappa - \kappa')\,, \tag{3.2.28}$$

$$\frac{1}{a}\, \tilde{\delta}(\kappa - \kappa')\, e^s(\kappa')\, e_s(\kappa') = \tilde{\delta}(\kappa - \kappa')\,. \tag{3.2.29}$$

Problem 3.2.4 Verify these formulae by direct calculation. [Hint: Use the condition of completeness of the basis (3.2.8)].

The connection between $f^s(k)$, $\Phi^{ss'}(k)$, $I^{ss'}$ and $f(\kappa)$, $\Phi(\kappa, \kappa')$, $\rho(\kappa)$ can be obtained in a similar manner. But it is simpler to do this using the expressions for the elastic and kinetic energies and the work of external forces in the representations of simple and complex structures

$$\begin{aligned} &\langle w | \Phi | w \rangle = \langle w_s | \Phi^{ss'} | w^{s'} \rangle\,, \\ &\langle \dot{w} | \rho | \dot{w} \rangle = \langle \dot{w}_s | I^{ss'} | w_{s'} \rangle, \quad \langle f | w \rangle = \langle f^s | w_s \rangle \end{aligned} \tag{3.2.30}$$

and the relations (3.2.23) and (3.2.27). For example, we have

$$\begin{aligned} \frac{1}{2\pi} \int \overline{f(\kappa)}\, w(\kappa)\, d\kappa &= \frac{1}{2\pi a} \int \overline{f(\kappa)}\, e^s(\kappa)\, \tilde{w}_s(\kappa)\, d\kappa \\ &= \frac{1}{2\pi a} \iint \overline{f(\kappa)}\, e^s(\kappa)\, \tilde{\delta}(\kappa - k)\, w_s(k)\, d\kappa dk \\ &= \frac{1}{2\pi} \int \overline{f^s(k)}\, w_s(k)\, dk\,. \end{aligned} \tag{3.2.31}$$

Hence, taking into account the arbitrariness of $w_s(k)$

$$f^s(k) = \frac{1}{a} \int \tilde{\delta}(k - \kappa)\, \overline{e^s(\kappa)}\, f(\kappa)\, d\kappa\,. \tag{3.2.32}$$

Similarly, we find

$$\Phi^{ss'}(k) = \frac{1}{a^2} \iint \tilde{\delta}(k - \kappa)\, \overline{e^s(\kappa)}\, e^{s'}(\kappa')\, \Phi(\kappa, \kappa')\, d\kappa d\kappa'\,, \tag{3.2.33}$$

$$I^{ss'} = \frac{1}{4\pi^2 a} \iint \overline{e^s(\kappa)}\, e^{s'}(\kappa')\, \rho(\kappa - \kappa')\, d\kappa d\kappa'\,. \tag{3.2.34}$$

Problem 3.2.5 For the cell, which has a center of symmetry, show that

$$\Phi^{ss'}(-k) = (-1)^{s+s'}\,\Phi^{ss'}(k)\,. \tag{2.3.35}$$

(Hint: In this case $e^s(-\kappa) = (-1)^s\,e^s(\kappa)$ and $\Phi(\kappa,\kappa') = \Phi(\kappa',\kappa)$.)

The inverse formulae have the form

$$f(\kappa) = e_s(\kappa)\,f^s(\kappa)\,, \tag{3.2.36}$$

$$\Phi(\kappa,\kappa') = \tilde{\delta}(\kappa-\kappa')\,e_s(\kappa)\,\tilde{\Phi}^{ss'}(\kappa)\,\overline{e_{s'}(\kappa')}\,, \tag{3.2.37}$$

$$\rho(\kappa-\kappa') = 2\pi\tilde{\delta}(\kappa-\kappa')\,e_s(\kappa)\,I^{ss'}\,\overline{e_{s'}(\kappa')}\,. \tag{3.2.38}$$

Let us now deduce two important relations. Multiplying (3.2.37) by $e^r(\kappa')$ and integrating with respect to κ', we have

$$\begin{aligned}&\int \Phi(\kappa,\kappa')\,e^{r'}(\kappa')\,d\kappa' \\ &\quad= e_s(\kappa)\,\tilde{\Phi}^{ss'}(\kappa)\int \tilde{\delta}(\kappa-\kappa')\,\overline{e_{s'}(\kappa')}\,e^r(\kappa')\,d\kappa'\,.\end{aligned} \tag{3.2.39}$$

In view of (3.2.21), applying Parseval's formula (2.2.10), we obtain for the integral in the right hand side

$$\int \tilde{\delta}(\kappa-\kappa')\,\overline{e_{s'}(\kappa')}\,e^r(\kappa')\,d\kappa' = a\sum_n \mathrm{e}^{-\mathrm{i}na\kappa}\int e_{s'}(x-na)\,e^r(x)\,dx\,. \tag{3.2.40}$$

But, due to the agreement about the continuation of the basis functions by zero, the integrand differs from zero only when $n=0$ and, in accordance with (3.2.5), is then equal to $\delta^r_{s'}$, Thus,

$$\int \tilde{\delta}(\kappa-\kappa')\,\overline{e_{s'}(\kappa')}\,e^r(\kappa')\,d\kappa' = a\delta^r_{s'}\,, \tag{3.2.41}$$

and the required relation is

$$\int \Phi(\kappa,\kappa')\,e^{s'}(\kappa')\,d\kappa' = a\tilde{\Phi}^{ss'}(\kappa)\,e_s(\kappa)\,. \tag{3.2.42}$$

Similarly, from (3.2.38) we find

$$\int \rho(\kappa-\kappa')\,e^{s'}(\kappa')\,d\kappa' = 2\pi a I^{ss'}e_s(\kappa)\,, \tag{3.2.43}$$

which agrees with (3.2.6).

The relations (3.2.42), (3.2.43) will be essential for the discussion in Sect. 3.7.

Thus above formulae establish a one-to-one correspondence between the descriptions of a periodic system in terms of the medium of simple and complex structures. Of course, there must be given the biorthogonal basis e^s, e_s in the ξ or κ representations which satisfies the condition of completeness (3.2.8). Observe that the selection of $e^s(\xi)$ in the form of polynomials, is by no means necessary—any other complete set of functions is possible. However, for the invariance of microdeformations with respect to translation, it is necessary in any case to let $e^0(\xi) = 1$ (compare with Problem 3.2.2).

3.3 Phenomenology

Let us proceed to a general study of the one-dimensional medium of complex structure. In Sects. 3.3–7 our aim will be to establish the governing laws, independent of specific model representations. Hence a purely phenomenological appoach has proved to be correct.

By definition, the displacements $u(x)$ and a set of quantities $\eta_p(x)$ characterizing the internal degrees of freedom (the argument t is as usual omitted) are the dynamical variables, determining the state of the system of complex structure. The index p takes the values from some finite or countable set. For example, in the case of an N-atomic chain, $p = 1, \ldots, N$-1. The physical significance of the quantity $\eta_p(x)$ is unimportant; in particular, they could be of non-mechanical nature.

The displacement $u(x)$ may be longitudinal, transverse or, in general, a three-dimensional vector. In the latter case, the vector index is omitted for simplicity.

The distinctive character of the displacement, in comparions with other dynamical variables, is connected with their law of transformation with respect to the tranlsation group. If u_0 is an infinitesimal translation, then

$$\delta u(x) = u_0, \quad \delta\eta_p(x) = 0\,. \tag{3.3.1}$$

Let us introduce temporary notations for dynamic variables $u(x)$, $\eta_p(x)$ and the corresponding generalized forces $q(x)$, $\mu^p(x)$

$$\begin{aligned} &w_0(x) \overset{\text{def}}{=} u(x), \quad w_p(x) \overset{\text{def}}{=} \eta_p(x)\,, \\ &f^0(x) \overset{\text{def}}{=} q(x), \quad f^p(x) \overset{\text{def}}{=} \mu^p(x)\,. \end{aligned} \tag{3.3.2}$$

Then, using the compact notations of Sect. 2.2, one is able to write the Lagrangian in the harmonic approximation in the form, which is invariant with respect to x, n (for quasicontinuum) and k representations:

$$L = \frac{1}{2}\langle \dot{w}_s | I^{ss'} | \dot{w}_{s'}\rangle - \frac{1}{2}\langle w_s | \Phi^{ss'} | w_{s'}\rangle - \langle \Phi^s | w_s\rangle + \langle f^s | w_s\rangle\,. \tag{3.3.3}$$

Here, $I^{ss'}$ and $\Phi^{ss'}$ are kinetic and potential energy operators. The quantities Φ^s account for the initial forces acting in the system. As the example of the system of rods with strain mentioned in Sect. 3.1 shows, the intitial forces in some problems may be essential. Excluding these special cases, we shall usually assume that $\Phi^s = 0$.

We emphasize here that the Lagranigan (3.3.3) does not describes the most general linear model. Later, we shall consider system, in which such effects as dissipation, time nonlocality and, in general, non-Hermiticity of the corresponding operator, are taken into account.

The equations of motion, obtained from the Lagranigian (3.3.3) have the form, under the assumption $\Phi^s = 0$,

$$\int I^{ss'}(x, x')\, \ddot{w}_{s'}(x')\, dx' + \int \Phi^{ss'}(x, x')\, w_{s'}(x')\, dx' = f^s(x) \tag{3.3.4}$$

or in the (k, ω) representation

$$\int [-\omega^2 I^{ss'}(k, k') + \Phi^{ss'}(k, k')]\, w_{s'}(k')\, dk' = f^s(k)\,. \tag{3.3.5}$$

A question naturally arises here, by which mechanism one can explain the occurrence of nonlocal inertial characteristics in the phenomenological theory. The concept of "associated masses," well known in hydrodynamics could serve well as one of the causes of nonlocality. The approximate consideration of the moment of inertia of transverse fibres in the dynamics of rods and plates, leads to the occurrence of a square of the velocity gradient in the kinetic energy (Rayleigh's model).

Inicidentally, we observe that from the point of view of generality a linear term in $\dot{w}_s$ of the form $\langle I^s | \dot{w}_s \rangle$ can be added to the Lagrangian (3.3.3). However, the author does not know of problems in which the consideration of such a term could lead to a nontrivial result (as distinct from the term $\langle \Phi^s | w_s \rangle$).

Let us now consider the properties of the energy operators.

Hermiticity. From the Lagrangian (3.3.3), it follows that

$$I^{ss'}(x, x') = I^{s's}(x', x), \quad \Phi^{ss'}(x, x') = \Phi^{s's}(x', x) \tag{3.3.6}$$

or in the k representation (for real values of k)

$$I^{ss'}(k, k') = \overline{I^{s's}(k', k)}, \quad \Phi^{ss'}(k, k') = \overline{\Phi^{s's}(k', k)}\,. \tag{3.3.7a}$$

In cases for which the analytical continuation of the functions $I^{ss'}(k, k')$ and $\Phi^{ss'}(k, k')$ in the complex region is considered, these relations are to be replaced by more general ones (cf. (2.2.24)), namely

$$I^{ss'}(k, k') = I^{s's}(-k', -k), \quad \Phi^{ss'}(k, k') = \Phi^{s's}(-k', -k)\,. \tag{3.3.7b}$$

Finiteness of Action at a Distance: As earlier, we shall assume that the action at a distance is bounded by a characteristic radius of interaction l (generally different for different operators). Consequences for the kernels of the operators $I^{ss'}$ and $\Phi^{ss'}$ are identical to those indicated in the Sect. 2.4 for the operator Φ.

Invariance with respect to Transformation Groups: In the above, we presented many examples of transformations with respect to which the invariance of the Lagrangian and equations of motions was postulated for physical reasons. In particular, a shift change of arguments, translation, rotation, homogenous deformation, pure bending, etc. were examples of such transformations.

All these transofrmations have the common feature that they form a group which depends continuously on a finite number of parameters ξ^a. Such groups are called continuous or Lie groups [3.2]. We shall be interested only in infinitesimal transformations.If we agree that the point $\xi^a = 0$ in the space of parameters corresponds to the identity transformation, then this means that the ξ^a are infinitesimal quantities.

Suppose that to each element (infinitesimal transformation) T of a Lie group G there corresponds a variation δw_s of the dynamical variables w_s. Usually only the variation δw_s, representable in the form of an inhomogenous linear function of dynamical variables w_s are considered, i.e., (recall that the summation over repeated indices is assumed)

$$\delta\omega_s = \zeta^b T_{bs}^{s'}\omega_{s'} + \xi_a \tau_s^a . \tag{3.3.8}$$

Here $T_{bs}^{s'}$ and τ_s^a are arbitary operators and functions, ζ^b and ξ_a are the infinitesimal parameters, that determine the transformation T from the group G.

Problem 3.3.1. Write down all the transformation groups indicated above in the from (3.3.8).

By definition, the invariance with respect to transformations of the group G means vanishing of the variation δL of the Lagrangian (3.3.3) when the variations of dynamic variables δw_s are given by the relations (3.3.8) with arbitary ζ^b, ξ_a. The variation is the linear with respect to δw_s part of the functional $L(w_s + \delta w_s) - L(w_s)$, with arbitary w_s.

Here, in order to make our considerations precise a series of remarks is necessary. We temporarily restrict ourselves to time-independent transformations, i.e. we assume that $\partial_t \delta\omega_s = \delta\dot{\omega}_s = 0$. This means that the kinetic energy in the Lagrangian is invariant under transformations of the form (3.3.8).

Furthermore, the Lagrangian (3.3.3) depends on the quantities $\Phi^{ss'}$, Φ^s and the external forces f^s. It is necessary to make an assumption about their behavior under transformations of the group G. For example, generally speaking, the external forces f^s must perform work during translation. Since the corresponding contribution into the Lagrangian is of no interest we could always assume that the forces f^s act on the body in such a way that the term $\langle f^s | \omega_s \rangle$ is invariant.

The operator $\Phi^{ss'}$ determines intrinsic properties of the medium and naturally may be considered invariant. However, the situation could become more complicated in the presence of strong initial forces Φ^s. For example, during rotation, it is necessary to distinguish the cases when the forces Φ^s are connected rigidly with the medium (as internal strains in the crystal) from those in which their origin is a field which is external with respect to the medium.

In such cases, not only will different rules of transformation of Φ^s occur but also a connection between Φ^s and $\Phi^{ss'}$ may appear. Later, some similar cases will be considered, but for the time being let us set for simplicity, $\Phi^s = 0$.

Thus, for the variation of the Lagrangian (3.3.3) we have

$$\delta L = -\frac{1}{2}\langle w_s + \delta w_s | \Phi^{ss'} | w_{s'} + \delta w_{s'}\rangle + \frac{1}{2}\langle w_s | \Phi^{ss'} | w_{s'}\rangle = -\langle w_s | \Phi^{ss'} | \delta w_{s'}\rangle = 0 . \quad (3.3.9)$$

Here the second order terms with respect to δw_s are neglected and the hermiticity of $\Phi^{ss'}$ is used.

Now δw_s is to be substituted by the expression (3.3.8) and the arbitrariness of ζ^b, ξ_a and w_s is to be taken into account. Then, one of the terms will have the structure $\langle w | \Phi T | w\rangle$. The operator ΦT is to be symmetrized, i.e., substituted by $1/2[\Phi T + (\Phi T)^+]$, since the antisymmetric part does not contribute. But, $(\Phi T)^+ = T^+ \Phi^+$ and $\Phi^+ = \Phi$. Hence, we find the required conditions on the operator $\Phi^{ss'}$:

$$\Phi^{ss''} T^{s'}_{bs''} + T^{+s}_{bs''} \Phi^{s''s'} = 0 , \quad (3.3.10)$$

$$\Phi^{ss'} \tau^a_{s'} = 0 . \quad (3.3.11)$$

From here, the conditions on kernels of the operator $\Phi^{ss'}$ in x, n (for quasicontinuum) or k representations follow in an obvious way. For example, in the x representation, the conditions (3.3.11) are written in the form

$$\int \Phi^{ss'}(x, x')\, \tau^a_{s'}(x')\, dx' = 0 . \quad (3.3.12)$$

Translational Invariance: For the group of translations, which is given by the relations (3.3.1), we have

$$\int \Phi^{00}(x, x')\, dx' = 0, \quad \int \Phi^{p0}(x, x')\, dx' = 0 , \quad (3.3.13)$$

or in the k representation

$$\Phi^{00}(k. k')|_{k'=0} = 0, \quad \Phi^{p0}(k, k')|_{k'=0} = 0 . \quad (3.3.14)$$

If the Hermitian symmetry of (3.3.7) and analyticity of $\Phi^{ss'}(k, k')$ in k and k' are taken into account, it follows from here that the matrix $\Phi^{ss'}(k, k')$ can be represented in the form

$$\Phi^{ss'}(k, k') = \begin{pmatrix} kk'\gamma(k, k') & -ik\chi^{+p}(k, k') \\ ik'\chi^{p}(k, k') & \Gamma^{pp'}(k, k') \end{pmatrix}, \quad (3.3.15)$$

where $\gamma(k, k')$, $\chi^p(k, k')$ and $\Gamma^{pp'}(k, k')$ are entire analytical functions of the variables k, k' (due to the boundedness of action at a distance), which are uniquely defined by $\Phi^{ss'}(k, k')$, and in accordance with (3.3.7a), (3.3.7b)

$$\begin{aligned}
&\gamma(k, k') = \gamma(-k', -k) = \overline{\gamma(k', k)}\,, \\
&\chi^{+p}(k, k') = \chi^{p}(-k', -k) = \overline{\chi^{p}(k', k)}\,, \\
&\Gamma^{pp'}(k, k') = \Gamma^{p'p}(-k', -k) = \overline{\Gamma^{p'p}(k', k)}\,,
\end{aligned} \tag{3.3.16}$$

the transition to complex conjugate quantities being correct only for real values of k.

In view of (3.3.15) and (3.3.2) the equations of motion (3.3.4) become now in the operator form

$$\begin{aligned}
&\rho\ddot{u} + j^{+p}\ddot{\eta}_p - \partial\gamma\partial u + \partial\chi^{+p}\eta_p = q\,, \\
&j^p\ddot{u} + I^{pp'}\ddot{\eta}_{p'} + \chi^p\,\partial u + \Gamma^{pp'}\eta_{p'} = \mu^p\,,
\end{aligned} \tag{3.3.17}$$

where

$$\rho \overset{\text{def}}{=} I^{00}, \quad j^p \overset{\text{def}}{=} I^{p0}\,, \tag{3.3.18}$$

and the operators j^{+p} and χ^{+p} are the Hermitian adjoints to the operators j^p and χ^p.

In the (k, ω) representation, equations (3.3.17) take the form

$$\begin{aligned}
&(-\,\omega^2\rho + k\gamma k)\,u - (\omega^2 j^{+p} + \mathrm{i}k\chi^{+p})\,\eta_p = q\,, \\
&(-\omega^2 j^p + \mathrm{i}\chi^p k)\,u + (-\omega^2 I^{pp'} + \Gamma^{pp'})\,\eta_{p'} = \mu^p\,,
\end{aligned} \tag{3.3.19}$$

where ρ, γ, ... are integral operators.

Rotational Invariance: The corresponding conditions will be considered in detail below for an example of the Cosserat medium model.

Internal Symmetry: Besides the fundamental conditions of translational and rotational invariance, which do not depend on the model concepts, additional conditions of invariance with respect to transformations of some groups G, which are due to the symmetries of the model may exist. Here two situations are possible. The transformations conserve the invariance of the Lagrangian and hence also the equations of motion (for example, transformations of independent variables) or conserve the equations of motions only (without considering the boundary conditions). The corresponding examples were given in sect. 3.1 and will also be considered later.

Local Approximation: Considerations similar to those in sect. 2.4 permit, in the case in which the dynamic variables change sufficiently slowly over distances of the order of l, to replace the integral equations of motion by approximate differential ones. In particular, in the zeroth approximation, (3.3.17) take the form

$$\begin{aligned}
&\rho_0(x)\,\ddot{u}(x) + j_0^p(x)\,\ddot{\eta}_p(x) - \partial\gamma_0(x)\,\partial u(x) - \partial\chi_0^p(x)\,\eta_p(x) = q(x)\,, \\
&j_0^p(x)\,\ddot{u}(x) + I_0^{pp'}(x)\,\ddot{\eta}_{p'}(x) + \chi_0^p(x)\;\partial u(x) + \Gamma_0^{pp'}(x)\,\eta_{p'}(x) = \mu^p(x)\,,
\end{aligned} \tag{3.3.20}$$

where the functions $\rho_0(x)$, $\gamma_0(x)$, ... are connected with the kernels of the corresponding operators by relations of the type (2.4.29) and (2.4.30).

Stability is equivalent to a requirement that the forms $\langle \dot{w}_s | I^{ss'} | \dot{w}_{s'} \rangle$ and $\langle w_s | \Phi_{ss'} | w_{s'} \rangle$ be positive definite for all nonzero $\dot{w}_s$ and w_s except translations and rotations.

For example, the system of masses, connected by springs shown in Fig. 3.4, is to be considered unstable, since under rotation of all the masses through the same angle, the length of the springs does not change and hence the elastic energy remains unchanged. Note that this transformation may be considered as an "angular translation." Therefore, if in the given system longitudinal displacement and angle of rotation of masses are taken as the generalized variables, then the system can be described in terms of the medium of simple structure with two degrees of freedom, stable with respect to any general displacement, except translation.

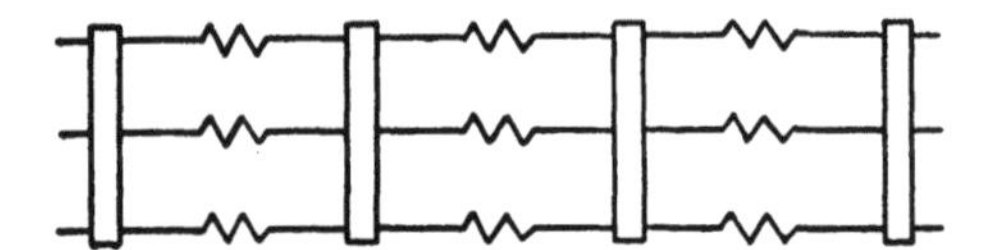

Fig.3.4. Unstable model

This example indicates that the definition of stability is to some extert a matter of convention.

Taking into account (3.3.15) and (3.3.16), one may write the expression for the elastic energy Φ in the form

$$2\Phi = \langle \partial u | \gamma | \partial u \rangle + 2\langle \eta_p | \chi^p | \partial u \rangle + \langle \eta_p | \Gamma^{pp'} | \eta_{p'} \rangle \,. \tag{3.3.21}$$

Hence the positive definitness of the operators γ and $\Gamma^{pp'}$ for all admissible ∂u and η_p is a necessary condition of stability. In fact, let us set, for example, $\partial u = 0$. Then the quantity $2\Phi = \langle \eta_p | \Gamma^{pp'} | \eta_{p'} \rangle$ must be strictly positive for all $\eta_p \neq 0$ since the rotation is excluded by the condition $\partial u = 0$ and the variables η_p are invariant with respect to translation.

These properties of the operators γ and $\Gamma^{pp'}$ will be used later.

Homogeneity: For a homogeneous medium, all the operators are, by definition, invariant with respect to the translation group.

Problem 3.3.2. Write in an explicit form the relations (3.3.8) and (3.3.10) for the group of infinitesimal translations. Show that (3.3.10) in the x representation, is equivalent to the condition of the dependence of the operator kernel on the difference of arguments only.

For a homogeneous medium $\Phi^{ss'}(k, k') = \Phi^{ss'}(k)\delta(k - k')$, where, in accordance with (3.3.15) and (3.3.16)

$$\Phi^{ss'}(k) = \begin{pmatrix} k^2\gamma(k) & -ik\chi^{+p}(k) \\ ik\chi^p(k) & \Gamma^{pp'}(k) \end{pmatrix}, \tag{3.3.22}$$

$$\gamma(k) = \gamma(-k) = \overline{\gamma(k)}, \quad \chi^{+p}(k) = \chi^p(-k) = \overline{\chi^p(k)},$$
$$\Gamma^{pp'}(k) = \Gamma^{p'p}(-k) = \overline{\Gamma^{p'p}(k)}, \tag{3.3.23}$$

the transition to complex conjugate quantities being allowed only for real values of k.

As was shown above repeatedly, for the homogeneous medium the equations of motion in the (k, ω) representation become algebraic. Let us for simplicity assume that the operators ρ and $I^{pp'}$ are constants and $j^p = 0$. Then (3.3.19) takes the form

$$[-\omega^2\rho + k^2\gamma(k)]\, u(k, \omega) - ik\chi^{+p}(k)\, \eta_p(k, \omega) = q(k, \omega),$$
$$ik\chi^p(k)\, u(k, \omega) + [-\,\omega^2 I^{pp'} + \Gamma^{pp'}(k)]\, \eta_{p'}(k, \omega) = \mu^p(k, \omega). \tag{3.3.24}$$

Quasicontinuum: In this case, the class of admissible functions is restricted by conditions presented in Sects. 2.2 and 2.3. Besides in the x- and k representations, the equations can be written in the discrete form as well.

Problem 3.3.3. Using (2.2.26), (3.2.10) and (3.1.7), show that $\Phi^{ss'}(k)$ is expressed in terms of the stiffnesses of elastic bonds $\Psi(n, \xi, \xi')$ in the following way

$$\Phi^{ss'}(k) = \frac{1}{a} \sum_n \iint \Psi(n, \xi, \xi')\, e^s(\xi)\, [e^{s'}(\xi) - e^{s'}(\xi')\, e^{-inak}]\, d\xi\, d\xi', \tag{3.3.25}$$

and hence, due to (3.3.22) and (3.2.7)

$$k^2\gamma(k) = \frac{2}{a} \sum_{n=1}^{\infty} (1 - \cos nak) \iint \Psi(n, \xi, \xi')\, d\xi\, d\xi',$$
$$ik\chi^p(k) = \frac{1}{a} \sum_n (1 - e^{-inak}) \iint \Psi(n, \xi, \xi')\, e^p(\xi)\, d\xi\, d\xi', \tag{3.3.26}$$
$$\Gamma^{pp'}(k) = \frac{1}{a} \sum_n \iint \Psi(n, \xi, \xi')\, e^p(\xi)\, [e^{p'}(\xi) - e^{p'}(\xi')\, e^{-inak}]\, d\xi\, d\xi'.$$

To conclude this section, let us consider briefly the question of the energy density and flux since the situation here is quite analogous to the case of simple structure. The energy balance equation still can be written in the integral form (2.5.10) or in the differential form (2.5.12). Multiplying the equations of motion (3.3.4) by $\dot{w}_s(x)$, we find

$$\tau(x) = \frac{1}{2}\, \dot{w}_s(x) \int I^{ss'}(x, x')\, \dot{w}_{s'}(x')\, dx',$$
$$\nu(x) = f^s(x)\, \dot{w}_s(x), \tag{3.3.27}$$
$$\varphi(x) + \partial S(x) = \dot{w}_s(x) \int \Phi^{ss'}(x, x')\, w_{s'}(x')\, ds'.$$

If we replace $w_s(x)$ by $u(x)$ and $\eta_p(x)$, then the density of elastic energy $\varphi(x)$ can be represented in the form of a sum $\varphi_0(x) + \varphi_1(x) + \varphi_2(x)$. Here $\varphi_0(x)$

depends only on $u(x)$ and is given by (2.5.22) with (2.5.7) taken into account where $\Phi(x, x') = \Phi^{00}(x, x')$, and, for definiteness, we may let $\xi = 0$. For $\varphi_1(x)$ and $\varphi_2(x)$, we have

$$\varphi_1(x) = \eta_p(x) \int \Phi^{p0}(x, x')\, u(x')\, dx' ,$$
$$\varphi_2(x) = \frac{1}{2}\, \eta_p(x) \int \Phi^{pp'}(x, x')\, \eta_{p'}(x')\, dx' . \tag{3.3.28}$$

From (3.3.1) and (3.3.13), it follows that these expressions are invariant with respect to translations.

Accordingly, $S(x) = S_0(x) + S_1(x) + S_2(x)$, where (cf. (2.5.17))

$$S_0(x) = -\int f_0(x, x')\, \dot{u}(x')\, dx' ,$$
$$S_1(x) = -\int f_1(x, x')\, \dot{u}(x')\, dx' , \tag{3.3.29}$$
$$S_2(x) = -\int f_2^p(x, x')\, \dot{\eta}_p(x')\, dx'$$

and (cf. (2.5.18) where $\xi = 0$)

$$f_0(x, x') = \int \theta(x' - x, x'' - x)\, \Phi^{00}(x', x'')\, \varepsilon(x', x'')\, dx'' ,$$
$$f_1(x, x') = \int \theta_-(x - x', x'' - x)\, \Phi^{0p}(x', x'')\, \eta_p(x'')\, dx'' ,$$
$$f_2^p(x, x') = \frac{1}{2} \int \theta_-(x' - x, x'' - x)\, \Phi^{pp'}(x', x'')\, \eta_{p'}(x'')\, dx'' . \tag{3.3.30}$$

Problem 3.3.4. By direct substitution, verify that expressions (3.3.28–30) satisfy the last equality in (3.3.27).

Problem 3.3.5. Show that, for homogeneous medium with the operator (3.3.22) one can set (cf. (2.5.6) and (2.5.25))

$$2\varphi(x) = \varepsilon(x)\, [\gamma(x) * \varepsilon(x)] + \eta_p(x)\, [2\chi^p(x) * \varepsilon(x) + \Gamma^{pp'}(x) * \eta_{p'}(x)] \tag{3.3.31}$$

and correspondingly

$$f_0(x, x') = \delta(x - x') \cdot [\gamma(x) * \varepsilon(x)]$$
$$\qquad - \frac{1}{2}\, \partial' \int \theta_-(x' - x, x'' - x)\, \gamma(x' - x'')\, \varepsilon(x'')\, dx'' ,$$
$$f_1(x, x') = -\, \delta(x - x') \cdot [\chi^{+p}(x) * \eta_p(x)]$$
$$\qquad - \partial' \int \theta_-(x' - x, x'' - x')\, \chi^{+p}(x' - x'')\, \eta_p(x'')\, dx'' ,$$
$$f_2^p(x, x'') = \frac{1}{2} \int \theta_-(x' - x, x'' - x)\, \Gamma^{pp'}(x' - x'')\, \eta_{p'}(x'')\, dx'' . \tag{3.3.32}$$

It is obvious that all the conclusions, concerning the degree of non-uniqueness of the energy density and flux presented in Sect. 2.5, are valid also in the case of the medium with complex structure. In particular, for periodical in time processes, the mean values of these quantities are determined uniquely.

3.4 Acoustical and Optical Modes of Vibration. General Solution and Green's Matrix

In this section, we consider a homogeneous medium with complex strcuture, which is described in the (k, ω) representation by equations (3.3.24). The study of free oscillations, i.e. solutions of the system (3.3.24) with zero right-hand side is of primary interest. Obviously, for the existence of nontrivial solutions it is necessary to require the fulfillment of the condition

$$\mathrm{Det}\begin{pmatrix} k^2\gamma(k) - \omega^2\rho & -ik\chi^{+p}(k) \\ ik\chi^p(k) & \Gamma^{pp'}(k) - \omega^2 I^{pp'} \end{pmatrix} = 0\,, \tag{3.4.1}$$

which can be interpreted as the disperison equation which connects the frequency ω and the wave number k. Its solution consists of N functions $\omega_j^2(k)$ $(j = 0, \dots .$ $N-1)$, which, due to the hermiticity of the operator Φ are even and real for real values of k.

From the conditions of stability, i.e. from the positive definitness of elastic energy, it follows that $\omega_j^2(k) > 0$ when $k \neq 0$ (Im $\{k\} = 0$). The case $k = 0$ requires a separate consideration. Then the dispersion relation separates into independent equations

$$\omega_0^2(0) = 0, \quad \mathrm{Det}(\Gamma^{pp'}(0) - \omega_j^2(0) I^{pp'}) = 0 \quad (j = 1, \dots, N-1). \tag{3.4.2}$$

The vanishing of the function $\omega_0^2(k)$ at the point $k = 0$ is the consequence of invariance of energy with respect to translation, as it is in the case of the medium of simple structure (Sect. 2.8). By assumption, this is the only admissible form of instability and, hence, for all other functions $\omega_j^2(0) > 0$. In Fig. 3.5, the curves $\omega_j(k)$ for an N-atomic chain are shown schematically.

By tradition, the curve $\omega_0(k)$ is called the acoustical mode of osciallations and

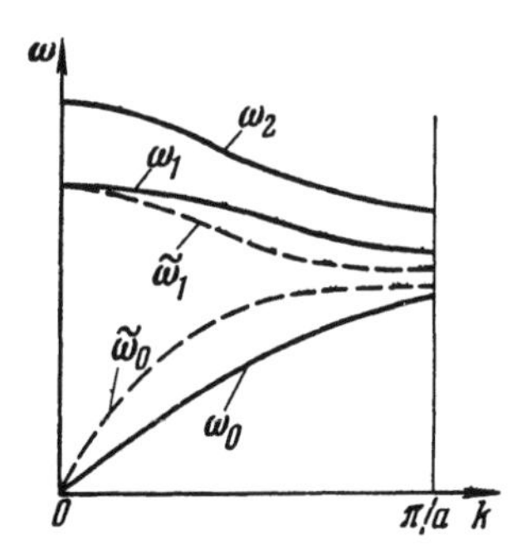

Fig.3.5. Acoustical and optical modes

the N-1 curves $\omega_j(k)$ are called optical modes. From what follows, it will be clear that for sufficiently small values of k, the acoustical mode corresponds to the low frequency oscillations of the centers of mass of the cells (the cell moves as a whole) and the optical modes to high frequency oscillations connected with the internal degrees of freedom (the centers of mass of cells are stationary). If the masses within the cell interact with each other much more strongly than with masses of neighboring cells, then the frequencies of optical oscillations are mainly determined by the internal forces and turn out to be close to frequencies of oscillations of an isolated cell. The influence of the neighboring cells will then not be large.

A few words about the origin of the term "optical oscillations." For ionic crystals (of the NACl type with oppositely charged atoms) long wave optical oscilliations are of importance when an interaction with light occurs, Usually, an electromagnetic wave interacts only with lattice oscillations having nearly the same frequencies and wavelengths. The lattice vibrations lie in the interval from 0 to 10^{13} s^{-1} and hence, the light waves with such frequencies have wavelengths longer than $3 \times 10^{10} \times 10^{-13} = 3 \times 10^{-3}$cm which is very large in comparison with the inter-atomic distance ($\sim 10^{-8}$ cm). Therefore, only long waves ($ak \sim 10^{-5}$) can interact with light.

The mechanism of interaction is connected with the fact, that in the presence of the relative displacements of oppositely charged atoms in the cell, dipole moments appear and cause the absorption of light. Hence, oscillations of such a type have been called optical. In the case of long wave acoutsic oscillations the dipole moment obviously does not appear and hence they do not interact with light.

Now, let us consider briefly the construction of a general solution of the system of equations (3.3.24), for which can be, as always, represented in the form of a sum of a general solution w_s^0 of the homogeneous system and a particular solution w_s' of the inhomogeneous system.

The vector w_s^0 has N components $w_s(k, \omega)$, where the agruments ω and k are connected by the dispersion relation (3.4.1). Since there exist N modes $\omega_j(k)$, we have N vectors $w^0(\omega_j)$ as the general solution of the homogeneous system. Then, as in the case of medium with simple structure, to the given value of ω_j there correspond $2N_*$ generally complex roots $k_m(\omega_j)$, where N_* is the number of interacting neighbors.

To construct the particular solution w_s' of the inhomogeneous system, let us introduce the Green's matrix $G_{ss}(k, \omega)$, defining it by the relation

$$G_{ss'}(k, \omega) = [-\omega^2 I^{ss'} + \Phi^{ss'}(k)]^{-1} \tag{3.4.3}$$

and by the additional conditions of radiation, identical to those indicated in Sect. 2.8.

Then, for the particular solution $w_s'(k, \omega)$ we have

$$w_s'(k, \omega) = G_{ss'}(k, \omega) f^{s'}(k, \omega) . \tag{3.4.4}$$

It is convenient to present the expression for $G_{ss'}(k, \omega)$ in the form

$$G_{ss'}(k, \omega) = \frac{g_{ss'}(k, \omega)}{g(k, \omega)}, \tag{3.4.5}$$

where

$$g(k, \omega) = \mathrm{Det}\,[-\omega^2 I^{ss'} + \Phi^{ss'}(k)] \tag{3.4.6}$$

and $g_{ss'}(k, \omega)$ is the corresponding algebraic cofactor. Analogously to the case of medium with simple structure, the properties of the Green's matrix $G_{ss'}(k, \omega)$ are in principle defined by the roots of $g(k, \omega)$.

It can be shown that independent fo the number N of degrees of freedom, the number of roots of the determinant $g(k, \omega)$ is equal to $2N_*$, there N_* is the number of interacting neighbors. This follows from the fact that the equations of motion in the discrete form and in the representation of simple structure can be written as an equation in finite differences, the order of which is equal to $2N_*$. From the general theory of such equations [3.3], it follows that the number of the independent fundamental solutions is equal to the order of the equation. But the roots of the determinant $g(k, \omega)$ exactly correspond to a choice of these fundamental solutions.

Problem 3.4.1. Prove this fact by direct computation. [Hint: It is sufficient to show that when $\omega = 0$, the equation $g(k, 0) = 0$ is equivalent to an equation of degree $2N^*$ with respect to $z = \exp(inak)$].

In conclusion, let us consider the energy flux $\langle S \rangle_t$, connected with an elementary solution of the form

$$w_s = A_s \, e^{i(kx - \omega t)}, \tag{3.4.7}$$

where the real values of k and ω belong to one of the oscillation modes. Calculation, analogous to those carried out for the case of simple structure, yield

$$\langle S \rangle_t = \frac{1}{2} \bar{A}_s I^{ss'} A_{s'} \omega^2(k) \frac{d\omega(k)}{dk}. \tag{3.4.8}$$

Problem 3.4.2. Obtain this expression using (3.3.29).

It is easy to see that the coefficient of $\omega'(k)$ is equal to twice the mean kinetic energy density. But the mean densities of kinetic and potential energies for the solution of the form (3.4.7) are equal and, hence, the energy is propagated with the group velocity, as it is in the simple structure. Let us emphasize that this is true only for the elementary wave (3.4.7).

An elementary wave with essentially complex k, obviously does not carry energy.

3.5 Long-Wave Approximation and Connection with One-Dimensional Analog of Couple-Stress Theories

When proceeding to approximate models, the reasoning carried out in Sect. 2.10 remains valid. In particular, to the simplest models of the medium with a weak spatial dispersion there corresponds a polynomial approximation of the functions $\gamma(k)$, $\chi^p(k)$, $\Gamma^{pp'}(k)$. This approximation, as earlier, can be obtained either by restricting to the first terms the expansions into the series in k (long-wave approximation) or by constructing the approximation using the first roots of the matrix $\Phi^{ss'}(k)$ the choice depending on the field of applicability of the approximate model.

Let us consider a simpler model of the long wave approximation, which, in the medium with internal degrees of freedom, approximately takes into account scaling effects caused, for example, by a discrete structure or by a finite radius of interaction. The model's field of applicability is the long wavelength region and, generally, such fields that change shoowly enough over distances of the order of a characteristic scaling parameter. In this case it is necessary to consider a homogeneous unbounded medium. According to the reasonings indicated in Sect. 2.10 it is necessary to use the "first roots" approximation for a correct approximate formulation of boundary problems or for descritpion of local inhomogeneities.

Let us assume for simplicity that $\chi^p(k)$ and $\Gamma^{pp'}(k)$ are even functions of k [for $\gamma(k)$, this is fulfilled automatically in accordance with (3.3.23)]. Then restricting ourselves to the first terms after the zeroth approximation, we have

$$\gamma(k) = \gamma_0 - \gamma_2 k^2, \quad \chi^p(k) = \chi_0^p - \chi_2^p k^2, \quad \Gamma^{pp'}(k) = \Gamma_0^{pp'} - \Gamma_2^{pp'} k^2 . \tag{3.5.1}$$

This corresponds to the replacing in the equations of motion in the x representation, the integral operators by differential operators of the second order. Taking into account (3.3.24), we obtain

$$\begin{aligned} &\rho\ddot{u} - \partial^2(\gamma_0 + \gamma_2\partial^2)\, u - \partial(\chi_0^p + \chi_2^p\partial^2)\, \eta_p = q\,, \\ &I^{pp'}\ddot{\eta}_{p'} + \partial(\chi_0^p + \chi_2^p\partial^2)\, u + (\Gamma_0^{pp'} + \Gamma_2^{pp'}\partial^2)\, \eta_{p'} = \mu^p\,. \end{aligned} \tag{3.5.2}$$

These equations constitute the one-dimensional analog of the equations of the couple-stress, multipolar, etc. theory of elasticity. More accurately, in order, for example, to obtain the one-dimensional equations of the usual couple-stress theory of elasticity [3.4,5], two additional assumptions are to be made: (*a*) there are only two degrees of freedom, i.e., $p = 1$; *b*) one has to restrict himself by the zeroth approximation in the first equation, i.e. to set $\gamma_2 = 0$ in it, and to keep the second approximation in the second equation.

While the first assumption corresponds to the choice of the simplest particular model and cannot give rise to objections, the second assumption is difficult to be substantiated; it seems to be incorrect from the physical point of view.

Actually the authors of couple-stress theories did not make such assumptions, since the corresponding models were constructed on a different basis—it was assumed that the elastic energy depends on first derivatives of all the dynamical variables. From the above it follows that this assumption is physically unfounded. In fact, due to the translational invariance, the elastic energy cannot depend on the displacement u itself and must depend on the displacement spatial derivatives only; in the zeroth approximation there occur the first derivatives only whereas in the higher approximations higher derivatives also appear. At the same time due to the condition of stability, the elastic energy must, in the zeroth approximation, depend on microdeformations and microrotations η_p themselves and only in the higher approximations—on their derivatives.

The consequences resulting from here will be considered in a general form in the next section and then illustrated by examples of concrete models.

3.6 Elimination of the Internal Degrees of Freedom in the Acoustic Region

In many problems, the frequencies of free or forced vibrations within the limits of acoustic frequencies are of primary interest. Thus, in crystals this limit is of the order 10^{-13} s^{-1}, i.e. considerably exceeds the frequencies of mechanical oscillations. In such cases, the system of equations (3.3.24) can be essentially simplified and reduced to a single equation for the displacement of the mass centers.

Let us first consider the equations of the zeroth approximation

$$(-\omega^2\rho + k^2\gamma_0)\,u - ik\chi_0^p\eta_p = q\,,$$
$$(-\omega^2 I^{pp'} + \Gamma_0^{pp'})\,\eta_{p'} + ik\chi_0^p u = \mu^p\,. \tag{3.6.1}$$

It was shown in Sect. 3.4 that, due to the conditions of stability, the determinant of the matrix $\Gamma_0 - \omega^2 I$ vanishes only for optical frequencies $\omega_j(0) \neq 0$. Hence, in the region of frequencies $0 \leqslant \omega < \min\{\omega_j(0)\}$, the matrix $A^0(\omega) = (\Gamma_0 - \omega^2 I)^{-1}$ exists. But from here follows also the existence of the Hermitian operator

$$A(k,\omega) = [\Gamma(k) - \omega^2 I]^{-1} \tag{3.6.2}$$

in some finite neighborhood of the origin of the (k, ω)-plane. This neighborhood is bounded by the curve $\tilde{\omega}_1(k)$ given by the equation

$$\mathrm{Det}[\Gamma(k) - \tilde{\omega}_1^2(k) I] = 0\,. \tag{3.6.3}$$

In the general case, the study of the boundary curve $\tilde{\omega}_1(k)$ presents certain difficulties. It can only be claimed that at $k = 0$, it is tangent to the lowest optical mode and $A(k, \omega)$ undoubtedly exists in the long wavelength part of the acoustic region. However, for some particular models, it will be shown below that under the additional condition of absolute stability, the boundary curve $\tilde{\omega}_1(k)$ has the form schematically shown in Fig. 3.5 by the dotted curve and nowhere crosses the upper boundary of the acoustic frequencies region. In the general case, the region of existance of the matrix $A(k, \omega)$ (containing the origin) will be called the admissible acoustical region.

The matrix $A(k, \omega)$ can be expressed in terms of $A^0(\omega)$ and the coefficients Γ_s of the expansion of the matrix $\Gamma(k)$

$$\Gamma(k) = \sum_{s\ 0}^{\infty} (ik)^s \Gamma_s . \tag{3.6.4}$$

Let $j(m) = \{j_1, j_2, \dots .\}$ be an arbitrary set of integers such that $1 \leqslant j_\nu \leqslant m$ and $\sum_\nu j_\nu = m$. By $A^{j(m)}(0)$ let us denote the following fringed product:

$$A^{j(m)} = A^0 \Gamma_{j_1} A^0 \dots A^0 \Gamma_{j_\nu} A^0 \dots A^0, \quad j_\nu \in j(m) . \tag{3.6.5}$$

Then, in the appropriate convergence circle the following representation of $A(k, \omega)$ is valid:

$$A(k, \omega) = \sum_{m=0}^{\infty} (ik)^m A^m(\omega), \quad A^m(\omega) = \sum_j A^{j(m)}(\omega) . \tag{3.6.6}$$

For the first terms of the series we have

$$A = A^0 + ikA^0\Gamma_1 A^0 - k^2(A^0\Gamma_2 A^0 + A^0\Gamma_1 A^0 \Gamma_1 A^0) + \dots$$

Note that in the static case ($\omega = 0$) the matrix $A(k)$ is expressed in terms of Γ_s and the numerical matrix Γ_0^{-1}.

Problem 3.6.1. Prove (3.6.6). (Hint: Use the identity

$$(B + \varepsilon C)^{-1} = B^{-1} + \varepsilon B^{-1}CB^{-1} + \varepsilon^2 B^{-1}CB^{-1}CB^{-1} + \dots ,$$

which is valid for small ε).

With the help of the matrix $A(k, \omega)$, it is possible, in the admissible acoustical region, to solve the second equation of the system (3.3.24) for η_p and then eliminate η_p in the first equation. The final result has the form

$$[-\omega^2 \rho + k^2 c(k, \omega)]\, u = q_* , \tag{3.6.7}$$

$$\eta_p = -\, ika_p(k, \omega)\, u + A_{pp'}(k, \omega)\, \mu^{p'} . \tag{3.6.8}$$

Here the notations

$$c(k, \omega) = \gamma(k) - \chi^{+p}(k)\, A_{pp'}(k, \omega)\, \chi^{p'}(k)\,,$$
$$a_p(k, \omega) = A_{pp'}(k, \omega)\, \chi^{p'}(k)\,,$$
$$q_* = q - \mathrm{i}k\mu, \quad \mu = -\,\overline{a_p(k, \omega)}\, \mu^p \tag{3.6.9}$$

are introduced.

In the admissible acoustic region, (3.6.7) and (3.6.8) are the exact equivalent of the system (3.3.24). In the zeroth approximation in ω and k they reduce to

$$(-\,\omega^2\rho + c_0k^2)\, u = q_*^0\,, \tag{3.6.10}$$

$$\eta_p = -\,\mathrm{i}a_p(0, 0)\, ku + A^0_{pp'}(0)\, \mu^{p'}\,. \tag{3.6.11}$$

Equation (3.6.10) coincides with the equation of one-dimensional elastic continuum characterized by the elastic constant c_0. It is essential that this constant is directly determined from macroscopic experiments, for instance for homogeneous deformation or propagation of long waves. Its connection with microparameters follows from (3.6.9)

$$c_0 = \gamma_0 - \chi_0^p A^0_{pp'}(0)\, \chi_0^{p'} \tag{3.6.12}$$

(the quantities γ_0, χ_0^p, $A^0_{pp'}(0)$ can be expressed by force constants of the micromodel).

The second term in the right hand side of (3.6.12) can be interpreted as a contribution of internal degrees of freedom to the effective elastic modulus c_0, neither this term nor γ_0 however are macroscopically measurable quantities (separately). In fact, according to (3.6.1), for a direct measurement, for example, of γ_0, it would be necessary to conduct an experiment with fixed internal degrees of freedom η_p; this is practically impossible: the variables u and η_p are connected for both the homogeneous deformation, and the free oscillations.

The right hand side of (3.6.10) contains the equivalent density of external forces q_*^0 which according to (3.6.9) consists of two items. The first is the average density of the external forces q, the second one is the derivative with respect to x (with a negative sign), of the density of micromoments μ^0, in an exact correspondence with the usual theory of elasticity (recall that the contribution of the moments to the volume forces is equal to minus the divergence of the tensorial moment density). The effective density of external forces q_* can be interpreted analogously in the general case of the system (3.6.7) as well.

Developing further the analogy with the macroscopic theory, it is natural to interpret the quantity

$$\sigma(k, \omega) = c(k, \omega)\, \varepsilon(k, \omega) \quad [\varepsilon(k, \omega) = \mathrm{i}ku(k, \omega)] \tag{3.6.13}$$

as stress. Then (3.6.7) in the (x, t)-representation takes the form

$$\rho\ddot{u}(x,t) - \partial\sigma(x,t) = q_*(x,t), \tag{3.6.14}$$

and to the relation (3.6.13) there corresponds the operator Hooke's law with an integral operator both in spatial coordinate and in time.

Transforming (3.3.31) with (3.6.8) taken into account, we find the following expression for the density of elastic energy

$$\varphi(x,t) = \frac{1}{2}\,\sigma(x,t)\,\varepsilon(x,t)\,. \tag{3.6.15}$$

Hence the stress σ can also be defined (in the admissible acoustic region) as a generalized force, which corresponds to the strain ε, in full agreement with the usual theory of elasticity.

It is necessary to emphasize that, as distinct from the medium of simple structure, the static and dynamic elastic moduli do not coincide in the present case, i.e. spatial as well as time dispersion take place. The latter is by no means connected with energy dissipation[1] and is a peculiar effect of polarization due to the contribution of the internal degrees of freedom.

In view of (3.3.23) and (3.6.2), it follows from (3.6.9) that $c(k,\omega)$ is an even function of k and ω. Therefore, in the case of weak dispersion (small values of k and ω) Hooke's law in the second approximation is written in the form

$$\sigma(k,\omega) = (c_0 + l^2 c_2 k^2 - \tau^2 c_2' \omega^2)\,\varepsilon(k,\omega)\,, \tag{3.6.16}$$

or in the (x, t) representation

$$\sigma(x,t) = (c_0 - l^2 c_2 \partial^2 + \tau^2 c_2' \partial_t^2)\,(x,t)\,, \tag{3.6.17}$$

where l, τ are characteristic spatial and time scale parameters.

Equation (3.6.14) can be written in the form

$$I\ddot{u}(x,t) = \partial(c_0 - l^2 c_2 \partial^2)\,\partial u(x,t) = q_*(x,t)\,, \tag{3.6.18}$$

where $I = \rho - \tau^2 c_2' \partial_t^2$ is an operator inertial characteristic of the medium.

Let us summarize: the possibility of reducing (for sufficiently small k and ω) the equations of the medium of complex structure to the equation of the medium of simple structure and to equations explicitly solved with respect to internal degrees of freedom has principal significance and, in particular, explains why, in the macroscopic theory, the displacement is the only essential variable. This is illustrated most clearly by the second equation of the system (3.6.1) which is algebraic with respect to the internal degrees of freedom which therefore can be eliminated. This is caused, after all, by the special character of the law of transformation (3.3.1).

---

[1]This is not in contradiction with the known Kramers-Kronig relations [3.6], because $c(k,\omega)$ as a function of ω has poles at the points $\omega_j(k)$.

Note that these points are not usually taken into account in couple-stress theories of elasticity. The question concerning the limiting transition to the classical theory of elasticity, either is not raised at all or is "solved" in the following way: the equations (3.5.2) coincide with the equations of the usual theory of elasticity if all the elastic constants except γ_0 tend to zero, i.e. the internal degrees of freedom can be ignored. This reasoning, being valid formally, actually does not solve the problem of limiting transition in such cases when the internal degrees of freedom and the corresponding elastic bonds really exist. Then it is necessary to eliminate the internal degrees of freedom from the equations but not to ignore them. The second term in (3.6.12), generally speaking, is of the same order as γ_0 and must be considered.

It is also hardly reasonable to interpret the quantity $\tau(x) = \gamma_0\varepsilon(x)$ as the analog of stress and to introduce, generally speaking, an infinite set of the so-called couple stresses. The fact is that as distinct from $\sigma(x)$, the couple stresses and $\tau(x)$ do not satisfy a conservation (i.e. equilibrium) law and, in the zeroth long-wavelength approximation, $\tau(x)$ is not transformed into the usual stress.

In conclusion, let us note that in the region of higher frequencies, it is the displacement which should naturally be excluded from the equations of motion (3.3.24) by writing them in the form [compare (3.6.7,8)]

$$[-\omega^2 I^{pp'} + \Gamma_*^{pp'}(k, \omega)]\,\eta_{p'} = \mu_*^p\,, \tag{3.6.19}$$

$$u = \frac{-ik\chi^{+p}}{\omega^2\rho - k^2\gamma(k)}\,\eta_p - \frac{1}{\omega^2\rho - k^2\gamma(k)}\,q\,. \tag{3.6.20}$$

Here

$$\Gamma_*^{pp'}(k, \omega) = \Gamma(k) + \frac{k^2\chi^p(k)\chi^{+p'}(k)}{\omega^2\rho - k^2\gamma(k)}\,, \tag{3.6.21}$$

$$\mu_*^p = \mu^p - \frac{-ik\chi^p(k)}{\omega^2\rho - k^2\gamma(k)}\,q\,, \tag{3.6.22}$$

and the equations are valid in the admissible optical region $\omega^2\rho - k^2\gamma(k) > 0$. The boundary curve $\tilde{\omega}_0(k)$ passes through the origin and has the form schematically shown in Fig. 3.5.

3.7 Equivalent Medium of Simple Structure

Let us now consider the equations (3.3.24) to be given (with the additional condition $|k| \leq \pi/a$) and to describe some homogeneous medium of complex structure. Let us investigate the possibility of putting this description into one-to-one correspondence with a model of homogeneous medium of simple structure without any restrictions on the region of admissible frequencies.

Thus, we assume that the medium of complex structure under consideration is a representation of homogeneous medium of simple structure constructed by the method of Sect 3.2, with the characteristics

$$\rho(x) = \rho_0, \quad \Phi(x, x') = \Phi(x - x') \tag{3.7.1}$$

or, in the κ representation

$$\rho(\kappa) = 2\pi\rho_0\delta(\kappa - \kappa'), \quad \Phi(\kappa, \kappa') = \Phi(\kappa)\,\delta(\kappa - \kappa'). \tag{3.7.2}$$

The relation (3.2.42) now takes the form

$$\tilde{\Phi}^{ss'}(\kappa)\, e_s(\kappa) = \frac{1}{a}\,\Phi(\kappa)\, e^{s'}(\kappa). \tag{3.7.3}$$

From (3.2.6) with $\rho(\xi) = \rho_0$, we have

$$e^{s'}(\kappa) = \frac{a}{\rho_0}\, I^{ss'} e_s(\kappa). \tag{3.7.4}$$

Substituting in (3.7.3) and setting as usual

$$\omega^2(\kappa) = \frac{1}{\rho_0}\,\Phi(\kappa), \tag{3.7.5}$$

we find the necessary conditions which must be satisfied by $e_s(\kappa)$

$$[\tilde{\Phi}^{ss'}(\kappa) - \omega^2(\kappa)\, I^{ss'}]\, e_s(\kappa) = 0. \tag{3.7.6}$$

Problem 3.7.1. Show that these conditions are also sufficient, i.e. if the basis e_s, $e^{s'}$ satisfies the relations (3.7.6) and (3.7.4), then the model is homogeneous in the representation of simple structure. (Hint: Use the formulae of Sect. 3.2 and verify that the relations (3.7.2) are fulfilled).

Problem 3.7.2. Show that the structure (3.3.22) of the matrix $\Phi^{ss'}(k)$ always permits to set $e^0(\xi) = 1$ and to satisfy the condition $\Phi(\kappa) = 0$ when $\kappa = 0$.

Thus, the existence of a biorthogonal basis satisying (3.7.6) and (3.7.4) is necessary and sufficient to be able to relate to a medium of complex structure (under the condition $k \in B$) an equvalent homogeneous medium of simple structure. But the solutions of the equation (3.7.6) exist if and only if $\omega^2(\kappa)$ satisfies the equation (cf. (3.4.1))

$$\mathrm{Det}[\tilde{\Phi}^{ss'}(\kappa) - \omega^2(\kappa)\, I^{ss'}] = 0. \tag{3.7.7}$$

Obviously, N solutions $\omega_j^2(\kappa)$ of this equation exist. We shall construct one of them in the following way. When $|\kappa| < \pi/a$, (3.7.7) coincides with the dispersion equation (3.4.1) and we set $\omega(\kappa) = \omega_0(\kappa)$ where $\omega_0(\kappa)$ is the acoustic mode. Analogously, in the region $\pi/a < |\kappa| < 2\pi a$ let us set $\omega(\kappa) = \omega_1(\kappa - 2\pi/a)$ where $\omega_1(k)$ is the first optical mode, etc. (See Fig. 3.6). It is easy to see that so defined $\omega(\kappa)$ satisfies the equation (3.7.7).

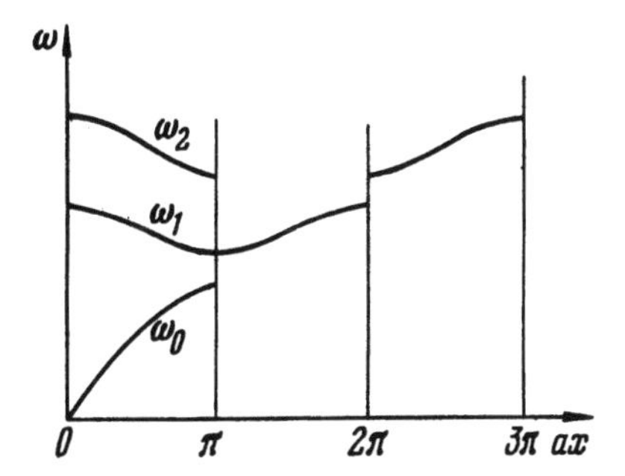

Fig.3.6. Discontinuous dispersion curve

It is essential that in general $\omega(\kappa)$ is discontinuous at the points $n\pi/a$ [in (3.7.6), its discontinuities are compensated by the corresponding discontinuities of $\Phi^{ss'}(\kappa)$]. Hence, in accordance with the results of Sect 2.7, the equivalent medium of simple structure has unbounded action at a distance (which corresponds to the discontinuities of $\omega(\kappa)$ according to the relation (2.7.19)) even in the case when the action at a distance in the medium of complex structure is bounded. It is not difficult to show that when $\omega^2(\kappa)$ is an anlytic or infinitely differentiable function and $N < \infty$, the action at a distance in the equivalent medium of simple structure decreases faster than any inverse power $|n|$. This follows from the properties of the coefficients of the Fourier series (2.4.35) for $\Phi(\kappa)$.

We emphasize that the unbounded action at a distance in the equivalent medium of simple structure is fictititious, since it is not connected with any physically measurable characteristics. The roots of the dispersion equation of the equivalent medium essentially coincide with the roots of the dispersion equation of the medium of complex structure, and hence the solutions also practically coincide as one would expect.

Thus we have established a principally important fact. Any quasi-continuum of complex structure can be considered as a representation of some homogeneous medium of simple structure. If the number of degrees of freedom is finite ($N < \infty$), then the medium of simple structure is also a quasi-continuum or a simple chain with the distance between particles $a_* = a/N$, i.e. $\kappa \in B_*$, where B_* is a circle of the radius a_*^{-1}. As $N \to \infty$, the medium of simple structure becomes a homogeneous elastic continuum.

The form of dispersion curves $\omega_j(k)$ allows one to distinguish the case of an "actual" medium of simple structure to which there corresponds a smooth $\omega(\kappa)$ and the case of "effective" medium with an infinite action at a distance of special type.

In the following section, we shall illustrate the above general results by a series of specific models, which are of interest in themselves.

3.8 Diatomic Chain

This model, shown in Fig. 3.1, due to its simplicity and clearness, deserves a detailed study.

In order to proceed to the collective cell variables from the quantities contained in the equation (3.1.1), let us introduce a biorthogonal basis, which in the given case, in accordance with (3.2.7), has the form (in the discrete representation)

$$e^0(j) = 1, \quad e^1(j) = \xi_j,$$
$$e_0(j) = \frac{\rho(j)}{a\rho}, \quad e_1(j) = \frac{\rho(j)\,\xi_j}{aI} \quad (j = 1, 2). \tag{3.8.1}$$

Here $\rho(j) = mj$ and (cf. (3.2.3))

$$\rho = \rho^{(0)} = a^{-1}\sum_j \rho(j) = a^{-1}(m_1 + m_2),$$
$$I = \rho^{(2)} = a^{-1}\sum_j \rho(j)\,\xi_j^2 = a^{-1}(m_1\xi_1^2 + m_2\xi_2^2) = -\rho\xi_1\xi_2. \tag{3.8.2}$$

Problem 3.8.1. Verify that the relations (3.2.5) with respect to the scalar product

$$(\rho, f) = \sum_j \rho(j)\, f(j) \tag{3.8.3}$$

are fulfilled.

Problem 3.8.2. Show that w_s and f^s calculated in accordance with the relations (3.2.10) coincide with u, η and q, μ, respectively (for the last ones–to within the factor a^{-1}), given by (3.1.8 and 9).

An expression for the matrix $\Phi^{ss'}(n)$ in terms of the stiffnesses of the elastic bonds $\Psi(n, j, j')$ can be found from (3.2.9), taking into account (3.1.7) and (3.1.6), and the corresponding one for the matrix $\Phi^{ss'}(k)$, is found from (3.3.25) and (3.2.26).

In particular, for the case of the interaction of two neighbors,

$$\Phi^{00}(k) = k^2\gamma(k) = 4\Psi a^{-1}(1 + \beta)\sin^2\frac{ak}{2},$$

$$\Phi^{10}(k) = \mathrm{i}k\chi(k)$$
$$= -2\Psi\sqrt{\frac{\lambda}{1 - \Delta^2}}\sin\frac{ak}{2}\left\{[(-\delta + \Delta(1 + \beta)]\sin\frac{ak}{2} + \mathrm{i}\cos\frac{ak}{2}\right\},$$

$$\Phi^{11}(k) = \Gamma(k)$$

$$= \Psi \frac{a\lambda}{1-\Delta^2}\left\{1 + \alpha - [1 - \beta - \Delta^2(1+\beta) + 2\delta\Delta]\sin^2\frac{ak}{2}\right\}. \quad (3.8.4)$$

Here, the relation $m_1 \xi_1 + m_2 \xi_2 = 0$ is taken into account and the notations

$$\Psi = \Psi(1,2,1), \quad \alpha\Psi = \Psi(0,1,2),$$
$$\beta\Psi = \Psi(1,1,1) + \Psi(1,2,2), \quad \delta\Psi = \Psi(1,2,2) - \Psi(1,1,1),$$
$$\Delta = \frac{m_2 - m_1}{m_2 + m_1}, \quad \lambda = \frac{4I}{\rho a^2}. \quad (3.8.5)$$

are introduced.

In the considered case the equations of motion (3.3.24) have the from $(|k| \leqslant \pi/a)$

$$[-\rho\omega^2 + k^2\gamma(k)]\,u(k,\omega) - ik\chi(-k)\,\eta(k,\omega) = q(k,\omega),$$
$$[-I\omega^2 + \Gamma(k)]\,\eta(k,\omega) + ik\chi(k)\,u(k,\omega) = \mu(k,\omega). \quad (3.8.6)$$

Problem 3.8.3. Show that, in order that forces not appear in the chain under homogeneous deformation (3.1.14), the fulfillment of the condition

$$\chi_0 + \Gamma_0 = 0 \quad \text{or} \quad \lambda(1+a)^2 = 1 - \Delta^2 \quad (3.8.7)$$

is necessary and sufficient. [Hint: In (3.1.16) pass to the k-representation or directly substitute $u_0(k) = -2\pi i\varepsilon_0\delta'(k)$, $\eta_0(k) = 2\pi\varepsilon_0\delta(k)$ in (3.8.6)].

It is interesting to note that only bonds between the nearest neighbors make a contribution to this condition, which is not a priori obvious. However if we assume that the second bonds are absent, then the relation (3.8.7) is immediately obtained by equating of forces of interaction among three neighboring particles under homogeneous deformation.

For the case of a homogeneous simple chain with interaction of two neighbors, it is obviously necessary to set $\Delta = \delta = 0$, $\alpha = 1$, $\lambda = 1/4$. Then

$$k^2\gamma(k) = 4\Psi a^{-1}(1+\beta)\sin^2\frac{ak}{2}, \quad ik\chi(k) = \frac{i}{2}\Psi\sin ak,$$
$$\Gamma(k) = \frac{1}{4}\Psi a\left[2 - (1-\beta)\sin^2\frac{ak}{2}\right]. \quad (3.8.8)$$

To the homogeneous simple chain there corresponds also the limiting transition $\alpha \to \infty$. In fact, the masses m_1 and m_2 in this case are rigidly connected, hence $\eta \to 0$ and the system (3.8.6) degeneratres into a single equation of the from (2.4.4) with $\Phi(k) = k^2\gamma(k)$ and $\rho = a^{-1}(m_1 + m_2)$.

For an arbitary diatomic chain, the dispersion equation (3.4.1) takes the from

$$\omega^2(k) = \omega_1^2[f_+(k) \mp \sqrt{f_-^2(k) + g(k)}]\,,$$

$$f_\pm(k) = \frac{\Gamma(k)}{2\Gamma_0} \pm \frac{I}{2\rho\Gamma_0} k^2\gamma(k), \quad g(k) = \frac{I}{\rho\Gamma_0^2} k^2\chi(k)\chi(-k)\,. \tag{3.8.9}$$

Here the minus (plus) sign before the root corresponds to the acoustic mode $\omega_0(k)$ (optical mode $\omega_1(k)$) and as easily seen, ω_1 is the optical frequency when $k = 0$, i.e. $\omega_1^2 = \Gamma_0/I$

From the stability requirement, we find the conditions (for $\mathrm{Im}\{k\} = 0$, $\mathrm{Re}\{k\} \neq = 0$)

$$f_+(k) > 0, \quad 0 < f_-^2(k) + g(k) < f_+^2(k), \quad \Gamma_0 > 0\,, \tag{3.8.10}$$

which imply restrictions on the admissible values of the parameters of the chain.

In the presence of interaction of two neighbors, taking into account (3.8.4), we have

$$f_\pm(k) = \frac{1}{2} - \frac{1 - \beta + 2\Delta\delta - (1 + \beta)[\Delta^2 \pm (1 - \Delta^2)]}{2(1 + \alpha)} \sin^2\frac{ak}{2}\,,$$

$$g(k) = \frac{1 - \Delta^2}{(1 + \alpha)^2} \sin^2\frac{ak}{2} \left\{[\delta - \Delta(1 + \beta)]^2 \sin^2\frac{ak}{2} + \cos^2\frac{ak}{2}\right\}\,,$$

$$\omega_1^2 = \frac{4\Psi}{m_1 + m_2} \cdot \frac{1 + \alpha}{1 - \Delta^2}\,. \tag{3.8.11}$$

For real k, in accordance with (3.3.23), $g(k) \geqslant 0$ and from (3.8.9) it follows that the acoustical and optical modes do not intersect, but generally speaking, can become tangent to each other at some point k_0, for which the radicand in (3.8.9) vanishes. Obviously, for this the conditions

$$\rho\Gamma(k_0) - Ik_0^2\gamma(k_0) = 0, \quad \chi(k_0) = 0\,, \tag{3.8.12}$$

are to be fulfilled, the second of them, due to the complex value of $\chi(k)$, being equivalent to two conditions.

In the presence of interaction of two neighbors, these conditions have the from

$$k_0 = \pi/a, \quad \delta = \Delta(1 + \beta), \quad \alpha = 1. \tag{3.8.13}$$

The boundary curve $\tilde{\omega}_1(k)$ of the admissible acoustic region for the diatomic chain, in accordance with (3.6.3), is given by the expression

$$\tilde{\omega}_1^2(k) = \omega_1^2 \frac{\Gamma(k)}{\Gamma_0}\,. \tag{3.8.14}$$

Taking into account (3.8.9), we have

$$\tilde{\omega}_1^2(k) - \omega_0^2(k) = \omega_1^2 [f_-(k) + \sqrt{f_-^2(k) + g(k)}] ,$$
$$\omega_1^2(k) - \tilde{\omega}_1^2 (k) = \omega_1^2 [-f_-(k) + \sqrt{f_-^2(k) + g(k)}] . \quad (3.8.15)$$

Thus $\tilde{\omega}_1(k)$ lies between the curves $\omega_0(k)$ and $\omega_1(k)$ without intersecting them, as shown in Fig. 3.5; however, it can become tangent to one of them at the point k_0, which satisfies the conditions

$$\rho\Gamma(k_0) - Ik_0^2\gamma(k_0) \gtrless 0, \quad \chi(k_0) = 0, \quad (3.8.16)$$

where the upper (lower) sign of inequality corresponds to the acoustical (optical) mode and the sign of equality corresponds to the case when all the three curves are tangent.

In the presence of interaction of two neighbors, the boundary curve $\omega_1(k)$ becomes tangent to the acoustic mode if

$$k_0 = \pi/a, \quad \delta = \Delta(1 + \beta), \quad \alpha > 1 . \quad (3.8.17)$$

Thus, for the diatomic chain, the system (3.8.6) can be replaced through almost the entire acoustic region by (3.6.7), containing $u(\kappa, \omega)$ only, and by (3.6.8) explicitly solved with respect to $\eta(k, \omega)$, where according to (3.6.9)

$$c(\kappa, \omega) = \gamma(k) - \frac{|\chi(k)|^2}{\Gamma(k) - \omega^2 I} . \quad (3.8.18)$$

Problem 3.8.4. Show that if the interaction of two neighbors is considered, then the effective moduli of the zeroth and the second approximations, which appear in (3.6.16,17), are equal to

$$c_0 = \Psi_a \frac{\alpha + \beta + \alpha\beta}{1 + \alpha}, \quad c_2' = \Psi a * \frac{1}{1 + \alpha},$$
$$c_2 = \Psi a \left[-\frac{1 + \beta}{12} + \frac{4 - 3[\Delta(1 + \beta) - \delta]^2}{12(1 + \alpha)} - \frac{1 - \beta - \Delta^2(1 + \beta) + 2\Delta\delta}{4(1 + \alpha)^2}\right], \quad (3.8.19)$$

if one lets $l = a$, $\tau = \omega_1^{-1}$.

Note that for stable bonds $\alpha > 0$, $\beta > 0$, the modulus c_0 is always positive and $\omega(k) \approx c_0 k$ for small values of k.

As it was pointed out in Sect. 3.6, in the region of high frequencies, the displacement can be eliminated from the equations of motion and the microstrain η is to be considered the fundamental variable, satisfying the equation (3.6.19) in which the indices p, p' are omitted. The boundary curve $\tilde{\omega}_0(k)$ of the admussible optical region is defined by the expression

$$\tilde{\omega}_0^2(k) = \omega_1^2 \cdot \frac{1}{\rho \Gamma_0} k^2 \gamma(k) . \tag{3.8.20}$$

Taking into account (3.8.9) and (3.8.15) we find the curious relations

$$\omega_1^2(k) - \tilde{\omega}_0^2(k) = \tilde{\omega}_1^2(k) - \omega_0^2(k), \quad \omega_0^2(k) \leqslant \tilde{\omega}_0^2(k) \leqslant \omega_1^2(k) , \tag{3.8.21}$$

from which it follows that in the case of diatomic chain, the admissible optical region covers all the optical frequencies.

Let us now consider a transition to the equivalent homogeneous simple chain. According to the results of Sect. 3.7 the dispersion curve $\omega(\kappa)$ coincides with $\omega_0(\kappa)$ for $|\kappa| < \pi/a$ and with $\omega_1(\kappa - 2\pi/a)$ for $\pi/a < |\kappa| \leqslant 2\pi/a$. The simplest way of obtaining the corresponding basis $e^s(\kappa)$, $e_s(\kappa)$ is to use (3.8.1), taking into account that for a homogeneous chain $m_1 = m_2$ and $\xi_2 = -\xi_1 = a_*/2$, where $a_* = a/2$ is the distance between the particles. In order to describe the basis in the κ representation, it is necessary to continue by the zero the functions of the discrete cell variable j and to use the formula (2.2.7) for $e^s(\kappa)$ (replacing k by $\kappa \in B_*$) and the formula (2.2.13) for the mass density $\rho(\kappa)$. We have

$$\begin{aligned} e^0(\kappa) &= 2a_* e_0(\kappa) = a_*(\mathrm{e}^{\mathrm{i}a_*\kappa} + \mathrm{e}^{2\mathrm{i}a_*\kappa}) , \\ e^1(\kappa) &= \frac{a_*}{2} e_1(\kappa) = \frac{a_*^2}{2} (-\mathrm{e}^{\mathrm{i}a_*\kappa} + \mathrm{e}^{2\mathrm{i}a_*\kappa}) . \end{aligned} \tag{3.8.22}$$

Let us also take into account that, in the given case, $\tilde{f}(\kappa)$ must correspond to the periodic continuation of $f(k)$, $k \in B$, to the circle B_* of the radius a_*^{-1}; in particular (cf. (3.2.21)),

$$\tilde{\delta}(\kappa) = \delta(\kappa) + \delta(\kappa \pm \pi/a_*), \quad \kappa \in B_* . \tag{3.8.23}$$

Problem 3.8.5. Verify that the basis (3.8.32) satisfies the conditions (3.7.6) and also the relations (3.2.28) and (3.2.29).

Using the formulae introduced in the end of Sect. 3.2, we can now establish a one-to-one correspondence between an arbitrary diatomic chain given by the equations (3.8.6), and a homogeneous chain of simple structure.

Problem 3.8.6. By direct calculation, verify the validity of the relations (3.2.33) and (3.2.37) between $\Phi^{ss'}(k)$ and $\Phi(\kappa, \kappa')$, taking into account that for the equivalent chain $\Phi(\kappa, \kappa') = \rho\omega^2(\kappa)\delta(\kappa - \kappa')$.

Problem 3.8.7. Show that to the diatomic chain with the characteristics (3.8.8) there corresponds an equivalent chain with two interacting neighbors.

To complete the analysis of the diatomic chain, let us investigate the complex dispersion curves $\Omega(\xi)$ which were introduced in Sect 3.1.8, when considering the simple chain.

Let

$$\Omega = \frac{m_1 + m_2}{4\Psi}\,\omega^2, \quad \zeta = \sin^2\frac{ak}{2}. \tag{3.8.24}$$

Then, for the case of interaction of two neighbors, the dispersion relation (3.8.9) with (3.8.11) taken into account takes in the variables Ω, ζ the form

$$a_{11}\Omega^2 + 2a_{12}\Omega\zeta + a_{22}\zeta^2 + 2a_1\Omega + 2a_2\zeta = 0\,, \tag{3.8.25}$$

where

$$\begin{aligned} a_{11} &= 1 - \Delta^2, \quad a_{12} = -(\beta - \delta\Delta), \quad a_{22} = \beta^2 - \delta^2\,, \\ a_1 &= -\frac{1+\alpha}{2}, \quad a_2 = \frac{\beta + \alpha + \alpha\beta}{2}. \end{aligned} \tag{3.8.26}$$

From the expression

$$D \stackrel{\text{Def}}{=} a_{11}a_{12} - a_{12}^2 = -(\delta - \Delta\beta)^2 \leqslant 0 \tag{3.8.27}$$

it follows that, except for degenerate cases, the dispersion curve (3.8.25) can be either a parabola ($D = 0$) or a hyperbola ($D < 0$).

The values $0 \leqslant \zeta \leqslant 1$ correspond on the cylinder $\mathscr{B}$, to the real axis $-\pi/a \leqslant k \leqslant \pi/a$ and hence the curve $\Omega(\zeta)$ must pass through the origin and intersect the strip $0 \leqslant \zeta \leqslant 1$ twice, taking positive values (the condition of stability). The lower segment of the curve is the acoustic mode and the upper one is the optical one.

Recall that to the extrema of the dispersion curve $\Omega(\zeta)$ when $\Omega \geqslant 0$ there correspond determining singular points of the "saddle" type; for a maximum the trajectories leave the directors and for a minimum they join them. Depending on the position of the extrema, a singular point will be located either on the real axis or on the imaginary axis ($-\infty < \zeta < 0$) or on the generator $\mathrm{Re}\{k\} = \pm\pi/a$ ($1 < \zeta < \infty$). From purely geometrical considerations, it follows that, under the fulfillment of the conditions of stability, there cannot be extrema (for $\Omega \geqslant 0$) on the semi-axis $-\infty < \zeta < 0$ and, hence, the determining singular points cannot appear on the imaginary axis of the cylinder $\mathscr{B}$.

Let us consider some examples. For the parabola ($\delta = \Delta\beta$) the conditions of stability lead to the inequalities

$$\alpha > -1, \quad \beta + \frac{1+\alpha}{2(1-\Delta^2)} > 1, \quad \beta > -\frac{\alpha}{1+\alpha}. \tag{3.8.28}$$

The dispersion curves of such a type are shown in Fig. 3.7. The case of the interaction of the nearest neighbors ($\beta = 0$, $\delta = 0$) is presented in Fig. 3.8 by the parabola with the axis $\Omega = 1 + \alpha/2(1 - \Delta^2)$, while, for the stable chains ($\alpha > 0$), the vertex of the parabola is situated to the right from the vertical $\zeta = 1$.

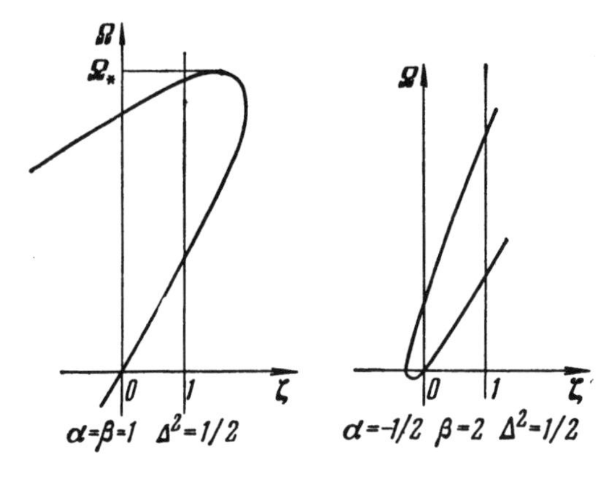

Fig. 3.7.

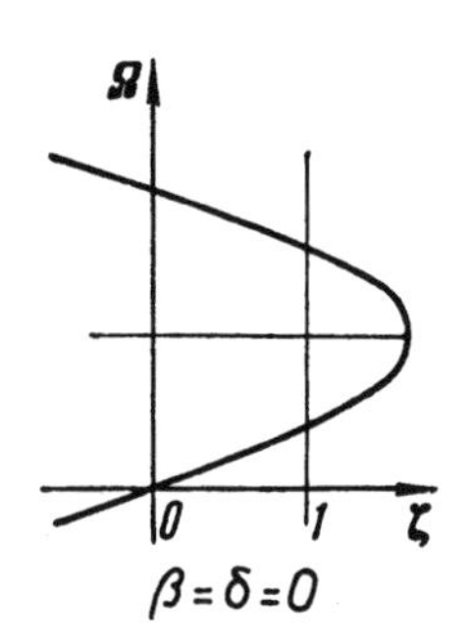

Fig. 3.8.

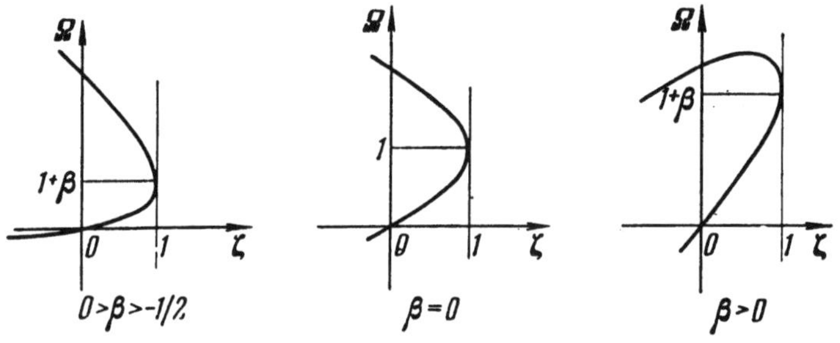

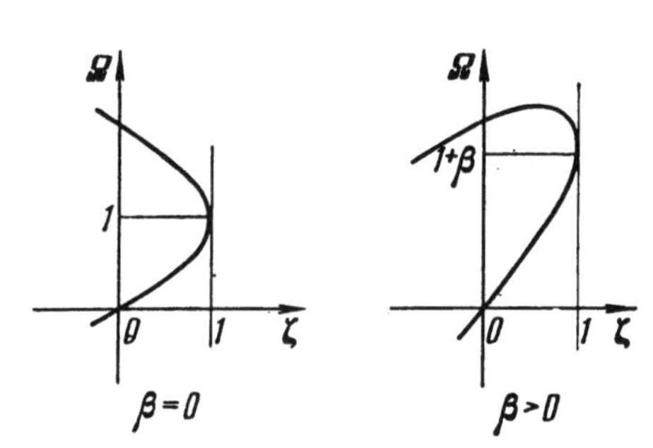

Fig. 3.9.

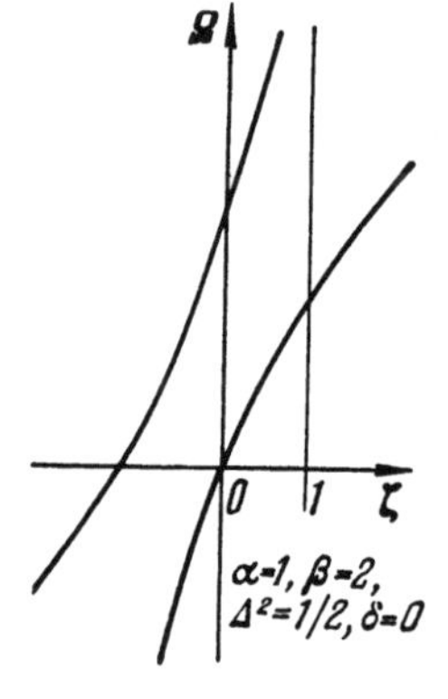

◀ Fig. 3.10.

Fig.3.7–10. Different types of dispersion curves

For the limiting case $\Delta = \delta = 0$, $\alpha = 1$ (homogeneous simple chain with the interaction of two neighbors in the representation of complex structure), the parabola is always tangent to the vertical $\zeta = 1$ (Fig. 3.9), the condition of stability requiring $\beta > -1/2$. In order to compare with the representation of the simple structure, it is necessary to pass to the variable $\zeta_* = \sin^2(a_* k/2 = \sin^2(ak/4)$ from the variable $\zeta = \sin^2(ak/2)$ This transformation is not unique and, apart from the required solution, gives also the solution which has no physical meaning [the parabola $\Omega(\zeta)$ is transformed into two parabolas in the plane Ω, ζ_*, one of which does not pass through the origin].

The example of the dispersion curve in the form of a hyperbola($D < 0$) is shown in Fig. 3.10.

3.9 The Cosserat Model

As shown in Sect. 3.1, the equations of motion for the Cosserat model coincide with the equations of motion of a diatomic chain and hence, in the (x, t) representation have the form

$$\rho\ddot{u} - \partial\gamma\,\partial u - \partial\chi^+\eta = q\,,$$
$$I\ddot{\eta} + \chi\,\partial u + \Gamma\eta = \mu\,, \tag{3.9.1}$$

but now u is the transverse displacement, η is the microrotation, μ is the density of the corresponding micromoments and I is the density of the moments of inertia of the particles. The other quantities have the same meaning as in the case of a diatomic chain.

The equations, written in this form, already take into account the condition of translational invariance. According to the new physical meaning of the variables, it is necessary to assume the fulfillment of the additional condition of invariance of the energy with respect to the rotation of the medium as a whole i.e. with respect to the transformation

$$u(x) \to u(x) + ax, \quad \eta(x) \to \eta(x) + a\,. \tag{3.9.2}$$

It coincides formally with the transformation of the homogeneous strain (3.1.14), but the requirement of invariance is now stronger, namely not only the equations of motion but also the elastic energy must be invariant.

In an attempt to formulate this requirement for an unbounded medium, a difficulty arises due to the fact that in this case, the energy is not defined by the displacements, increasing linearly as $|x| \to \infty$. Therefore, we shall first consider a bounded medium and then we pass to the limiting case.

In order to obtain conditions on the operators γ, χ and Γ, arising from the necessity of the invariance of energy with respect to rotation, we can use the general formulae (3.3.10), (3.3.11), obtained for an arbitrary group of transformations. Comparison of (3.9.2) with (3.3.8) shows that in the given case

$$\xi_a = 0,\; \zeta^b \equiv \zeta = a,\; \tau_s = (x, 1)\,. \tag{3.9.3}$$

Hence, only the condition (3.3.11) is to be taken into account (or (3.3.12), which is equivalent); here it takes the form

$$\int \Phi^{00}(x, x')\, x'\, dx' + \int \Phi^{01}(x, x')\, dx' = 0\,,$$
$$\int \Phi^{10}(x, x')\, x'\, dx' + \int \Phi^{11}(x, x')\, dx' = 0\,. \tag{3.9.4}$$

In view of (3.9.1) we find

$$\int [\gamma(x, x') + \chi^+(x, x')]\, dx' = \text{const}\,,$$
$$\int [\chi(x, x') + \Gamma(x, x')]\, dx' = 0 \tag{3.9.5}$$

or in the k representation

$$\gamma(k, 0) + \chi^+(k, 0) = \text{const}\cdot\delta(k)\,,$$
$$\chi(k, 0) + \Gamma(k, 0) = 0\,. \tag{3.9.6}$$

As for a bounded medium, $\gamma(k, k')$ and $\chi^+(k, k')$ are analytical functions of k, k', the constant of integration is to be set equal to zero.

We assume that these conditions are fulfilled also in the limiting case of unbounded medium. This corresponds to an extension of the definition of the functional of energy such that it will possess the required properties with respect to transformations (3.9.2), i.e. to a special regularization of the functional.

We can now proceed to a case of fundamental interest, i.e. to the case of a homogeneous (and hence, unbounded) Cosserat medium, for which

$$\gamma(k, k') = \gamma(k)\, \delta(k - k'), \quad \chi(k, k') = \chi(k)\, \delta(k - k')\,,$$
$$\Gamma(k, k') = \Gamma(k)\, \delta(k - k'), \quad \chi^+(k) = \chi(-k)\,. \tag{3.9.7}$$

The equations of motion in the (k, ω) representation take the form

$$[-\rho\omega^2 + k^2\gamma(k)]\, u(k, \omega) - \mathrm{i}k\chi^+(k)\, \eta(k, \omega) = q(k, \omega)\,,$$
$$[-I\omega^2 + \Gamma(k)]\, \eta(k, \omega) + \mathrm{i}k\chi(k)\, u(k, \omega) = \mu(k, \omega) \tag{3.9.8}$$

and, formally, coincide with the equations (3.8.6) for the diatomic chain; however, in the general case the restriction $|k| \leqslant \pi/a$ is not assumed.

It is easy to see that the conditions (3.9.6) take the form

$$\gamma_0 + \chi_0 = 0, \quad \chi_0 + \Gamma_0 = 0\,, \tag{3.9.9}$$

where, as usual, the index zero corresponds to the value of the function when $k = 0$.

Problem 3.9.1. Show that for a discrete model these conditions are equivalent to the conditions (3.1.18).

Problem 3.9.2. For the Cosserat model, the quantity $\Omega = \partial u$ has the meaning of macrorotation of an element of the medium (as distinct from the microrotation η). Show that the conditions (3.9.9) are equivalent to the condition that in the long wavelength approximation the density of energy depends only on the difference $\Omega - \eta$ which is invariant with respect to the transformation (3.9.2).

Comparison of (3.9.9) with (3.8.7) shows that the conditions of invariance with respect to rotation are stronger than the condition of homogeneous deformation, although, formally, the same transformations (3.1.14) and (3.9.2) correspond to them. We shall see below that the conditions (3.9.9) lead to an important distinction between the dispersion laws for the chain and the Cosserat model.

We emphasize once again the essentially different origin of the conditions (3.9.9) and (3.8.7). The first ones are valid for any Cosserat model, being independent of the model's special structure as a consequence of invariance of the energy with respect to fundamental group of rotations.

The second ones are a consequence of invariance only of the equations of motion with respect to the group of internal symmetry, which corresponds to a choice of a specific model of a chain. Note that, if one proceeds only from the invariance of the equation of motion (3.9.8) with respect to the group of rotations, then the first of the conditions (3.9.9) is lost since the first equation of the system (3.9.8) is identically invariant with respect ot rotation.

For the Cosserat model, the invariance of the equations of motion with respect to pure bending

$$u_0(x) = ax^2, \quad \eta_0(x) = 2ax \tag{3.9.10}$$

is the analog of the requirement of invariance with respect to homogeneous deformation,

Problem 3.9.3. Show that this requirement leads to the additional condition (by a prime derivatives in k are denoted)

$$\chi_0' + \Gamma_0' = 0\,, \tag{3.9.11}$$

which is equivalent to the condition (3.1.20) concerning the force constants.

The system of masses (with finite moments of inertia) connected with each other by elastic rods described in Sect. 3.1. may serve as a micromodel of the Cosserat medium. The force constants appearing in (3.1.10) can be found if one writes the connection between the transverse displacement of the rod u, the angle of rotation η, the rod transverse force q and the moment μ, acting on the rod,

$$\begin{pmatrix} u \\ \eta \end{pmatrix} = A \begin{pmatrix} q \\ \mu \end{pmatrix} \text{ or } \begin{pmatrix} q \\ \mu \end{pmatrix} = A^{-1} \begin{pmatrix} u \\ \eta \end{pmatrix}, \tag{3.9.12}$$

where the matrix A is given by the formulae of the strength of materials.

For the case when neighboring masses are connected by homogeneous rods

$$\Phi^{00}(1) = -\Psi_*, \quad \Phi^{01}(1) = -\Phi^{10}(1) = -\frac{1}{2} a\Psi_*,$$

$$\Phi^{11}(1) = \frac{1}{6}(1 - \nu) a^2 \Psi_*,$$

$$\Psi_* = \frac{12EJ}{a^3(1 + 2\nu)}, \quad \nu = \frac{6EJ}{\kappa a^2 GF}. \tag{3.9.13}$$

Here EJ and GF are stiffnesses of the rod in bending and shear respectively, κ is a dimensionless coefficient which depends on the shape of the cross-section. The other force constants are found from the symmetry conditions (3.1.11), the translations invariance (3.1.12) and the rotational invariance (3.1.18). In the k representation

$$\Phi^{00}(k) = k^2\gamma(k) = 4\Psi_* a^{-1} \sin^2 \frac{ak}{2},$$

$$\Phi^{10}(k) = ik\chi(k) = -i\Psi_* \sin ak,$$

$$\Phi^{11}(k) = \Gamma(k) = \Psi_* a\left[1 + \frac{2}{3}(\nu - 1) \sin^2 \frac{ak}{2}\right]. \tag{3.9.14}$$

It is directly verified that the given model satisfies the condition of pure bending (3.1.20) or (3.9.11).

Comparison of (3.9.14) with (3.8.4) shows that, for the Cosserat model under consideration to be mathematically identical with the diatomic chain with the interaction of two neighbors, the latter must obey the additional conditions

$$(1 + \alpha)(1 + \beta) = 1, \quad \lambda = (1 + \beta)^2(1 - \Delta^2), \quad \delta = \Delta(1 + \beta) \tag{3.9.15}$$

and, under the assumption of the equality of inertial characteristics of both models it is necessary to set

$$\Psi_* = \Psi(1 + \beta), \quad \frac{2\nu + 1}{3} = \frac{\lambda}{(1 + \beta)^2} - \frac{1 - \beta}{1 + \beta}. \tag{3.9.16}$$

Note that the conditions (3.9.15) can also be obtained by applying the requirements (3.8.7 and 11) to the chain.

From the conditions (3.9.15), it follows that two independent dimensionless parameters are left for the equivalent chain, for which α and Δ can be taken, for example. Then for the identity of the two models it is necessary and sufficient to set

$$\lambda = \frac{1-\Delta^2}{(1+\alpha)^2}, \qquad \frac{2\nu+1}{3} = \frac{\alpha^2-\Delta^2}{(1+\alpha)^2}. \tag{3.9.17}$$

The conditions of stability (3.8.10) for the Cosserat model yield

$$\lambda = \frac{4I}{\rho a^2} > a, \quad 2\nu + 1 > 0 \tag{3.9.18}$$

or in terms of the parameters α and Δ

$$\Delta^2 < 1, \quad \alpha^2 - \Delta^2 > 0. \tag{3.9.19}$$

On the other hand, if we consider Δ^2 and α to be the parameters of the chain, then the additional restriction $\Delta^2 > 0$ arises resulting from the reality of Δ as is seen from (3.8.5). For the Cosserat model this limitation, of course, does not exist. Furthermore, it follows from the first condition of (3.9.15), that at least one of the parameters α or β is negative, i.e. at least one of the bonds between the nearest or between the second neighbors in the chain is unstable.

Thus, although the transverse vibrations of the Cosserat model with the interaction of nearest neighbors are formally identical to the particular case of longitudinal vibrations of a diatomic chain with the interaction of two neighbors, the parameters of the latter take non-real values–negative α or β and possibly imaginary values of Δ. This is not surprising if we recall that the energy of the equivalent chain should not change during homogenous deformation, which corresponds to the rotation for the Cosserat model.

If we do not compare the Cosserat model with the chain, then rod bonds are to be considered unpaired. In fact, only by this could it be explained why in the presence of interaction of nearest neighbors, additional roots of the dispersion equation appear, which are typical for the chain with interaction of two neighbors.

This phenomenon of the additional roots has also to be taken into account when constructing the general solution and the Green's matrix $G_{ss'}(k, \omega)$.

As an example, let us present the static Green's matrix $G_{ss'}(k)$ constructed for the interaction of nearest neighbors according to the formulae (3.4.5), (3.4.6)

$$G_{ss'}(n) = \frac{a}{6EJ} \sum_{m=0,1,-1} A_{ss'}^{(m)} \left(|n+m|^3 - |n+m|\right),$$

$$A_{ss'}^{(0)} = \begin{pmatrix} \frac{a^2(2+\nu)}{6} & 0 \\ 0 & 1 \end{pmatrix}, \quad A_{ss'}^{(\pm 1)} = \begin{pmatrix} \frac{a^2(1-\nu)}{12} & \pm\frac{a}{4} \\ \mp\frac{a}{4} & -\frac{1}{2} \end{pmatrix}. \tag{3.9.20}$$

Apart from the term $|n+m|$, typical for the static Green's function of the chain with interaction of nearest neighbors, there is also the term $|n+m|^3$, caused by the fact that for the acoustic mode in the long-wavelength region $\omega_0^2(k) \sim k^4$ i.e. when $\omega = 0$, the fourfold root $k = 0$ exists. For the general case of the diatomic chain with the interaction of two neighbors $\omega_0^2(k) \sim k^2$ and the structure of the Green's matrix is analogous to (3.9.10), but $|n+m|^3$ is replaced by the term proportional to $\exp[iak_1(n+m)]$, where k_1 is the corresponding root of the determinant $g(k)$.

Let us now consider the long wavelength approximation for the Cosserat model, both in the acoustic and optical regions.

Problem 3.9.4. Verify that a consequence of the conditions (3.9.9) is that $\omega_0^2(k) \sim k^4$ and that the elastic constant c_0 vanishes, c_0 being defined by the expression (3.6.12) (cf. also (3.8.19)) and show that this is impossible for a chain with stable bonds.

Problem 3.9.5. Show that in the acoustical region of frequencies the equation of zeroth approximation for the Cosserat micromodel considered above coincides, as it should, with the equation of transverse vibrations of the rod

$$\rho \frac{\partial^2 u}{\partial t^2} + EJ \frac{\partial^4 u}{\partial x^4} = q_* . \tag{3.9.21}$$

Thus, for sufficiently long wavelengths and low frequencies, $\omega(k) \sim k^2$, this is to be taken into account in a concrete construction of longwave approximations. In the second approximation

$$c(k, \omega) = (c_{20}k^2)_0 + (c_{02}\omega^2 + c_{40}k^4)_1 + (c_{22}k^2\omega^2 + c_{60}k^6)_2 , \tag{3.9.22}$$

where the brackets indicate the order of approximation. For the model considered above, taking into account (3.8.18) and (3.9.14),

$$c_{20} = EJ, \quad c_{02} = -I, \quad c_{40} = -\frac{E^2J^2}{\kappa GF},$$

$$c_{22} = \frac{2EJ}{\kappa GF} I, \quad c_{60} = \frac{13EJa^4}{360} . \tag{3.9.23}$$

The substitution of (3.9.22) in (3.6.7) yields the equation of the second approximation for the Cosserat model, which in the considered case has the form

$$\left[\rho \frac{\partial^2 u}{\partial t^2} + EJ \frac{\partial^4 u}{\partial x^4}\right]_0 + \left[\frac{E^2J^2}{\kappa GF} \frac{\partial^6 u}{\partial x^6} - I \frac{\partial^4 u}{\partial x^2 \partial t^2}\right]_1 + \left[\frac{13EJa^4}{360} \frac{\partial^8 u}{\partial x^8} - \frac{2EJ}{\kappa GF} I \frac{\partial^6 u}{\partial x^4 \partial t^2}\right]_2 = q_* . \tag{3.9.24}$$

In the zeroth approximation (of order k^4), the contribution is made only by the bending stiffness and the mass density. The first approximation ($\sim k^6$) considers in addition the shear stiffness and the moments of inertia of the masses. The discreteness parameter appears only in the second approximation $\sim k^8$.

It is interesting to compare the approximate equations of the Cosserat model with the known improved Timoshenko theory of vibrations of a rod [3.7]. With this in mind, let us recall that (3.6.7) can be presented in different equivalent forms; this can be achieved by multiplying by an operator, which has no zeros or poles in the admissible acoustic region. As such an operator, let us choose $[\omega^2 I - \Gamma(k)/\chi(k)]$, which, in accordance with (3.9.9), coincides with the identity operator in a small neighborhood or the origin of the k, ω plane. Then the eguation of the second approximation takes the form

$$\left[\rho\frac{\partial u^2}{\partial t^4} + EJ\frac{\partial^4 u}{\partial x^4}\right]_0 - \left[\left(\frac{\rho EJ}{\kappa GF} + I\right)\frac{\partial^4 u}{\partial x^2\,\partial t^2}\right]_1 + \left[\frac{EJa^4}{240}\frac{\partial^8 u}{\partial x^8} + \frac{\rho a^2}{12}\left(\frac{a^2}{15} + \frac{EJ}{\kappa GF} + I\right)\frac{\partial^6 u}{\partial x^4\,\partial t^2} + \rho I\left(\frac{a^2}{12EJ} + \frac{1}{\kappa GF}\right)\frac{\partial^4 u}{\partial t^4}\right]_2 = q_{**}\,. \qquad (3.9.25)$$

where q_{**} is evidently connected with q_* by (3.6.7) or (3.9.24).

When $a = 0$ all the terms in the bracket $[\;]_2$, except the last one, vanish and (3.9.25) coincides with Timoshenko's equation. This is connected with the fact, that Timoshenoko's equation corresponds to the homogeneous rod without microstructure, which is the limiting case of the model investigated above. Generally, the zeroth and first approximations can be considered as universal ones (to within equivalent transformations) in the sense that they do not depend on the specific choice of the Cosserat model, which satisfies the conditions (3.9.9). At the same time, the second approximation already contains some information about the microstructure of the model.

In the region of optical frequencies the microrotation η becomes the basic variable, and the equation of motion must be written in a form analogous to (3.6.19). Now, when proceeding to the long-wavelength approximation, ω cannot be considered to be a small quantity. Preserving the terms of the zeroth and second order in k, we find

$$\left[I\frac{\partial^2\eta}{\partial t^2} + \frac{24EJ}{a^2(\nu+2)}\eta\right]_0 - \left[EJ\left(1 - \frac{6}{\nu+2} + \frac{12\lambda^2}{\nu+2}\right)\frac{\partial^2\eta}{\partial x^2}\right]_1 = \mu_*\,. \qquad (3.9.26)$$

In the zeroth approximation, the masses perform independent rotational vibrations with an optical frequency ω_1. Only in the next approximation do these vibrations become coupled and can to propagate.

In conclusion, let us construct the complex dispersion curves for the Cosserat model. In the given case, the coefficients of the dispersion equation (3.8.25) have the form

$$a_{11} = \frac{\lambda(1+2\nu)}{3}, \quad a_{12} = \frac{2(1-\nu)}{3} - \lambda, \quad a_{22} = 4,$$

$$a_1 = -1, \quad a_2 = 0, \quad \Omega = \frac{\rho a^1}{8EJ}\omega^2. \tag{3.9.27}$$

From the expression

$$D = a_{11}a_{22} - a_{12}^2 = 4\lambda - \left[\lambda + \frac{2(1-\nu)}{3}\right]^2 \tag{3.9.28}$$

it follows that in the region of stable values of λ and ν, dispersion curves in the form of hyperbolae ($D < 0$), parabolae ($D = 0$) and ellipses ($D > 0$) are possible. The last of these was not possible for the diatomic chain with the interaction of two neighbors because this would correspond to an imaginary value of the parameter Δ.

The dispersion curves are tangent to the axis ζ at the origin (twofold roots ζ, which correspond to fourfold roots k at zero) and intersect twice the region $0 \leqslant \zeta \leqslant 1$ when $\Omega > 0$, forming the acoustical (lower) and the optical (upper) modes.

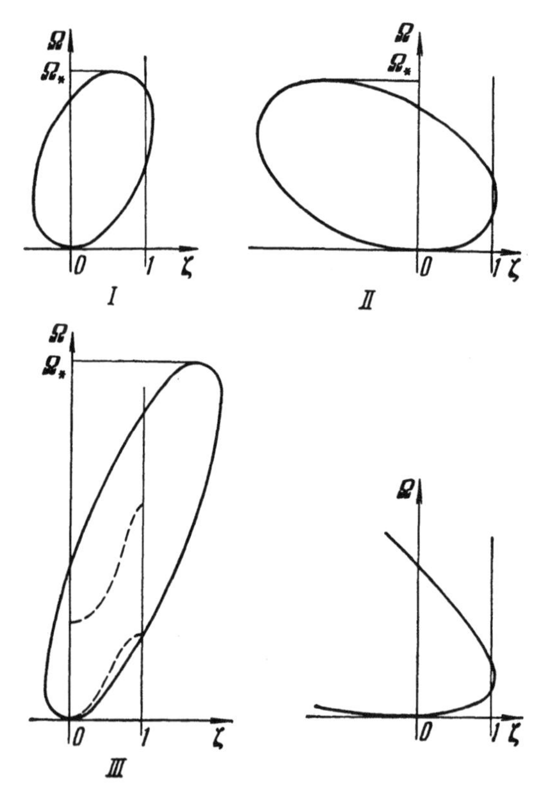

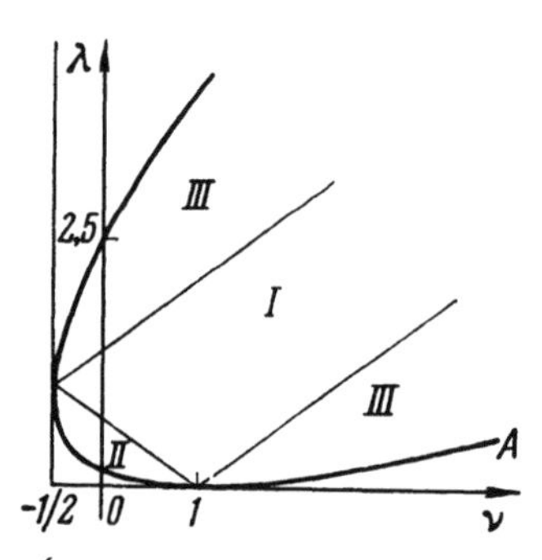

▲ **Fig.3.11.** Regions corresponding to different types of complex dispersion curves

Fig.3.12. Typical complex dispersion curves

The plane of the parameters λ, ν which is divided into a number of regions which correspond to different types of the complex dispersion curves is shown in Fig. 3.11.

First of all the region of stable values of parameters $\lambda > 0$, $\nu > -1/2$, see (3.9.18), can be distinguished. This itself is divided by the parabola $D = 0$ into two regions. The internal one, divided into subregions I, II, III, corresponds to dispersion curves of the form of ellipses while the external one corresponds to hyperbolae and points of the boundary curve A to parabolae.

In the subregion *I*, the dispersion curves (complex and usual ones) have a maximum in the interval $0 < \zeta < 1$. Subregions II and III correspond to the maximum, displaced to the intervals $-\infty < \zeta < 0$ and $1 < \zeta < \infty$

In Fig. 3.12, typical examples of complex dispersion curves $\Omega(\zeta)$ are shown: ellipses and hyperbolae (for the latter ones, only one mode is shown since the second one is situated in the region $\Omega < 0$). Moreover, the usual dispersion curves $\omega(k)$ are shown by dotted curves.

3.10 Notes*

The phenomenological model of the medium of complex structure was first considered by E. and F. Cosserat [B2.7]. About couple stress theories see Section *B* of the Bibliography, in particular [B2.2].

The theory of crystals of complex structure (crystals with basis) can be found in classical monographs [B5.2,10]. The nonlocal theory of media of complex structure was developed in [B3.23].

*The citations refer to the bibliography to be given in [1.12]

4. Nonstationary Processes

Nonstationary processes in media with space dispersion are investigated in detail. A new energy method in wave propagation is developed. The energy density of a wave packet is considered as a basic physical quantity. The evolution of the wave packet is described in terms of energy characteristics which are functionals of the energy density. Examples of such characteristics are the energy itself, the coordinate of "the center of mass" of the energy density, as well as its mean velocity, width, etc. The advantages of such an approach are as follows: 1) In applications, it is sufficient to consider only a few of these characteristics, and these satisfy equations simpler than the original wave equations (a consequence of the laws of conservation). 2) When developed systematically, the corresponding mathematical framework is analogous to quantum-mechanical formalism, thus enabling one to draw on this well-developed technique. Examples of applications, in particular to scattering problems, are considered.

4.1 Green's Functions of the Generalized Wave Equation

We shall study nonstationary solutions of the equation of motion of an (unbounded) medium of simple structure

$$Lu \stackrel{\text{def}}{=} (\rho\partial_t^2 + \Phi)\, u = q; \tag{4.1.1}$$

in the general case, the homogeneity of the medium and the invariance of the elastic energy operator Φ with respect to translation are not assumed.

Equation (4.4.1) can be considered as the generalized wave equation which coincides with the usual one when $\Phi = c\partial^2$.

In order to solve (4.1.1), it is convenient to introduce appropriate Green's functions. Let us first assume that $q = 0$ and investigate the Cauchy problem for the homogenous equation

$$Lu = 0, \quad u(x, 0) = u^0(x), \quad \dot{u}(x, 0) = \dot{u}^0(x)\,. \tag{4.1.2}$$

Let us introduce the Green's function $D(x, x', t)$ of the Cauchy problem, which satisfies the conditions

$$LD = 0, \quad D(x, x', 0) = 0, \quad \dot{D}(x, x', 0) = \delta(x - x'). \tag{4.1.3}$$

Its existence and uniqueness for a sufficiently "well-behaved" operators Φ are shown by the usual methods of functional analysis. Observe that $D(t)$ as a solution of the equation $LD = 0$ can be continued onto the negative semiaxis of time as an odd function, i.e. $D(-t) = -D(t)$.

It is easy to see that the solution of the Cauchy problem (4.1.2) can now be represented in the form

$$u(x, t) = \int \dot{D}(x, x', t)\, u^0(x')\, dx' + \int D(x, x', t)\, \dot{u}^0(x')\, dx' \tag{4.1.4}$$

or in a compact form

$$u(t) = \dot{D}(t)\, u^0 + D(t)\, \dot{u}^0 . \tag{4.1.5}$$

Similarly, for the solution of the inhomogenous ($q \neq 0$) equation (4.1.1), but with zero initial conditions, we introduce the Green's function $G^r(x, x', t - t')$, which obeys the condition

$$\begin{aligned} &LG^r = I , \\ &G^r(x, x', t - t') = 0 \quad \text{when} \quad t < t' , \end{aligned} \tag{4.1.6}$$

where I is the unit operator with the kernel $\delta(x - x')\, \delta(t - t')$. The Green's function G^r is commonly known as the retarded Green's function.

The corresponding solution has the form

$$u(x, t) = \iint G^r(x, x', t - t')\, q(x', t')\, dx' dt' \tag{4.1.7}$$

or in a compact form

$$u(t) = \int G^r(t - t')\, q(t')\, dt' . \tag{4.1.8}$$

For forces $q(x, t)$, which differ from zero only when $t > 0$, obviously, the solution $u(x, t)$ also differs from zero only when $t > 0$ (the result cannot precede the cause), i.e. the solution in the form (4.1.7) satisfies the principle of causality.

The general solution of (4.1.1) is obtained as the superposition of (4.1.5) and (4.1.8) and has the form

$$u(t) = \dot{D}(t)\, u^0 + D(t)\, \dot{u}^0 + \int G^r(t - t')\, q(t')\, dt' . \tag{4.1.9}$$

It is essential that the Green's functions $D(t)$ and $G^r(t)$ are not independent and are connected by the relation

$$\theta(t) D(t) = G^r(t)\, \rho , \tag{4.1.10}$$

where $\theta(t)$ is the Heavyside function, see (2.3.32), and the right-hand side is a product of the operator $G^r(t)$ and the operator of multiplication by $\rho(x)$; i.e. the operator with the kernel $G^r(x, x', t - t')\, \rho(x')$.

Problem 4.1.1. Verify the equality (4.1.10) [Hint: substitute G^r from (4.1.10) into (4.1.6)].

Along with the retarded Green's functions $G^r(t)$, it is sometimes also convenient to use an advanced Green's function

$$G^a(t) \stackrel{\text{def}}{=} G^r(-t) \tag{4.1.11}$$

and a symmetric Green's function

$$G(t) \stackrel{\text{def}}{=} 1/2[G^r(t) + G^a(t)]\,. \tag{4.1.12}$$

Let us write the following relations, which appear as consequences of (4.1.10) and will be useful in future:

$$\operatorname{sgn}(t)D(t) = 2G(t)\,\rho, \quad \theta(-t)D(t) = -\,G^a(t)\,\rho\,, \tag{4.1.13}$$

$$D(t) = 2\operatorname{sgn}(t)G(t)\,\rho = [G^r(t) - G^a(t)]\,\rho\,, \tag{4.1.14}$$

$$G^r(t) = 2\theta(t)G(t) = G(t) + 1/2D(t)]\,\rho^{-1}. \tag{4.1.15}$$

As usual, we shall consider the introduced Green's functions not only in the t, but also in the ω representation. Due to the evenness of $G(t)$ and oddness of $D(t)$, their Fourier images $G(\omega)$ and $D(\omega)$ are correspondingly an even real and a purely imaginary odd function, respectively. From (4.1.15) it follows that (up to the a multiplicative factor), they are the real and imaginary parts of the retarded Green's function $G^r(\omega)$. Due to the first relations (4.1.13) and (4.1.14), $D(\omega)$ and $G(\omega)$ are not independent. From physical considerations, it is obvious that under the fulfillment of the conditions of stability ($\rho > 0$ and Φ is assumed to be positive definite), the function $D(t)$ is bounded. Then the function $G^r(\omega)$ is analytic in the upper halfplane and $D(\omega)$ and $G(\omega)$ are connected by the relation (Appendix B)

$$D(\omega) = \frac{2i}{\pi} \fint \frac{G(\omega')}{\omega - \omega'}\,\rho\, d\omega'\,. \tag{4.1.16}$$

This connection between the real and imaginary parts of the retarded Green's function $G^r(\omega)$ in field theory are called the dispersion relations. From the above, it follows that they are direct consequences of the principle of causality.

Note that the Green's functions permit an interesting energy interpretation. Obviously, the conservation law for the total energy $E(t)$ has the form

$$\dot{E}(t) = \langle q(t)\,|\,\dot{u}(t)\rangle\,. \tag{4.1.17}$$

Let us assume the initial conditions to be zero and the wave process $u(x, t)$ to be caused only by an action of the force $q(x, t)$. Then for $E = E(\infty)$, we have

$$E = (q\,|\,\dot{u})\,, \tag{4.1.18}$$

where the parenthes denote the scalar product, including integration in the spatial as well as in the time variable (in the limit $-\infty < t < \infty$)

$$(u|v) \stackrel{\text{def}}{=} \iint u(x,t)\, v(x,t)\, dx\, dt$$
$$= \frac{1}{(2\pi)^2} \iint \overline{u(k,\omega)}\, v(k,\omega)\, dk\, d\omega\,. \tag{4.1.19}$$

Let us also introduce the invariant notation for the bilinear functional, defined by the operator A,

$$(u|A|v) \stackrel{\text{def}}{=} \iiiint u(x,t)\, A(x,x',t,t')\, v(x',t')\, dx dx'\, dt dt'$$
$$= (2\pi)^2 \iiiint \overline{u(k,\omega)}\, A(k,k',\omega,\omega')\, v(k',\omega')\, dk\, dk' d\omega d\omega'\,. \tag{4.1.20}$$

Substituting in (4.1.18) the expression for $\dot{u}$ from (4.1.17), we have

$$E = (q|\dot{G}^{\mathrm{r}}|q)\,. \tag{4.1.21}$$

Obviously, only the component of $G^{\mathrm{r}}(t)$ which is even in time, i.e. $1/2\dot{D}(t)\rho^{-1}$, makes a contribution to this expression. Thus, finally

$$E = 1/2(q|\dot{D}\rho^{-1}|q)\,. \tag{4.1.22}$$

This expression can be taken as a definition of the Green's function $D(t)$, and all properties of the latter can be deduced from it.

In what follows up to Sect. 4.9, we shall investigate a homogeneous medium for which $\Phi(x,x') = \Phi(x-x')$ and $\rho =$ constant. As usual, let

$$\omega^2(k) = \Phi(k)/\rho\,. \tag{4.1.23}$$

Then the Green's functions introduced above can be expressed explicitly in terms of $\omega(k)$.

Problem 4.1.2. Verify that in the (k,t) and (k,ω) representations

$$D(k,t) = \frac{\sin \omega(k)t}{\omega(k)}\,, \tag{4.1.24}$$

$$D(k,\omega) = \frac{\pi i}{\omega(k)}\,\{\delta[\omega-\omega(k)] - \delta[\omega+\omega(k)]\}\,. \tag{4.1.25}$$

In the (x,t) representation, $D(x,t)$ and $\dot{D}(x,t)$ are written as integrals

$$D(x,t) = \frac{1}{2\pi}\int \frac{\sin \omega(k)\,t}{\omega(k)}\, e^{ixk}\, dk\,, \tag{4.1.26}$$

$$\dot{D}(x,t) = \frac{1}{2\pi}\int \cos \omega(k)\,t \cdot e^{ixk}\, dk\,, \tag{4.1.27}$$

the investigation of which will be presented in the next section.

Explicit expressions for $D(x, t)$ can be obtained in two simple cases. For the usual wave equation $\omega(k) = v_0|k|$, where $v_0^2 = c_0/\rho$ and the function

$$D(x, t) = \frac{1}{2v_0}[\theta(x + v_0 t) - \theta(x - v_0 t)] \tag{4.1.28}$$

is a rectangular impulse with the amplitude $1/2v_0$, which expands with velocity v_0 (Fig. 4.1). The Green's function

$$\dot{D}(x, t) = 1/2[\delta(x + v_0 t) + \delta(x - v_0 t)] \tag{4.1.29}$$

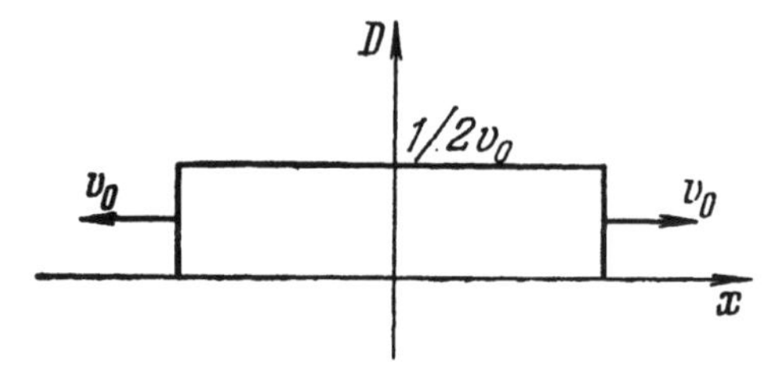

Fig.4.1. Expanding rectangular impulse

consists of two δ-functions, which scatter in the opposite directions.

For the chain with interaction of the nearest neighbors, according to (2.7.3)

$$\omega(k) = \omega_{\max}\left|\sin\frac{ak}{2}\right|. \tag{4.1.30}$$

Using the well-known integral representation for the Bessel functions, we find

$$D(n, t) = \frac{1}{a}\int_0^t J_{2n}(\omega_{\max} t)\, dt\,, \tag{4.1.31}$$

$$\dot{D}(n, t) = \frac{1}{a} J_{2n}(\omega_{\max} t)\,. \tag{4.1.32}$$

Problem 4.1.3. Show that for long wavelengths and not too large t, the expressions (4.1.31) and (4.1.32) are equivalent to (4.1.28) and (4.1.29).

The expressions for the Green's functions G, G^r and G^a can be obtained by taking into account their connection with the Green's function D in accordance with (4.1.10) and (4.1.13), but the straightforward method, directly connected with the definition of these functions is more instructive.

From (4.1.6) in the (k, ω) representation, we find formally

$$\rho G^r(k, \omega) = \frac{1}{\omega^2(k) - \omega^2}\,.$$

However, this expression still does not uniquely define $G(k, t)$, since the rule for passing around the poles is not specified. To different contours there correspond to changes in the Green's function by solutions of the homogenous equation $D(k, t)$. But, it is exactly this contribution which distinguishes G, G^r, G^a, according to the previous discussion.

Since, by definition, $G(t)$ is an even function of t, the only possible expression for it is

$$G(k, t) = \frac{1}{2\pi\rho} \fint \frac{1}{\omega^2(k) - \omega^2} e^{-i\omega t} d\omega , \tag{4.1.33}$$

where the integral is understood in the sense of its principal value.

From the condition $G^r(t) = 0$ when $t < 0$, it follows that we may write

$$G^r(t) = \lim_{\varepsilon\to 0} G^r_\varepsilon(t), \quad G^r_\varepsilon(t) \overset{\text{def}}{=} G^r(t)\, e^{-\varepsilon t} , \tag{4.1.34}$$

where $\varepsilon > 0$. The function $G^r_\varepsilon(t)$ satisfies (4.1.6) in which it is necessary to replace ∂_t by $\partial_t + \varepsilon$. But in the ω representation, $\partial_t + \varepsilon$ corresponds to $-i\omega + \varepsilon$ and hence, as $\varepsilon \to 0$

$$\rho G^r(k, \omega) = \frac{1}{\omega^2(k) - \omega^2 - 2i\varepsilon\omega} . \tag{4.1.35}$$

Thus, the poles are displaced in the lower half plane or, equivalently, are to be passed around from above. Symbolically this rule for deforming the contour is written as

$$G^r(k, t) = \frac{1}{2\pi\rho} \int \frac{1}{\omega^2(k) - (\omega + i0)^2} e^{-i\omega t} d\omega . \tag{4.1.36}$$

In the corresponding expression for $G^a(k, t)$, the sign before $i\cdot 0$ is to be changed.

In what follows, it will sometimes be convenient to represent the equation of motion, which is of second order in time for the scalar function $u(x, t)$, in the form of first order equation, but with respect to a two-component vector function $w(x, t)$ by setting

$$w_1 = u, \quad \text{and} \quad w_2 = \dot{u} . \tag{4.1.37}$$

Then (4.1.1) takes the form

$$(\rho\hat{I}\partial_t + \hat{J}\hat{H})\, w = f , \tag{4.1.38}$$

where

$$\hat{I} = \begin{pmatrix} 1 & 0 \\ 0 & 1 \end{pmatrix}, \quad \hat{J} = \begin{pmatrix} 0 & -1 \\ 1 & 0 \end{pmatrix}, \quad \hat{H} = \begin{pmatrix} \Phi & 0 \\ 0 & \rho \end{pmatrix}, \quad f_1 = 0, \quad f_2 = q . \tag{4.1.39}$$

Note that $\hat{J}^2 = -\hat{I}$, and the operator matrix $\hat{H}$ determines the total energy $E(t)$ (the sum of kinetic and elastic energies)

$$E(t) = 1/2 \langle w | \hat{H} | w \rangle = \frac{\rho}{2} \langle w_2 | w_2 \rangle + 1/2 \langle w_1 | \Phi | w_1 \rangle . \qquad (4.1.40)$$

It is easy to verify that the solution of (4.1.38) may be written in the form

$$w(t) = \hat{D}(t)\, w^0 + \hat{G}^r(t-t') f(t')\, dt' , \qquad (4.1.41)$$

where the matrix Green's functions are given by the expressions

$$\hat{D}(t) = \begin{pmatrix} \dot{D}(t) & D(t) \\ \ddot{D}(t) & \dot{D}(t) \end{pmatrix}, \quad \rho \hat{G}^r(t) = \theta(t)\, \hat{D}(t) . \qquad (4.1.42)$$

The advanced Green's function $\hat{G}^a(t)$ is defined by the expression

$$\rho \hat{G}^a(t) = -\,\theta\,(-t)\, \hat{D}(t) , \qquad (4.1.43)$$

where from the connection between the Green's functions follows (cf. Eq. (4.1.14))

$$\hat{D}(t) = \rho[\hat{G}^r(t) - \hat{G}^a(t)] . \qquad (4.1.44)$$

Obviously, the Green's function $\hat{D}(t)$ satisfies the homogeneous equation

$$(\rho \hat{I} \partial_t + \hat{J}\hat{H})\, \hat{D} = 0 \qquad (4.1.45)$$

and the initial condition $\hat{D}(0) = \hat{I}$, cf. (4.1.37). From the definition of $\hat{D}(t)$ it follows that it propagates in time any solution $w(t)$ of the homogeneous equation. Hence for two arbitrary times t and t', we have

$$w(t) = \hat{D}(t-t')\, w(t') . \qquad (4.1.46)$$

Green's functions $\hat{G}^r(t-t')$ and $\hat{G}^a(t-t')$ possess similar properties, but for times $t > t'$ and $t < t'$, respectively. In connection with this, in field theory Green's functions are often called propagators.

The formalism of Green's functions introduced here permits us to proceed to a detailed analysis of the solutions themselves.

4.2 Investigation of the Asymptotics Behavior

In many problems, the asymptotic of behavior the solutions $u(x, t)$ as $t \to \infty$ is of fundamental importance. The most simple way of obtaining it, is by the stationary phase method [4.1] used for the investigation of the asymptotic behavior of the integral

$$I(t) = \int_{k_1}^{k_2} f(k)\, e^{i\omega(k)t}\, dk\,, \tag{4.2.1}$$

where $f(k)$ and $\omega(k)$ are sufficiently smooth functions, while $f(k)$ does not have roots between k_1 and k_2.

The main contribution in the asymptotic behaviour of $I(t)$ is made by the neighborhoods of the ends of the segment $[k_1, k_2]$ and of the points at which the function $\omega(k)$ is stationary, i.e. $\omega'(k) = 0$. Moreover, the contribution of stationary points, as a rule, has the order $t^{-1/2}$ and is more significant than the contribution ($\sim t^{-1}$) of the ends of the segment.

More accurately, if $\omega'(k) \neq 0$ when $k_1 \leq k \leqslant k_2$, then

$$I(t) = \frac{1}{it}\left[\frac{f(k_2)}{\omega'(k_2)}\, e^{i\omega(k_2)t} - \frac{f(k_1)}{\omega'(k_1)}\, e^{i\omega(k_1)t}\right] + O\left(\frac{1}{t^2}\right). \tag{4.2.2}$$

If $\omega'(k_0) = 0$, $\omega''(k_0) \neq 0$ and k_0 is the only internal stationary point on the segment $[k_1, k_2]$, then

$$I(t) = \sqrt{\frac{2\pi}{|\omega''(k_0)|\, t}}\, f(k_0)\, e^{i[\omega(k_0)t + \frac{\pi}{4}\operatorname{sgn}\omega''(k_0)]} + O\left(\frac{1}{t}\right). \tag{4.2.3}$$

If there are several stationary points, each one makes a similar contribution and for a stationary point which coincides with one of the ends of the segment $[k_1, k_2]$, it is necessary to introduce a factor 1/2. Finally, let the stationary point have the order m, i.e. $\omega'(k_0) = \dots = \omega^{(m)}(k_0) = 0$ and $\omega^{(m+1)}(k_0) \neq 0$. Then such a point makes a contribution of the order $t^{-1/(m+1)}$ to the asymptotic behavior.

Now, let there be a stationary point of first order when $f(k)$ has an n-fold root. Then $I(t) \sim t^{-(n+1)/2}$. And if $f(k)$ has a simple real pole, then the asymptotics begins with a constant [4.1,2].

Let us consider the asymptotics of the solution of Cauchy problem for (4.1.2) in the case of a homogeneous medium. For simplicity we restrict ourselves to the solution of the form $u = \dot{D}u^0$, because the asymptotic behavior of the term $u = D\dot{u}^0$ may be obtained in an analogous way. Taking into account (4.1.6), we have

$$u(x, t) = \frac{1}{4\pi}\int u^0(k)\, e^{ixk}\, [e^{i\omega(k)t} + e^{-i\omega(k)t}]\, dk\,. \tag{4.2.4}$$

The investigation of the asymptotic behavior of $u(x, t)$ with fixed x does not have any special physical meaning, because it depends on the behaviour of the initial distribution $u^0(k)$ in the neighborhood of stationary points of the function $\omega(k)$ or the ends of the segment of integration. In fact, the vanishing of $u^0(k)$ in a small neighborhood of these points does not influence the physical properties of the solution, particularly, its energy properties, but significantly changes the asymptotics.

In connection with this, we shall consider the asymptotic behavior of the solution in a moving coordinate system by setting $x = vt$. Then for the observer, who is located at the origin of the moving coordinate system, the asymptotic behavior is determined by the points k_0 which satisfy the equation

$$\pm\,\omega'(k_0) = v = \frac{x}{t}\,. \tag{4.2.5}$$

Using (4.2.3), in which $\omega(k)$ should be replaced by $\pm\,\omega(k) - vk$, we find for x, which satisfies (4.2.5),

$$u(x, t) \simeq \sqrt{\frac{1}{2\pi|\omega''(k_0)|t}}\,\mathrm{Re}\,\{u^0(k_0)\,\mathrm{e}^{\mathrm{i}[\omega(k_0)t - k_0x + \frac{\pi}{4}\,\mathrm{sgn}\,\{\omega''(k_0)\}]}\} \tag{4.2.6}$$

where $k_0 = k_0(v)$ according to (4.2.5).

In particular, the asymptotic behavior of $D(x, t)$ is obtained from (4.2.6) when $u^0(k_0) = 1$.

Problem 4.2.1. Show that

$$D(x, t) \simeq \frac{1}{2v_0} + \sqrt{\frac{1}{2\pi|\omega''(k_0)|t}}\,\frac{\sin\,[\omega(k_0)\,t - k_0x + \frac{\pi}{4}\,\mathrm{sgn}\,\{\omega''(k_0)\}]}{\omega(k_0)} \tag{4.2.7}$$

where $v_0 = |\omega'(0)|$ is the velocity of propagation of long-wavelength waves.

Note that the additive constant $1/2v_0$ in (4.2.7) is unimportant because only the spatial and time derivatives of the displacement $u(x, t)$ have physical meaning.

If v in (4.2.6) is given all admissible values, for which (4.2.5) has real solutions, then we obtain the region of those $x = vt$, where the given asymptotic behavior is valid. Outside this moving region, the solution decreases at least as fast as than t^{-1}.

Strictly speaking, there exists a distinguished velocity $v_0 = |\omega'(0)|$, for which the function $\pm\,\omega(k) + v_0(k)$ usually has a stationary point of second order (the first and the second derivatives vanish). The corresponding asymptotic behavior of the order $t^{-1/3}$ permits one to estimate the asymptotic maximum of the solution. However, it is necessary to emphasize that this maximum depends on the value $u^0(k)$ at one point $k = 0$, as distinct from the asymptotics (4.2.6), where k^0, as a solution of (4.2.5) runs over all the possible values (except $k = 0$). In connection with this, the energy localized in neighborhood of a solution maximum, moving with the velocity $v = v_0$, tends to zero as $t \to \infty$. Therefore, from the energy point of view, it is the asymptotics (4.2.6) of the order $t^{-1/2}$ that has the main significance, but the stationary point of the second (and higher) order determines the evolution of the wave in the neighborhood of the maximum.

Thus, at large t and in the admissible region of values of x, the solution $u(x, t)$ has the form of wave packet with the local wave number $k_0(x/t)$ and the frequency $\omega_0(x/t)$ and slowly varying amplitude proportional to $|\omega''(k_0)t|^{-1/2}u^0(k_0)$. Moreover, from (4.2.5) it follows that the local wave number $k_0(x/t)$ moves in space with the group velocity $v = \omega'(k_0)$, i.e. remains constant as $x = vt$.

Of certain interest is the question as to what extent the packet is localized in space. The example of Green's function for the chain shows that the packet can be well localized for arbitrary t. In fact, $J_{2n}(\omega_{max}t)$ decreases exponentially as $|n| \to \infty$. This permits to evaluate the effective width of the packet from the energy considerations.

Since $\dot{u} \sim u' \sim t^{-1/2}$, the amplitude of the energy density is of order t^{-1}. But the total energy is conserved and therefore, the width of the packet is of order t i.e., asymptotically the packet diffuses with a constant velocity. A more detailed study of the spreading of the packet will be carried out in the following sections.

Let us examine as an illustration some problems which have significance in applications.

Suppose that the particle at the origin was subject to an impulse, i.e. $\dot{u}^0 \sim \delta(x)$. In the continuum without dispersion, according to (4.1.29), there appear δ-function impulses of velocity which scatter in opposite directions, the perturbation propagating with finite velocity v_0.

The presence of dispersion changes the picture. Let us study the chain as a concrete example. The perturbation propagates throughout the whole chain instansaneously, but its tails fade exponentially. This effect of instantaneous action at a distance is connected with the fact that the inertia of the bonds is neglected. Otherwise the disturbance tails should propagate with a finite velocity. The main disturbance at the initial times is nothing but two diverging smeared out δ-functions, which diffuses in the course of time. It is essential that all the energy of the disturbance is localized in them and they diverge with a finite velocity. All this depends very weakly on the inertial properties of the bonds changing even substantially, such an idealization therefore is justified.

An analogous picture takes place in many problems involving collisions. Let, for example, a collision with a subsequent cohesion of two equal semi-chains occur. It is easy to see that if we introduce the difference between displacements of neighboring atoms as a new variable, then the problem is identical to the one considered above.

A series of problems for semi-bounded ($n \geq 0$) chains with nearest-neighbor interaction may also be reduced to this scheme. Let a particle with velocity v and a large mass hit the boundary atom so that for $t > 0$, the displacement of the boundary atom will be vt. It is convenient to introduce the new variable $\tilde{u} = u$

$-vt$ for which when $t > 0$, the boundary condition is $\tilde{u}(0, t) = 0$. Continuing $\tilde{u}(n, t)$ in the region $n < 0$, as an odd function we obtain the Cauchy problem considered earlier for the unbounded chain with the condition $\tilde{u}(0, t) = 0$.

Another example is the Cauchy problem for a free semichain when the boundary condition is the vanishing of the force acting on the boundary atom. On passing to the unbounded chain this condition is equivalent to $u(0, t) = u(-1, t)$ and the solution can be obtained by the even continuation of $u(n, t)$. For example, let the initial velocity $\dot{u}^0$ of m-th atom of the semichain be prescribed. Then the solution

$$u(n, t) = \dot{u}^0 [D(n - m, t) + D(n + m + 1, t)], \tag{4.2.8}$$

satisfies, as easily seen, the required condition. Here the second term describes the pulse, reflected from the end of the chain.

4.3 Decomposition into Packets and Factorization of Wave Equations

It is well known that a solution $u(x, t)$ of the wave equation can always be uniquely represented as a superposition of waves $u_+(x - v_0t)$ and $u_-(x + v_0t)$ running in opposite directions. A similar decomposition of the solution $u(x, t)$ of the generalized wave equation for unbounded homogeneous medium

$$\rho\partial_t^2 u + \Phi u = 0 \tag{4.3.1}$$

is of interest. As we shall see, this will permit us to obtain essential additional information about the properties and characteristics of the solution.

Let us represent the solution as a superposition of harmonic waves, each of which is propagated in the homogeneous medium with constant velocity. Then it is natural to set

$$u(x, t) = u_+(x, t) + u_-(x, t), \tag{4.3.2}$$

where the wave packets $u_+(x, t)$ and $u_-(x, t)$ consist of harmonic waves, propagating in the positive or negative x-direction, respectively.

However, we need to be more accurate here. Associated with a harmonic wave in the medium are two velocities—the phase velocity and the group velocity. Later we shall be mainly interested in energy characteristics of wave processes and, hence, we shall consider the group velocity only. Obviously, the decomposition into packets does not have physical meaning in a neighborhood of points on the dispersion curve, for which the group velocity is equal to zero. Therefore, for the moment, let us assume that in the k representation the initial values and, hence also the solutions are concentrated in the region where the group velocity does not vanish. The case of localization of the solutions in a neighborhood of the dispersion curve extrema will be studied separately in Sect. 4.7.

Let us remember that by definition $\omega(k)$ is the positive root of $\omega^2(k)$ and, moreover, $\omega(k) = \omega(-k)$ since the function $\Phi(k)$ is even. Let $\tilde{\omega}_+(k) = \omega(k)\mathrm{sgn}\{k\}$ and $\tilde{\omega}_-(k) = -\tilde{\omega}_+(k)$ be the odd roots of $\omega^2(k)$. Then, as is easily seen,

$$u_\pm(x, t) = \frac{1}{2\pi}\int \exp[\mathrm{i}xk - \mathrm{i}t\omega_\pm(k)]\, u_\pm^0(k)\, dk \tag{4.3.3}$$

are the required wave packets, while $u_\pm^0(k)$ are their Fourier-images at $t = 0$.

Later it will be convenient to use an operator notation. Let the real symmetric operator $\Omega = \Omega^+$, such that $\rho\Omega^2 = \Phi$, correspond in the x representation to multiplication by the function $\omega(k)$ in the k representation. Similarly, an imaginary operator corresponds to the odd function $\tilde{\omega}_k(k)$ in the x representation. In this connection let us define the operator $\tilde{\Omega}$ with the kernel $\mathrm{i}\tilde{\omega}_+(k)$, which is real in the x representation. Obviously, $\tilde{\Omega}$ is antisymmetric (in contrast to Ω).

Let us also introduce the function

$$j(k) \stackrel{\text{def}}{=} \mathrm{i}\ \mathrm{sgn}\{k\} = -j(-k), \quad j^2(k) = -1\,, \tag{4.3.4}$$

which defines an operator j which is real in the x representation and which commutes with all operators having difference kernels, in particular with Ω; it has the properties

$$j^2 = -1, \quad j^+ = -j\,. \tag{4.3.5}$$

Problem 4.3.1. Show that in the x representation the operator j has the kernel $[\pi(x - x')]^{-1}$ and coincides with the well-known Hilbert transform [4.3].

Then $\tilde{\Omega} = j\Omega$ and (4.3.3) takes the form

$$u_\pm = D_\pm u_\pm^0, \quad D_\pm \stackrel{\text{def}}{=} \mathrm{e}^{\mp j\Omega t}\,, \tag{4.3.6}$$

where the exponential operator is, as usual, defined by the Taylor series.

The last relations suggest that the decomposition (4.3.2) permits one more interpretation, which appears to be most useful for different generalizations. In fact, taking into account that $\Phi = -\rho(j\Omega)^2$ and ∂_t commutes with j and Ω, let us decompose the operator $\rho\partial_t^2 + \Phi$ into factors, linear in ∂_t, and write (4.3.1) in the form

$$(\partial_t - j\Omega)(\partial_t + j\Omega)\, u = 0\,. \tag{4.3.7}$$

Obviously, a general solution of the factorized equation is representable in the form (4.3.2) where u_+ and u_- satisfy the equations of first order in time.

$$(\partial_t + j\Omega)\, u_+ = 0\,,$$
$$(\partial_t - j\Omega)\, u_- = 0\,, \qquad (4.3.8)$$

which is equivalent to (4.3.6). Now, of course, u_+ and u_-, are also solutions of (4.3.1).

Let U be the space of all solutions of (4.3.1) over which the operators j and Ω^{-1} are defined. Then for any $u \in U$, the representation (4.3.2) is unique and

$$u_+ = p_+ u, \quad u_- = p_- u\,, \qquad (4.3.9)$$

where

$$p_+ \stackrel{\text{def}}{=} 1/2(1 + j\Omega^{-1}\partial_t), \quad p_- \stackrel{\text{def}}{=} 1/2(1 - j\Omega^{-1}\partial_t)\,. \qquad (4.3.10)$$

Problem 4.3.2. Show that p_+ and p_- satisfy in U the identities

$$p_+^2 = p_+,\ p_-^2 = p_-,\ p_+ p_- = p_- p_+ = 0,\ p_+ + p_- = 1 \qquad (4.3.11)$$

and, hence, are projection operators which decompose U into a direct sum of subspaces U_+ and U_-.

In what follows, we shall investigate only such solutions of (4.3.1) for which the total energy

$$E = 1/2\rho\langle \dot{u} | \dot{u}\rangle + 1/2\,\langle u | \Phi | u\rangle \qquad (4.3.12)$$

is finite. By assumption, the energy dissipation is absent, so that E is independent of time. The spaces U, U_+ and U_- are now Hilbert spaces with respect to the scalar product $\langle u | v\rangle$, induced by the kinetic energy.

We choose to adopt the following terminology. The above-introduced solutions u_+ and u_- with finite energy will be called packets[1], and a solution of the form (4.3.2) a superposition of two packets. The packet possesses a remarkable property: its kinetic and potential energies are equal and, hence, are conserved. In fact,

$$\langle u_+ | \Phi | u_+\rangle = \rho\langle u_+ | -(j\Omega)^2 | u_+\rangle = \rho\langle j\Omega u_+ | j\Omega u_+\rangle = \rho\langle \dot{u}_+ | \dot{u}_+\rangle \qquad (4.3.13)$$

and similarly for u_-.

Problem 4.3.3. Verify directly that the kinetic energy of the packet is conserved (Hint: The operator $j\Omega$ is antihermitian and the corresponding exponential one is unitary).

[1]Usually, any solution localized in the x, k, or ω space is called the wave packet, without the assumption that the packet contains harmonic waves, propagating in one direction.

Problem 4.3.4. Show that the packets u_+ or u_- are orthogonal with respect to the total energy, i.e. the energy of interaction is equal to zero:

$$\rho\langle \dot{u}_+ | u_- \rangle + \langle u_+ | \Phi | u_- \rangle = 0 . \tag{4.3.14}$$

Thus, the total energy of each packet is equal to twice the kinetic (potential) energy

$$E_+ = \rho\langle \dot{u}_+ | \dot{u}_+ \rangle, \quad E_- = \rho\langle \dot{u}_- | \dot{u}_- \rangle \tag{4.3.15}$$

and for the superposition of packets

$$E = E_+ + E_- , \tag{4.3.16}$$

i.e. the subspaces U_+ and U_- are orthogonal with respect to the total energy.

Note that for the superposition $U = U_+ + U_-$ of packets, the kinetic and potential energy, generally speaking, are not conserved separately, i.e. energy transfer takes place. Let us illustrate this by a simple example. Let two packets run towards each other and the condition of symmetry be fulfilled, according to which $u_-(x, t) = -u_+(x, t)$ (Fig. 4.2) at some moment; for their superposition $u(x, t) = 0$ (but, of course, $\dot{u}(x, t) \neq 0$) and hence all the energy is converted into the kinetic energy.

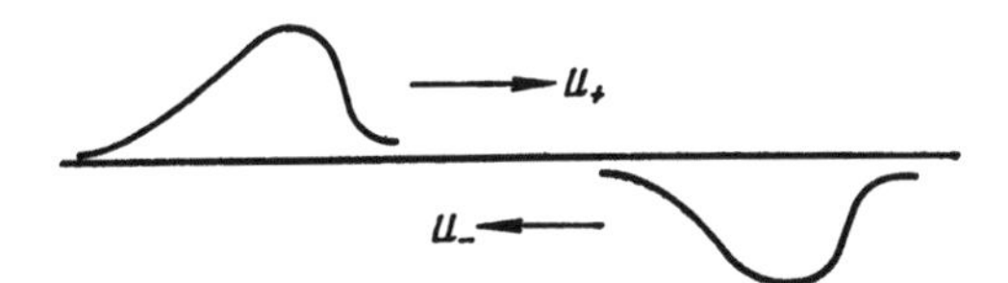

Fig.4.2. Symmetric packets

Let us investigate the decomposition into packets by the example of a chain. In Sect 4.1 we saw that for the usual wave equation, the Green's function $\dot{D}(x, t)$ in accordance with (4.1.29) consists of two scattering δ-functions, which may serve as the simple example of packets (but with infinite energy). The corresponding solution of (4.1.32) for the chain is described by Bessel functions. It is interesting to see what is the analog of a single δ-function here.

Problem 4.3.5. Decompose the solution (4.1.32) into packets $\dot{D}_+$ and $\dot{D}_-$ (as distinct from (4.1.32), they are not expressed in the (n, t) representation in terms of special functions).

The evolution of the total energy density $e_+(n, t)$ of the packet $\dot{D}_+(n, t)$ is shown in Fig. 4.3.

Let us now study the decomposition into packets from another point of view. Let $w(x, t)$ be the two-dimensional vector-function introduced in Sect. 4.1 with components $w_1 = u$ and $w_2 = \dot{u}$. Then $w(x, t)$ satisfies the equation

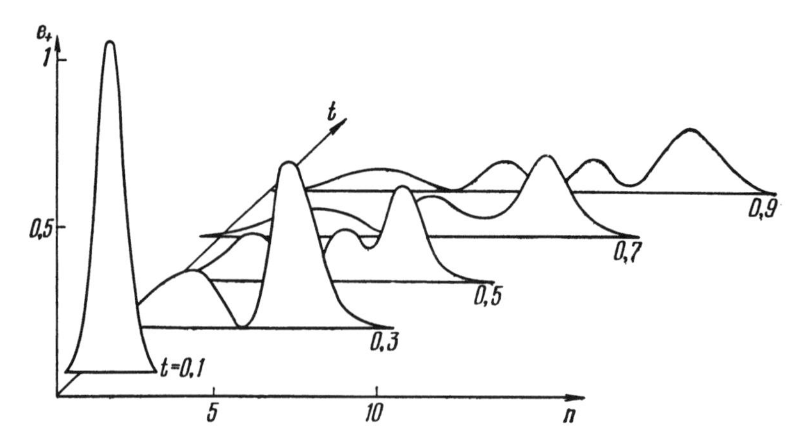

Fig.4.3. Evolution of the energy density $e_+(n,t)$

$$\rho \hat{J}\dot{w} = \hat{H}w\,, \tag{4.3.17}$$

where the matrices $\hat{J}$ and $\hat{H}$ are defined by the expressions (4.1.39). As pointed out above, the operator $\hat{H}$ has the meaning of an operator for the total energy and may be represented in the form

$$\hat{H} = \hat{T} + \hat{\Phi}, \quad \hat{T} \stackrel{\text{def}}{=} \begin{pmatrix} 0 & 0 \\ 0 & \rho \end{pmatrix}, \quad \hat{\Phi} \stackrel{\text{def}}{=} \begin{pmatrix} \Phi & 0 \\ 0 & 0 \end{pmatrix}, \tag{4.3.18}$$

where $\hat{T}$, $\hat{\Phi}$ are the operators of kinetic and elastic energy.

Problem 4.3.6. Show that to the decomposition in packets there corresponds the representation of the vector w in the form

$$w = w_+ + w_-, \quad w_\pm \stackrel{\text{def}}{=} \hat{P}_\pm w\,, \tag{4.3.19}$$

where the matrix projection operators are defined by the expressions

$$\hat{P}_\pm = 1/2 \begin{pmatrix} 1 & \pm j\Omega^{-1} \\ \pm j\Omega & 1 \end{pmatrix} \tag{4.3.20}$$

and identically satisfy relations similar to (4.3.11), as distinct from the operators $p_\pm$, which satisfy (4.3.11) only on solutions of (4.3.1).

Problem 4.3.7. Verify that w_+ and w_- satisfy the equations

$$j\dot{w}_+ = \Omega w_+, \quad j\dot{w}_- = -\Omega w_-\,. \tag{4.3.21}$$

It is essential that, as distinct from (4.3.17), the equations for w_+ and w_- contain the scalar (proportional to $\hat{I}$) operators j and Ω and the systems (4.3.21) split into equations of the first order in time with respect to individual components of the vectors w_+, w_-. Moreover, as is easily verified, due to the relations $\hat{P}_+ w_- = \hat{P}_- w_+ = 0$, there is only one independent equation in each system and it coincides with the corresponding equation (4.3.8).

The convenience of introducing the matrix projection operators P_+ and P_- is connected with the fact that, as distinct from p_+ and p_-, they do not contain the operator ∂_t and are defined not only on w but also on w_0. In fact, it is directly verified that $\hat{P}_+$ and $\hat{D}$ commute and hence

$$w_+ = \hat{P}_+ w = \hat{P}_+ \hat{D} w^0 = \hat{D}\hat{P}_+ w^0 = \hat{D}_+ w^0_+ \,, \tag{4.3.22}$$

$$\hat{D}_+ \stackrel{\text{def}}{=} \hat{P}_+ \hat{D} = \hat{D}\hat{P}_+ = \begin{pmatrix} \dot{D}_+ & D_+ \\ \ddot{D}_+ & \dot{D}_+ \end{pmatrix} \tag{4.3.23}$$

where, due to (4.3.6) $D_+ = \exp(-j\Omega t)$. The relations for w_- are analogous.

It is now possible to decompose directly the initial disturbance $w^0 = (u^0, \dot{u}^0)$ into packets. In particular, in the k representation

$$u^0_\pm(k) = 1/2\,[u^0(k) \pm \frac{j(k)}{\omega(k)}\, \dot{u}^0(k)] \,. \tag{4.3.24}$$

It is convenient to write also the energy operator in the vector notations. Then, from the relations

$$\langle w_+ |\, \hat{T} \,| w_+ \rangle = \langle \hat{P}_+ w_+ |\, \hat{T} \,|\, \hat{P}_+ w_+ \rangle = \langle w_+ |\, \hat{P}_+^\dagger \hat{T} \hat{P}_+ \,| w_+ \rangle \tag{4.3.25}$$

it follows that for the packet, the kinetic energy operator is

$$\hat{T}_+ = \hat{P}_+^\dagger\, \hat{T} \hat{P}_+ \,, \tag{4.3.26}$$

and similarly the potential energy operator is

$$\hat{\Phi}_+ = \hat{P}_+^\dagger\, \hat{T} \hat{P}_+ \,. \tag{4.3.27}$$

Here

$$\hat{P}_+^\dagger = 1/2 \begin{pmatrix} 1 & j\Omega \\ -j\Omega^{-1} & 1 \end{pmatrix} \tag{4.3.28}$$

is the Hermitian conjugate of the operator $\hat{P}_+$.

Problem 4.3.8. Verify that $\hat{T}_\pm = \hat{\Phi}_\pm$.

Let us summarize these results. The state of the medium is characterized by the solution $u(x, t)$ of the generalized wave equation (4.3.1) or equivalently by the vector $w(x, t)$ which satisfies the system (4.3.12), equivalent to (4.3.1). Any state is uniquely represented in the form of a superposition of two packets, which possess the following properties:

a) Each packet consists of harmonic waves having the same directions of their group velocities.
b) Each packet satisfies an equation of evolution type, i.e. of first order in time.
c) Kinetic and elastic potential energies of each packet are equal and conserved.
d) The packets do not interact, i.e. their total energies are additive.[2]

Of these properties, the first should not be considered as fundamental. In fact, it cannot be retained when passing to inhomogeneous media since, in that case, the elementary harmonic waves are not solutions. The other three properties have an invariant meaning and may be taken as a basis of any reasonable generalization.

From the above it follows that for the factorization of the wave equation and decomposition in packets, two operations are essential. They are: construction of the operator Ω, i.e. the extraction of a positive self-adjoint root from the operator Φ, which is done uniquely, and the construction of the operator j, which commutes with Ω and satisfies the conditions (4.3.5). However these conditions do not yet define j uniquely. In fact, along with $j(k) = \mathrm{i}\,\mathrm{sgn}\{k\}$, a different $j_*(k)$ might have been introduced, as shown in Fig. 4.3. In the considered case, the uniqueness is a chieved only if one takes into account the additional condition (a) for the directions of propagation of harmonic waves.

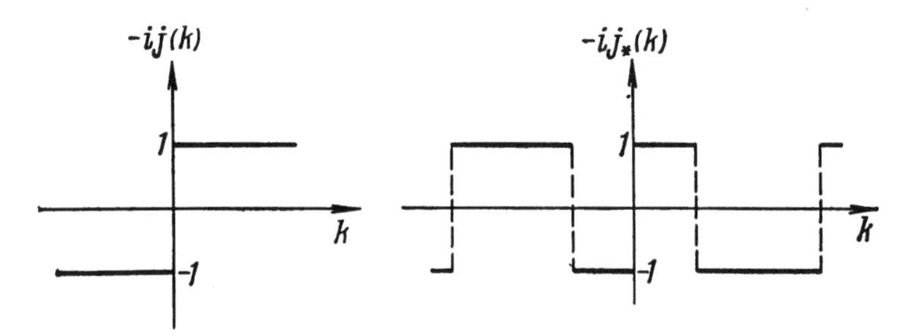

Fig.4.4. Operators j and j^*

4.4 Energy Method and Quantum-Mechanical Formalism

Let us now consider a wave packet from a new point of view. As a basic quantity which characterizes the packet and plays the role of measure, we shall consider the energy density of the packet. We repeatedly emphasized above that in a nonlocal theory, the elastic energy density $\varphi(x)$ (unlike the kinetic energy density) may be defined in different ways and, generally speaking, none of them should be given preference, either from a physical or from a mathematical point of view.

[2]In Sect 4.4, we shall see that this property of addivity covers also the density of the total energy.

For the packet, another situation occurs. The equality of kinetic and potential energies is its characteristic property. Therefore, for the packet, it is natural also to assume the elastic energy density $\varphi(x, t)$ to be equal to the kinetic energy density $1/2\rho\dot{u}^2(x, t)$.

Problem 4.4.1. Verify that this condition is satisfied by the expression

$$\varphi = \frac{\rho}{2}(j\Omega u)^2, \tag{4.4.1}$$

which, in the long-wavelength approximation, (i.e. with the law of dispersion $\omega(k) \sim k$), coincides with the usual elastic energy density.

Then for the total energy density $e(x, t)$ of the packet, we have

$$e(x, t) = \rho\dot{u}^2(x, t), \quad \int e(x, t)\,dx = E. \tag{4.4.2}$$

Thus, $e(x, t)$, by definition, is the energy measure of the distribution of the packet along the x-axis.

We can also introduce the total energy density $e(x, t)$ for the superposition of packets $u = u_+ + u_-$, setting

$$e \stackrel{\text{def}}{=} 1/2\rho\dot{u}^2 + 1/2\rho(j\Omega u)^2. \tag{4.4.3}$$

By such a definition, the total energy density of the superposition $u = u_+ + u_-$ of packets is additive, i.e. is equal to the sum of the densities for each of the packets $e = e_+ + e_-$, but this property is not fulfilled for the densities of kinetic and potential energy separately.

Problem 4.4.2. Verify the additivity property for the energy density (4.4.3).

Together with $e(x, t)$, it is convenient to introduce the spectral density of energy $e(k, t)$, defining it by analogy with the relation (4.4.3).[3]

$$\begin{aligned} e(k, t) &\stackrel{\text{def}}{=} \frac{\rho}{4\pi}|\dot{u}(k, t)|^2 + \frac{\rho}{4\pi}|\tilde{\omega}(k)\,u(k, t)|^2 \\ &= \frac{\rho}{4\pi}|\dot{u}(k, t)|^2 + \frac{1}{4\pi}\Phi(k)|u(k, t)|^2, \\ \int e(k, t)\,dk &= E. \end{aligned} \tag{4.4.4}$$

Taking into account (4.1.5), (4.1.24) and (4.3.3), we find that the spectral density of energy is independent of time and additive with respect to packets.

[3]Here, as an exceptional case, the agreement about notations is not adhered to; $e(k,t)$ is not the Fourier-image of $e(x,t)$.

$$e(k) = \frac{\rho}{4\pi} \dot{u}^0(k)|^2 + \frac{1}{4\pi} \Phi(k)|u^0(k)|^2 = \frac{\rho}{2\pi} |\dot{u}^0_+(k)|^2 + \frac{\rho}{2\pi} |\dot{u}^0(k)|^2 . \quad (4.4.5)$$

In the following we shall study the characteristics of packets which are continuous (generally, nonlinear) functionals of the energy density and, hence, energetically stable. As an example of quantities which do not possess this property, we mention functionals linear with respect to fields, in particular, the "total momentum"

$$P = \rho \int \dot{u}(x, t)\, dx = \rho \dot{u}(k)|_{k=0} = \rho \dot{u}^0(k)|_{k=0} \quad (4.4.6)$$

(here we used $\omega(k) = 0$ when $k = 0$). In fact, changing $\dot{u}^0(k)$ in an arbitrarly small neighborhood of the point $k = 0$, one can change P to any value while $e(k)$ changes negligibly and the energy E remains constant.

In this and the next sections, we restrict ourselves to the case of one packet setting for definiteness $u_- = 0$; the index $+$ will be omitted.

Of the main points of interest there will be packet evolution characteristics averaged with respect to the energy density: coordinate and velocity of the center of mass, the packet width and velocity with which it changes, asymmetry of the packet, etc. and also the correlations between these quantities.

For example, the coordinate $\langle x \rangle$ of the center of gravity of the energy density is given by the expression

$$\langle x \rangle = \frac{1}{E} \int x e(x, t)\, dx , \quad (4.4.7)$$

and for the average frequency $\langle \Omega \rangle$ we have

$$\langle \Omega \rangle = \frac{1}{E} \int \omega(k)\, e(k)\, dk . \quad (4.4.8)$$

For a unified description of all similar quantities it is convenient to present them in another form. With this in mind, let us introduce the normalized function

$$\psi(x, t) \stackrel{\text{def}}{=} \sqrt{\frac{\rho}{E}}\, \dot{u}(x, t), \quad \langle \psi | \psi \rangle = 1 , \quad (4.4.9)$$

and operators for the coordinate x and of frequency Ω. The first of these has kernels $x\, \delta(x - x')$ and $2\pi i \delta'(k - k')$ and the second one has kernels $\Omega(x - x')$ and $\omega(k)\, \delta(k - k')$. Then (4.4.7) and (4.4.8) can be rewritten in the form

$$\langle x \rangle = \langle \psi | x | \psi \rangle, \quad \langle \Omega \rangle = \langle \psi | \Omega | \psi \rangle \quad (4.4.10)$$

and interpreted as mean values of operators x and Ω in the state ψ.

All the other characteristics of the packet can be expressed analogously, as the mean values of operators of the corresponding physical quantities. For an arbitrary operator y with kernels $y(x, x')$ and $y(k, k')$ let

$$\langle y\rangle \stackrel{\text{def}}{=} \langle\psi|y|\psi\rangle = \iint \psi(x, t)\, y(x, x')\, \psi(x', t)\, dx\, dx'$$
$$= \frac{1}{2\pi}\iint \psi^*(k, t)\, y(k, k')\, \psi(k', t)\, dk\, dk' \,. \tag{4.4.11}$$

With such a definition, a mean value is, generally speaking, no more a functional of the energy density $e(x, t)$ or of the spectral density $e(k)$. This property is conserved only for operators, which are diagonal ones in k or x representations, respectively, as we had seen in the examples of operators of coordinate and frequency[4]. Incidentally, in the case of homogeneous medium, we shall as a rule, deal with such operators.

Since $\psi(x, t)$ is real, the operators have real kernels in the x representation, while the contribution to the mean value $\langle y\rangle$ is made only by the symmetric component of the kernel $y(x, x')$ or the Hermitian component of the kernel $y(k, k')$. Therefore, without loss of generality, one may assume that the characteristics of the packet are described by Hermitian operators $y = y^+$.

The approach demonstrated above suggests a far reaching analogy with the formalism of quantum mechanics. The state of the system is described by a normalized function $\psi(x, t)$ which satisfies an evolution equation of Schrödinger type

$$j\partial_t\psi = \Omega\psi \,, \tag{4.4.12}$$

where the frequency operator Ω plays the role of the Hamiltonian and the operator j plays the role of the imaginary unit. Obviously, the solution has the form

$$\psi(x, t) = \mathrm{e}^{-tj\Omega}\,\psi^0(x) \quad \text{or} \quad \psi(k, t) = \mathrm{e}^{-it}\tilde{\omega}_+(k)\,\psi^0(k) \,. \tag{4.4.13}$$

The normalized energy density

$$e(x, t) = \psi^2(x, t) \tag{4.4.14}$$

or spectral density of the energy

$$e(k) = |\psi(k, t)|^2 = |\psi^0(k)|^2 \tag{4.4.15}$$

corresponds to the quantum mechanical probability density, and to quantum mechanical observables there correspond the operator characteristics of the packet.

[4] Any operator with symmetric kernel $y(x,x')$ can be transformed to a diagonal form. But in that representation $\langle y\rangle$ is a linear functional of the corresponding "spectral" density of energy. This permtis to consider (4.4.11) as a natural generalization of (4.4.7) and (4.4.8).

There are also differences. Firstly, in the given model, there is no characteristic quantity which would have the dimension of action. In connection with this when carrying out the anology with quantum mechanics, the correspondence is established between quantities of different dimensions (for example, frequency—energy). Secondly, and more important, the function $\psi(x, t)$ is real and, as has been pointed out, the role of the imaginary unit is played by the operator j with the properties (4.3.5).

In the approach under consideration, states are determined by time-dependent ψ-functions, and to physical characteristics of the packet there correspond time-independent Hermitian operators. Leaving invariant the average values of the operators, one can, however, transfer the time-dependence from ψ-functions to the operators. In fact, setting

$$y_t \stackrel{\text{def}}{=} e^{j\Omega t} y e^{-j\Omega t}, \quad \psi^0 \stackrel{\text{def}}{=} e^{j\Omega t} \psi(t), \tag{4.4.16}$$

we have

$$\langle y_t \rangle \stackrel{\text{def}}{=} \langle \psi^0 | y_t | \psi^0 \rangle = \langle \psi(t) | y | \psi(t) \rangle = \langle y \rangle . \tag{4.4.17}$$

In quantum mechanics this corresponds to the transition from the Schrödinger representation to the Heisenberg one. In the following, the index t will point out that the operator is considered in the Heisenberg representation.

From (4.4.16) follows

$$\dot{y}_t = [j\Omega, y_t], \tag{4.4.18}$$

where the brackets, as usual, denote the commutator of operators,

$$[y, z] \stackrel{\text{def}}{=} yz - zy . \tag{4.4.19}$$

The evolution equation (4.4.18) in the Heisenberg representation is equivalent to (4.4.12) for the ψ function in the Schrödinger representation. As distinct from quantum mechanics, the operator j in (4.4.18) cannot be factored out because, in general, it does not commute with y_t.

If the operator y in the x representation is a difference one and hence is multiplicative in the k representation, then there is a conservation law for it: $\dot{y}_t = 0$ and $y_t = y$, i.e. the Heisenberg and Schrödinger representations for the operator y coincide. In particular, the frequency operator Ω possesses this property.

A few words about the operator Ω: when k is small, three basic laws of dispersion $\omega(k)$ are to be distinguished. For longitudinal oscillations of a chain and rod or for transverse oscillations of a string, $\Phi(k) \sim k^2$ and $\omega(k) \sim k$. Equation (4.4.12) is the analog of the equations of quantum mechanics for massless particles (photon, neutrino, acoustic phonon). For transverse oscillations of a rod $\Phi(k) \sim k^4$, $\omega(k) \sim k^2$, corresponding to the law of dispersion of the Schrödinger equation for nonrelativistic particles. Finally, for longitudinal vibrations of a rod, or for transverse vibrations of a string attached to an elastic foundation $\Phi(k) \sim a_0 + a_2k^2$ and $\omega(k) \sim \omega_0 + \alpha k^2$; this corresponds to the Klein-Gordon equation for a relativistic particle. In the last two cases the dispersion curve has an extermum when $k = 0$, and therefore we temporarily postpone their investigation.

Thus, in what follows, wave packets will be analyzed from the energy point of view. Let us note several important features of such an approach. We leave aside a detailled space-time description of solutions of the equations of motion and are interested only in the main energy characteristics of packets. As will be shown, for these characteristics, it is possible to obtain generalized conservation laws and to calculate the energy characteristics from the initial conditions, without solving the equations of motion. It is essential that within the framework of the energy method, the well-developed formalism of elementary quantum mechanics can be used effectively.

4.5 Characteristics of the Evolution of a Packet

Let us proceed to the detailed study of the energy characteristics of a packet. In the k representation, the operator $i\partial/\partial k$ is put into correspondence with the coordinate operator and hence

$$\langle x \rangle = \frac{i}{2\pi} \int \psi^*(k,t) \frac{\partial}{\partial k} \psi(k,t)\, dk\,. \tag{4.5.1}$$

Taking into account (4.4.13) and introducing the group velocity operator v, which in the k representation corresponds to the multiplication by $v(k) = \tilde{\omega}'_+(k)$, we find[5]

$$\langle x \rangle = \langle v \rangle\, t + \langle x \rangle_0 \tag{4.5.2}$$

where $\langle x \rangle_0$ is the coordinate of the center of gravity of the packet when $t = 0$. The quantity $\langle v \rangle$ is independent of time and is determined by the initial conditions only

$$\langle v \rangle = \langle \psi | v | \psi \rangle = \langle \psi^0 | v | \psi^0 \rangle\,. \tag{4.5.3}$$

Thus, the center of gravity of the packet moves with a constant velocity $\langle v \rangle$ as a free particle. We shall write this in the form of a basic conservation law for the packet

$$\frac{d}{dt} \langle v \rangle = \frac{d^2}{dt^2} \langle x \rangle = 0\,. \tag{4.5.4}$$

[5] Recall that it is assumed here $u = u_+$. (i.e. $v = v_+$). For the packet u_-, obviouly, $v_- = -v_+$, since $v_-(k) = \tilde{\omega}'_-(k) = -\tilde{\omega}'_+(k)$.

In the Heisenberg representation, the conservation law is written as

$$\frac{d}{dt} v_t = \frac{d^2}{dt^2} x_t = 0 , \tag{4.5.5}$$

whence it follows

$$v_t = v_0 = v, \quad x_t = vt + x_0 , \tag{4.5.6}$$

where v is the velocity operator in the Schrödinger representation.

Problem 4.5.1. Verify (4.5.5) using (4.4.18).

The conservation law (4.5.5) is, obviously, the analog of the conservation law for the momentum of a free particle in quantum mechanics. In the given case, it is a result of the homogenity of the medium, i.e. of the invariance of its properties with respect to the translation group. We shall see that in the case of an inhomogenous medium in the right-hand side part of (4.5.4) or (4.5.5), there appears a "force", which acts on the packet.

The operator of wave number k, which has the kernel $|k|\delta(k-k')$ in the k representation determines the mean wavelength $2\pi\langle k\rangle^{-1}$ of the packet. If the packet is localized in a small neighborhood of the point k_0, then $\langle k\rangle \simeq k_0$ and $\langle\Omega\rangle \simeq \omega(k_0)$. It is also obvious that the energy operator is proportional to the unit operator with coefficient of proportionality E.

Let us now consider the packet characteristics of higher orders. With this in mind, for the arbitrary operator y, let us set

$$\delta y \stackrel{\text{def}}{=} y - \langle y\rangle \tag{4.5.7}$$

and introduce the correlator (yz) and the correlation function $K(yz)$ of two generally noncommuting operators y and z:

$$2(yz) \stackrel{\text{def}}{=} \delta y\delta z + \delta z\delta y, \quad K(yz) \stackrel{\text{def}}{=} \langle (yz)\rangle . \tag{4.5.8}$$

The quantity

$$\Delta x \stackrel{\text{def}}{=} \sqrt{K(xx)} \tag{4.5.9}$$

can be interpreted as the effective width of the packet, and (xx)-as an operator for the square of the packet width.

Similarly, for an arbitrary operator y the dispersion

$$(\Delta y)^2 \stackrel{\text{def}}{=} K(yy) \tag{4.5.10}$$

characterizes the deviation of the corresponding quantity from its mean value.

In quantum mechanics, Δy is interpreted as the uncertainty of the observable y in the state ψ.

The correlator (yz) and the correlation function $K(yz)$ for different operators y and z, characterize the degree of connection between the corresponding quantities. The physical sense of $K(yz)$ will be clear from examples.

Problem 4.5.2. Verify that

$$(yz)_t = (y_t z_t)\,. \tag{4.5.11}$$

Substituting (4.5.6) in (4.5.8), we find

$$K(xv) = tK_0(vv) + K_0(xv)\,, \tag{4.5.12}$$

$$(\Delta x)^2 = K(xx) = t^2K_0(vv) + 2tK_0(xv) + K_0(xx)\,, \tag{4.5.13}$$

where the index "zero" denotes the values at $t = 0$ and instead of $K_0(vv)$ we could also write $K(vv)$, because $v_t = v$.

At large t, the width Δx of the packet spreads out with a constant velocity $\Delta v = \sqrt{K(vv)}$ and, hence, the operator

$$(vv) = (vv)_t = \frac{d^2}{dt^2}(xx)_t \tag{4.5.14}$$

has the meaning of the square of the asymptotic spreading out velocity of the packet.

The conservation law (4.5.5) leads to the equations

$$\frac{d^2}{dt^2}(xv)_t = 0, \quad \frac{d^3}{dt^3}(xx)_t = 0 \tag{4.5.15}$$

and similar equations for $K(xv)$ and $K(xx)$. If desired, they can be also interpreted as the conservation laws for the corresponding characteristics of the packet.

From (4.5.13), it is seen that Δx has its minimum value at the time

$$t_* = -\frac{K_0(xv)}{K_0(vv)}\,, \tag{4.5.16}$$

while t_* is uniquely determined by the condition $K_*(xv) = 0$. The packet has the minimum width when the position and velocity correlation is zero.

The expression (4.5.13) is now written in the form

$$(\Delta x)^2 = (t - t_*)^2(\Delta v)^2 + (\Delta x)^2{}_*\,. \tag{4.5.17}$$

Thus, to each packet there can be ascribed the absolute age

$$t - t_* = \frac{K(xv)}{K(vv)} ; \tag{4.5.18}$$

when $t < t_*$, the packet is focusing up to the minimum width $(\Delta x)_*$ and then, when $t - t_* > 0$, it spreads out with its asympototic velocity Δv. The direction of the evolution (focusing or spreading out) is determined by the sign of $K(xv)$. It is obvious that in a dispersive medium, the time interval in which the packet is focusing, is finite and at sufficiently large times, the packet always spreads out.

The physical sense of the correlation function $K(xv)$, becomes especially clear, if one represents the packet as a combination of particles, which are moving in the same direction, but with different velocities. Let the coordinates of all the particles practically coincide when $t = t_*$. Then when $t - t_* > 0$ ($t - t_* < 0$), the particles with higher velocities have larger (smaller) coordinates.

For the relative width of a packet at an arbitrary time t, we have

$$\frac{(\Delta x)^2}{(\Delta x)_*^2} = 1 + \frac{K^2(xv)}{K_*(xx)\,K(vv)} . \tag{4.5.19}$$

If we introduce a characteristic time $T = (\Delta x)_*/\Delta v$, during which the dispersion $(\Delta x)_*^2$ doubles, and measure the age of the packet in units of T, then from (4.5.19) it follows that

$$\Delta x \simeq \frac{t - t_*}{T} (\Delta x)_* . \tag{4.5.20}$$

Note that for packets of small width Δk, the asymptotic velocity of spreading out $\Delta v \simeq |\omega''(\langle k \rangle)|\, \Delta k$ is small and proportional to Δk. If $\langle k \rangle$ coincides with a point of inflection of $\omega(k)$, then $\Delta v \sim (\Delta k)^2$.

The packet evolution characteristics of the next level may be obtained, if we introduce the triple correlator

$$(xyz) \overset{\text{def}}{=} S\, \delta x\, \delta y\, \delta z \tag{4.5.21}$$

and the corresponding correlation function

$$K(xyz) \overset{\text{def}}{=} \langle (xyz) \rangle , \tag{4.5.22}$$

where $S\,xyz$ is the symmetrized product of operators x, y, z.

It is not difficult to show that the quantity

$$K(xxx) = t^3 K(vvv) + 3t^2 K_0(xvv) + 3t K_0(xxv) + K_0(xxx) \tag{4.5.23}$$

characterizes the degree of the packet-form asymmetry. For a symmetrical packet (in the x representation), this quantity is equal to zero and is positive (negative) for a packet extended to the right (left) with respect to the packet center of gravity. Asymptotically as $t \to \infty$, the packet form asymmetry is determined by the quantity $K(vvv)$.

Problem 4.5.3. Let the energy density $e(x)$ have a form of a triangle whose vertex projection divides the base in the ratio $b : a$ (b corresponds to the left side). Show that

$$K(xxx) = \frac{1}{270}(a - b)(2a^2 + 5ab + 2b^2). \tag{4.5.24}$$

Analogously, n-tuple correlators and correlation functions may be introduced. In statistics, higher central moments of the probability distribution function correspond to them.

Correlators of the operators x and v are the main characteristics of evolution of the packet. Let (n, m) be the correlator of n operators x and m operators v. Then the equation

$$\frac{\partial^{n+1}}{\partial t^{n+1}}(n, m)_t = 0 \tag{4.5.25}$$

is a direct consequence of the conservation law.

The corresponding correlation functions $K(n, m)$ of the $(n + m)$-th order are polynomials of the n-th degree in t, while the coefficients of such a polynomial are the values at $t = 0$ of correlation functions of the same order. Therefore the evolution characteristics of different orders are not correlated with each other and could be calculated independently and directly from the initial data.

Thus, the possibility of the description of the evolution characteristics of a packet without solving the equations of motion is, from a practical point of view, an important result of the packet analysis. For example, for the calculation of the width of a packet at any moment of time, in accordance with (4.5.13), it is enough to know ($K_0(vv)$, $K_0(xv)$ and $K_0(xx)$, which are determined by the initial conditions.

4.6 Superposition of Packets

Let us now proceed to a more general case, in which a state is the superposition of two packets $u_+(x, t)$ and $u_-(x, t)$ moving in opposite directions. Let E be the total energy of the packets, determined by (4.3.15) and (4.3.16). Analogous to the case of one packet, let us introduce the ψ-functions

$$\psi_1 \stackrel{\text{def}}{=} \sqrt{\frac{\rho}{E}}\,\dot{u}_+, \quad \psi_2 \stackrel{\text{def}}{=} \sqrt{\frac{\rho}{E}}\,\dot{u}_- \tag{4.6.1}$$

and consider them as two components of the vector ψ, which characterizes the state.

Problem 4.6.1. Show that the vector $\psi = (\psi_1, \psi_2)$ and the vector $w = (u, \dot{u})$ are connected by the relation.

$$\begin{pmatrix}\psi_1\\ \psi_2\end{pmatrix} = \frac{1}{2}\sqrt{\frac{\rho}{E}}\begin{pmatrix}-j\Omega & 1\\ j\Omega & 1\end{pmatrix}\begin{pmatrix}w_1\\ w_2\end{pmatrix}. \tag{4.6.2}$$

[Hint: use (4.3.20) or (4.3.24)]

From (4.3.8), it follows that the vector ψ satisfies the equation

$$\partial_t\begin{pmatrix}\psi_1\\ \psi_2\end{pmatrix} = j\Omega\begin{pmatrix}-1 & 0\\ 0 & 1\end{pmatrix}\begin{pmatrix}\psi_1\\ \psi_2\end{pmatrix}, \tag{4.6.3}$$

which is equivalent to (4.3.17) for w.

Problem 4.6.2. Verify that (4.6.3) can be obtained from (4.3.17) with the help of the transformation (4.6.2), which diagonalizes the matrix $\hat{J}\hat{E}$.

From (4.3.15) and (4.3.16), it follows that the vector ψ is normalized:

$$\langle\psi|\psi\rangle \overset{\text{def}}{=} \langle\psi_1|\psi_1\rangle + \langle\psi_2|\psi_2\rangle = 1, \tag{4.6.4}$$

while the quantities

$$\mu_i \overset{\text{def}}{=} \langle\psi_i|\psi_i\rangle, \quad \mu_1 + \mu_2 = 1 \tag{4.6.5}$$

determine the relative energy weight of each of the packets. The quantities μ_i will be considered as given parameters of the state.

With each physical quantity, we associate a Hermitian operator, which is represented as a 2×2-matrix with operator components. For a Hermitian operator $\mathfrak{Y}$ with the matrix y_{ij} $(i, j = 1, 2)$, the conditions

$$y_{ij} = y_{ji}^{+}, \tag{4.6.6}$$

where the adjointness sign is applied to the operator elements of the matrix, must be fulfilled.

The mean value of the operator $\mathfrak{Y}$ in the state ψ is defined by the expression

$$\langle\mathfrak{Y}\rangle \overset{\text{def}}{=} \langle\psi|\mathfrak{Y}|\psi\rangle = \sum_{ij}\langle i|y_{ij}|j\rangle. \tag{4.6.7}$$

Here the following notations are introduced:

$$\langle i|\ldots|j\rangle \overset{\text{def}}{=} \langle\psi_i|\ldots|\psi_j\rangle. \tag{4.6.8}$$

In the case of a superposition of packets, the evolution characteristics of a single packet (Sect. 4.5) are replaced by matrix ones. For example, the operator $\hat{x}$ of the coordinate of the resultant center of gravity has the matrix $x_{ij} = x\delta_{ij}$ and

$$\langle \hat{x} \rangle = \langle 1|x|1 \rangle + \langle 2|x|2 \rangle = \mu_1 \langle x_1 \rangle + \mu_2 \langle x_2 \rangle , \tag{4.6.9}$$

where the coordinates of the packets' centers gravity are denoted by $\langle x_i \rangle$. The appearance of the weight factors in the relation

$$\langle i|x|i \rangle = \mu_i \langle x_i \rangle \tag{4.6.10}$$

is connected with the fact that the mean values $\langle x_i \rangle$ were calculated with the help of the normalized ψ-function.

Analogous to (4.6.9), the additivity property for the resultant mean values holds for all previously introduced first-order characteristics (i.e. linear functionals of energy density). In particular, these include the wave number, velocity, frequency, etc. It is obvious that the corresponding matrix operators are diagonal.

For the superposition of packets, the kinetic energy operator $\hat{T}$ and the potential energy operator $\hat{\Phi}$ can serve as examples of nondiagonal operators. Additivity does not hold for these.

Problem 4.6.3. Using (4.3.18) and (4.6.2), show that

$$\hat{T} = \frac{1}{2} E \begin{pmatrix} 1 & 1 \\ 1 & 1 \end{pmatrix}, \quad \hat{\Phi} = \frac{1}{2} E \begin{pmatrix} 1 & -1 \\ -1 & 1 \end{pmatrix}. \tag{4.6.11}$$

At the same time, the additivity property is fulfilled for the total energy operator $\hat{H} = \hat{T} + \hat{\Phi}$ which is proportional to the unit matrix.

Let us proceed to characteristics of higher orders, i.e. nonlinear functionals of the energy density. It is already clear that such characteristics are not additive.

The correlator $(\delta\hat{y}\delta\hat{z})$, correlation function $\mathrm{K}(\delta\hat{y}\delta\hat{z})$, and dispersion $(\Delta y)^2$ are defined as in the case of a single packet.

For example, for the width $\Delta\hat{x}$ of he superposition of packets we find after simple calculations that

$$(\Delta\hat{x})^2 = \mu_1(\Delta x_1)^2 + \mu_2(\Delta x_2)^2 + \mu_1\mu_2(\langle x_1 \rangle - \langle x_2 \rangle)^2 , \tag{4.6.12}$$

where Δx_i is the width of an individual packet. Here the additivity property is violated by the last term, which characterizes the scattering of packets.

From (4.5.2) and (4.5.13), it follows that asymptotically, as $t \to \infty$

$$\Delta\hat{x} \simeq t\Delta\bar{v} , \tag{4.6.13}$$

where the spreading-out velocity $\Delta\bar{v}$ depends on the velocities of spreading-out, Δv_i, of the individual packets and the scattering velocity $\langle v_1 \rangle - \langle v_2 \rangle$:

$$(\Delta\hat{v})^2 = \mu_1(\Delta v_1)^2 + \mu_2(\Delta v_2)^2 + \mu_1\mu_2(\langle v_1\rangle - \langle v_2\rangle)^2\,. \tag{4.6.14}$$

The quantity

$$\xi(t) \stackrel{\text{def}}{=} \frac{|\langle x_1\rangle - \langle x_2\rangle|}{\Delta x_1 + \Delta x_2} \tag{4.6.15}$$

characterizes the degree of overlap of the packets. For $\xi \ll 1$ the packets practically coincide and for $\xi \gg 1$ they are practically uncoupled. At $t \to \infty$, in view of (4.5.12) and (4.5.13), we obtain

$$\xi_\infty = \frac{|\langle v_1\rangle| + |\langle v_2\rangle|}{\Delta v_1 + \Delta v_2}\,. \tag{4.6.16}$$

Obviously, the decomposition into packets can have meaning only if the order of magnitude of ξ_∞ is not less than unity. Let us evaluate the quantity ξ_∞ for the packets, which are localized in the k representation in the neighborhood of points k_1 and k_2. We have

$$\xi_\infty \simeq \frac{|\omega'(k_1)| + |\omega'(k_2)|}{\Delta k_1|\omega''(k_1)| + \Delta k_2|\omega''(k_2)|}\,. \tag{4.6.17}$$

where Δk_i is the width of the i-th packet in the k representation.

It follows from here that sufficiently narrow packets are always asympototically uncoupled (recall that we are considering the case when $\omega'(k) \neq 0$ in the region where the packets are concentrated).

In the following section we shall consider cases for which the decomposition into packets is not justified.

4.7 Solutions Localized in the Neighborhood of Extrema of the Dispersion Curve

Up to now, we assumed that the region in which the initial conditions $u^0(k)$ and $\dot{u}^0(k)$ are prescribed, contains no extrema of the dispersion curve $\omega(k)$. Let us now consider the case when the initial conditions and hence also the solution are localized in a small neighborhood Δk of the extrema points.

If point of extremum k_0 does not coincide with the origin, then $\langle v\rangle \sim (\Delta k)^2$ and $\Delta v \sim \Delta k$. In view of (4.6.16), we see that $\xi_\infty \sim \Delta k$ and hence the packets do not uncouple. Thus, in this case, a decomposition into packets does not have any meaning.

Let $k_0 = 0$; then $\langle v\rangle \sim \Delta k$, $\Delta v \sim \Delta k$ and $\xi_\infty \sim 1$ i.e. the problem of uncoupling of the packets is left open. However, there are more serious reasons which indicate that in the given case the decomposition into packets is not expedient.

In order to see this, let us return to the solution representation (4.3.6). Let us first consider in the long-wavelength (local) approximation, an equation of the type of transverse oscillations of the rod when $\omega(k) \sim k^2$ for small values of k. Then $j(k)\,\omega(k) \sim k/k/$ and the corresponding operator $j\Omega$ (due to the factor $/k/$) has an essentially nonlocal character in the local approximation (as distinct from the case $j(k)\,\omega(k) \sim k$, which was studied in previous sections). As a result, the solution and the potential energy density, written in the forms (4.3.6) and (4.4.1), do not satisfy the natural requirement of transition to the local theory for small values of k. A similar situation occurs for longitudinal vibrations of a rod, attached to an elastic foundation (equation of the Klein-Gordon type).

Although the decomposition into packet is formally possible in the given case, it has not meaning for the reasons pointed out. Therefore, while studying solutions localized in the neighborhood of extrema of the dispersion curve, we shall not connect the energy method with the decomposition into packets, which run in the opposite directions.

In contrast to (4.4.1) assume for the potential energy density

$$\varphi = \frac{\rho}{2}(\Omega u)^2\,, \tag{4.7.1}$$

this expression possesses the required property of locality in the long-wavelength approximation. Correspondingly, we have for the total energy density

$$e = \frac{1}{2}\,\rho[(\dot{u}^2 + (\Omega u)^2]\,. \tag{4.7.2}$$

For the convenience of presentation of the energy characteristics, let us introduce a normalized complex function $\psi(x, t)$ connected with the solution $u(x, t)$, of the equation (4.3.1) by the relation

$$\psi = \sqrt{\frac{\rho}{2E}}(\dot{u} - \mathrm{i}\Omega u)\,. \tag{4.7.3}$$

It is easy to see that the normalized total energy density is equal to

$$e(x, t) = \psi^*(x, t)\,\psi(x, t)\,, \tag{4.7.4}$$

and the function ψ satisfies the Schrödinger equation

$$\mathrm{i}\partial_t\psi = \Omega\psi\,, \tag{4.7.5}$$

so that

$$\psi = \mathrm{e}^{-\mathrm{i}t\Omega}\psi^0\,, \tag{4.7.6}$$

As earlier, we shall associate with each energy characteristic of a solution the mean value of the corresponding Hermitian operator. Note that, as distinct from (4.1.11), the mean value of an operator y for the complex function (state) $\psi(x, t)$ has the form

$$\langle y \rangle = \langle \psi | y | \psi \rangle = \iint \psi^*(x, t)\, y(x, x')\, \psi(x', t)\, dx\, dx'$$
$$= \frac{1}{2\pi} \iint \psi^*(k, t)\, y(k, k')\, \psi(k', t)\, dk\, dk' \qquad (4.7.7)$$

and Hermitian operators with complex kernals $y(x, x')$ are also admissible. This extension of the class of operators, which is formally connected with the complexity of the ψ-function, comes about because the state (as distinct from the packet) is now given by two independent real functions $u(x, t)$ and $\dot{u}(x, t)$. For the same reason the spectral energy density

$$e(k) = \psi^*(k, t)\, \psi(k, t) = \psi^{0*}(k)\, \psi^0(k)\,, \qquad (4.7.8)$$

is, generally speaking, no longer an even function of k and the mean value of the operator with the kernal

$$y(k, k') = y(k)\, \delta(k - k')\,, \qquad (4.7.9)$$

where $y(k) = -y(-k)$, is not equal to zero.

In other respects, the construction of energy characteristics is carried out according to the same rules as in Sect 4.5. For the coordinate of the center of gravity, for example, we have

$$\langle x \rangle = \langle v \rangle\, t + \langle x \rangle_0\,, \qquad (4.7.10)$$

where the kernal of the velocity operator v is now defined by the expression

$$v(k) = \omega'(k) \qquad (4.7.11)$$

[for a packet $v_\pm(k) = \tilde{\omega}'_\pm(k)$].

Problem 4.7.1. Show that when $u^0 = 0$ (or $\dot{u}^0 = 0$) the center of gravity is stationary, i.e. $\langle v \rangle = 0$.

We see that the given model, considered from the energy point of view, is mathematically equivalent to the quantum mechanical model, described by the Schrödinger equation (4.7.5) with the Hamiltonian Ω (for $\hbar = 1$). This permits us to use the quantum mechanical formalism more completely, than in Sect. 4.5.

By a standard method which is presented in textbooks on quantum mechanics (see, for example [4.4]), one can obtain the inequality for the dispersion of two arbitrary Hermitian operators y and z.

$$(\Delta y)^2 (\Delta z)^2 \geqslant K^2(yz) + \frac{1}{4} |\langle [y, z] \rangle|^2\,, \qquad (4.7.12)$$

which is usually written in the weaker form (uncertainty relation)

$$(\Delta y)(\Delta z) \geqslant \frac{1}{2} |\langle [y, z] \rangle| . \tag{4.7.13}$$

In particular, for the operators x and v we have

$$(\Delta x)(\Delta v) \geqslant (\Delta x)_*(\Delta v) \geqslant \frac{1}{2} |\langle v' \rangle| , \tag{4.7.14}$$

where $(\Delta x)_*$ is the minimum width and $v' = \mathrm{i}\,[x, v]$ is an operator with the kernel $v'(k) = \omega''(k)$. For $\omega \sim k^2$, we find

$$(\Delta x)(\Delta k) \geqslant 1 , \tag{4.7.15}$$

i.e. in this case k plays the role of momentum, adjoint to the coordinate x.

Problem 4.7.2. Varify that, for the packets, introduced in Sect. 4.3–5, the uncertainty relation (4.7.13) degenerates, since $\langle [y, z] \rangle \equiv 0$ (Hint: The commutator $[u, z]$ of two real self-adjoint operators is antiself-adjoint).

Problem 4.7.3. Show that the second inequality in (4.7.14) is transformed into an equality only for a state $\psi(k, t)$ proportional for $t = t_*$ to $\exp[-\alpha\omega(k)]$, which corresponds to the Gausian function if the law of dispersion is $\omega \sim k^2$.

A non-Hermitian operator, defined by the kernel $v^{-1}(k)\partial/\partial k$, may be associated with two Hermitian operators, which have the dimension of time

$$\tau = \mathrm{i}\,[x, v^{-1}], \quad \bar{\tau} = (xv^{-1}) . \tag{4.7.16}$$

For the second of these $\langle \bar{\tau} \rangle = t + \langle \bar{\tau} \rangle_0$ i.e. it differs from the time t only by a shift. The operator τ enters the uncertainty relation

$$(\Delta x)(\Delta v^{-1}) \geqslant \frac{1}{2} |\langle \tau \rangle| . \tag{4.7.17}$$

Problem 4.7.4. Show that if T is the characteristic time during which the packet moves through a distance Δx equal to its width, then

$$T \geqslant T_* \geqslant \Delta t_{\min} = \alpha |\langle \tau \rangle| , \tag{4.7.18}$$

where T_* corresponds to the minimum width Δx_* of the packet and $\alpha \sim (\Delta k)^{-1}$.

4.8 The Case of External Forces

Let us study a wave process, which, as distinct from the above, is caused by an action of the force $q(x, t)$. Let us assume that $q(x, t)$ tends to zero sufficiently rapidly as $|t| \to \infty$; for example, $q(t)$ may be a function with finite suppopt. Then the solution of the equation

$$\rho\ddot{u} + \Phi u = q \tag{4.8.1}$$

must assymptotically transform into two scattering packets u_+ and u_-. Our task will be to find the characteristics of these packets in the form of functionals of q.

Let us introduce the projection operators

$$\pi_+ = \frac{1}{2}(1 - j_t j), \quad \pi_- = \frac{1}{2}(1 + j_t j), \tag{4.8.2}$$

where the operator j acts on the spatial variable x (or on k) and is defined by the expression (4.3.4); the operator j_t has the same properties but acts on the variable t (or ω).

$$j_t(\omega) \overset{\text{def}}{=} \mathrm{i}\ \mathrm{sgn}\ \{\omega\}, \quad j_t^2(\omega) = -1. \tag{4.8.3}$$

It is easy to see that $\pi_+(k,\omega)$ coincides with the characteristic function in the first and third quadrants of the (k,ω)-plane and $\pi_-(k,\omega)$ coincides with it in the second and fourth quadrants.

Taking into account that the projectors $\pi_\pm$ commute with the operators ∂_t^2 and Φ, let us rewrite (4.8.1) in the form of a system of two independent equations

$$\rho\ddot{u}_+ + \Phi u_+ = q_+, \quad \rho\ddot{u}_- + \Phi u_- = q_-. \tag{4.8.4}$$

As $t \to \infty$, $q_\pm(t) \to 0$ (as easily verified) and hence these equations transform into homogenuous ones. It is essential that when acting on the solutions of the homogenuous equations, the operators $\pi_\pm$ coincide with the operators $p_\pm$ introduced above see (4.3.10), and therefore may be considered as their natural extension.

Problem 4.8.1. Show that the operators $p_\pm$ and $\pi_\pm$ coincide when acting on solutions of the homogenuous equation (4.3.1).

Obviously, the energies E_+ and E_- of the packets are equal to the work done by the external forces q_+ and q_-, respectively. We have

$$\begin{aligned} E_+ &= \int dt\, \langle q_+(t)\,|\,\dot{u}_+(t)\rangle\, dt = \iint dt\, dt'\, \langle q_+(t)\,|\,\dot{G}^{\mathrm{r}}(t-t')\,|\,q_+(t')\rangle \\ &= \frac{1}{2\rho}\iint dt\, dt'\, \langle q_+(t)\,|\,\dot{D}(t-t')\,|\,q_+(t')\rangle, \end{aligned} \tag{4.8.5}$$

where G^{r} and D are the Green's functions, introduced in Sect 4.1 and connected by the relation (4.1.10).

For what follows the notations introduced in (4.1.19) and (4.1.20) will be convenient:

$$(u|u) = \int dt\, \langle u(t)|u(t)\rangle = \frac{1}{2\pi}\int d\omega\, \langle \overline{u(\omega)}|u(\omega)\rangle\,,$$

$$(u|A|u) = \iint dt\, dt' \langle u(t)|A(t,t')|u(t')\rangle$$

$$= \frac{1}{2\pi}\iint d\omega\, d\omega'\, \langle \overline{u(\omega)}|A(\omega,\omega')|u(\omega')\rangle\,, \tag{4.8.6}$$

which are invariant with respect to the t and ω-representations. In particular, for the difference operator $A(t-t')$ in the ω representation,

$$(u|A|u) = \frac{1}{2\pi}\int d\omega\, \langle \overline{u(\omega)}\,|A(\omega)|u(\omega)\rangle\,. \tag{4.8.7}$$

Thus for the energy of the packets we have

$$E_\pm = \frac{1}{2\rho}(q_\pm|\dot{D}|q_\pm)\,, \tag{4.8.8}$$

and the total energy E is equal to the sum of E_+ and E_- since, as $t\to\infty$, the packets are orthogonal in energy.

Let v be a time-independent Hermitian operator with a difference kernel $v(x-x')$, for example, the velocity operator. Let us evaluate its average value

$$\langle v_+\rangle = \frac{1}{E_+}\langle u_+|v|u_+\rangle\,. \tag{4.8.9}$$

In general, $\langle v_+\rangle$ depends on time, but, as $t\to\infty$, approaches a constant value, namely the average value $\langle v_+\rangle_\infty$ for the packet u_+. We shall be interested only in this limiting value which is a characteristic of the packet u_+, when the external force $q_+(t)$ has already ceased to work. If $q(t)=0$ for $t>t_0$ then obviously $\langle v_+\rangle_0 = \langle v_+\rangle_\infty$. In future the index ∞ will be dropped, and we shall understand by $\langle v_+\rangle$ its asymptotic value.

In order to find $\langle v_+\rangle$, let us take the scalar product of both sides of the first (4.8.4) with the function $v\dot{u}_+$. Taking into account that v is independent of time and commutes with Φ we find

$$\frac{1}{2}\frac{d}{dt}(\rho\langle \dot{u}_+|v|\dot{u}_+\rangle + \langle u_+|v\Phi|u_+\rangle) = \langle q_+|v|\dot{u}_+\rangle \tag{4.8.10}$$

which is the obvious anolog of the energy conservation law. Integrating with respect to t from $-\infty$ to $+\infty$ and taking into account that the kinetic and potential energy densities are equal for the free packet, we have

$$\rho\langle \dot{u}_+|v|\dot{u}_+\rangle = (q_+|v\dot{G}^{\mathrm{r}}|q_+) = \frac{1}{2\rho}(q_+|v\dot{D}|q_+) \tag{4.8.11}$$

or

$$\langle v_+ \rangle = \frac{1}{2\rho}(f_+ | v\dot{D} | f_+) , \tag{4.8.12}$$

where normalized forces are introduced

$$f_+ = \frac{1}{\sqrt{E_+}} q_+ . \tag{4.8.13}$$

Of course, the introduced expressions are meaningful if $E_+ \neq 0$, i.e. $u_+ \neq 0$.

Thus we have obtained the required algorithm for the calculation of all asymptotic characteristics of a packet under the action of external forces. For example

$$\langle x_+ \rangle \simeq t \langle v_+ \rangle . \tag{4.8.14}$$

4.9 Weakly Inhomogeneous Medium

Let us proceed to a consideration of inhomogenuous media. In the general case an inhomogenity of mass and inhomogenity of elastic bonds are both possible. The operator L in the generalized wave equation (4.1.1) can be represented in the form

$$L = L_0 + L_1 , \tag{4.9.1}$$

where L_0 is an operator for a homogeneous medium, and L_1 is an operator which describes a pertubation.

An effective solution of the corresponding equation is possible only under special assumptions concerning the form of the operator L_1. Let us ennumerate some of them.

There are models of local defects which permit exact solutions and which are interesting for applications. These models will be considered below.

Two cases are well known, for which the construction of approximate solutions is justified. Firstly, the case when characteristics of the medium changes sufficiently slowly in comparison with wavelengths under investigation so that one can consider the medium to be approximately homogeneous. Secondly, the case when the perturbation L_1 may be considered small in comparison with L_0. Both of these cases are described in detail in textbooks on quantum mechanics. To the first one corresponds the quasiclassical approximation, to the second–pure perturbation theory. Here we restrict outselves to a brief consideration of the second case, which illustrates the main features of the energy method.

Let us assume, for simplicity, that the mass density $\rho(x) = \rho_0$ — constant and in the equation

$$\rho_0 \partial_t^2 u + \Phi u = 0 \tag{4.9.2}$$

the operator Φ has the form

$$\Phi = \Phi_0 + \varepsilon\Phi_1 , \tag{4.9.3}$$

where Φ_1 is a perturbation with a small parameter ε.

Let us construct an analogue of the factorization (4.3.7) by setting

$$\frac{1}{\rho_0}\Phi = -\Omega^2, \quad \Omega^+ = -\Omega \tag{4.9.4}$$

(the corresponding operator j is included in Ω). Under natural assumptions (Sect. 4.3), the existence and uniqueness of the operator Ω can be demonstrated by methods of the spectral theory of operators. From the above it follows that the factorization is justified if for $\varepsilon = 0$ the solution is concentrated in a region containing no extrem points of the dispersion curve.

Let us set

$$\Omega = \Omega_0 + \varepsilon\Omega_1 , \tag{4.9.5}$$

where

$$\Omega_0^2 = -\frac{1}{\rho_0}\Phi_0, \quad \Omega_0^+ = -\Omega_0, \quad \Omega_1^+ = -\Omega_1 . \tag{4.9.6}$$

Substitution in (4.9.4), taking into account (4.9.3), leads to the equation for Ω_1

$$\Omega_0\Omega_1 + \Omega_1\Omega_0 = -\frac{1}{\rho_0}\Phi_1 \tag{4.9.7}$$

or, in the k representation,

$$\Omega_0(k)\,\Omega_1(k, k') + \Omega_1(k, k')\,\Omega_0(k') = -\frac{1}{\rho_0}\Phi_1(k, k') . \tag{4.9.8}$$

The formal solution is written in the from

$$\Omega_1(k, k') = -\frac{\Phi_1(k, k')}{\rho_0[\Omega_0(k) + \Omega_0(k')]} . \tag{4.9.9}$$

This expression has a singularity at the point $k' = -k$ and for the determination of the operator Ω_1 it is necessary to establish the rule for passing around the pole when integrating with respect to k'. We simplify the problem assuming that the operators Φ_0 and Φ are invariant with respect to translation.

Problem 4.9.1. Show that in this case the representation, compare (2.4.23),

$$\Phi_1(k, k') = kk' c_1(k, k'), \tag{4.9.10}$$

is valid and the integral

$$\int \Omega_1(k, k')\, u(k')\, dk'$$

may be understood in the sense of its principal value.

It is expedient to distinguish two basic cases of the behavior of the kernel of the pertubation operator of elastic models

$$c(x, x') = c_0(x - x') + \varepsilon c_1(x, x'). \tag{4.9.11}$$

For a local inhomogenity in the neighborhood of the point x_0, the pertubation c_1 (x, x') is a finite function, whose support is a neighborhood of the point (x_0, x_0) in the plane x, x'. Correspondingly, in the k representation, the kernel c_1 (k, k') is an entire analytic function. As the simplest model, a point defect in quasicontinuum can serve for which

$$c_1(x, x') = \delta(x - x_0)\, \delta(x' - x_0), \tag{4.9.12}$$

$$c_1(k, k') = \frac{1}{2\pi} \mathrm{e}^{-\mathrm{i}x_0(k-k')}, \quad k, k' \in B. \tag{4.9.13}$$

In the second case, the characteristics of the medium are changing periodically or randomly. The perturbation $c_1(x, x')$ is a periodic or almost periodic function of the variable $x + x'$ in the (x, x')-plane

$$c_1(x, x') = \sum_n b_n(x - x')\, \mathrm{e}^{\mathrm{i}\alpha_n(x+x')}, \tag{4.9.14}$$

where the $\alpha_n \neq 0$ are arbitrary constants for the almost periodic function and $\alpha_n \sim n$ $(n \neq 0)$ for periodic functions. In the simplest case

$$c_1(x, x') = b(x - x') \cos \alpha[(x + x')],$$

$$c_1(k, k') = \frac{1}{2} b(k + k)\, [\delta(k' - k - 2\alpha) + \delta(k' - k + 2\alpha)], \tag{4.9.15}$$

where $b(x)$ is a function which decreases sufficiently rapidly for $|x| > l$. In particular, in the local approximation one can set $b(x) = \delta(x)$.

Taking into account that $\Omega_0 = \mathrm{i}\omega(k)\mathrm{sgn}\{k\}$ and introducing the phase velocity

$$v_f(k) = \left|\frac{\omega(k)}{k}\right| = v_f(-k), \tag{4.9.16}$$

let us rewrite (4.9.9) in the form

$$\Omega_1(k, k') = \frac{ikk'c_1(k, k')}{kv_f(k) + k'v_f(k')} \tag{4.9.17}$$

where $c_1(k, k')$, for example, is given by one of the expressions above. In the local approximation or in the model of quasicontinuum without dispersion, $v_f(k) = \text{const}$, and we may also take, compare (2.3.30),

$$c_1(k, k') = c_1(k - k') . \tag{4.9.18}$$

We now can transfer the general scheme of decomposing into packets and calculating their energy characteristics, developed in the previous sections, to the case of a weakly inhomogeneous medium. Let us consider some new effects which arise in the presence of inhomogenity.

As we know, velocity conservation law (4.5.5) for the packet center of gravity is valid for the homogeneous medium, since it is an analog of the mementum conservation law. In the considered case, it takes a more general form.

In the Heisenberg representation, the equation of motion for the arbitrary operator y_t has the form, compare (4.4.18),

$$\frac{d}{dt} y_t = [\Omega, y_t] = [\Omega_0, y_t] + \varepsilon[\Omega_1, y_t] . \tag{4.9.19}$$

Setting

$$v_t \overset{\text{def}}{=} \frac{d}{dt} x_t \tag{4.9.20}$$

and taking into account that

$$[\Omega_0, [\Omega_0, x]] = 0 ,$$

we obtain

$$\frac{d}{dt} v_t = \varepsilon f_t \tag{4.9.21}$$

where the 'external force' f_t operator is defined by the expression

$$f_t = [\Omega_0, [\Omega_1, x_t]] + [\Omega_1, [\Omega_0, x_t]] . \tag{4.9.22}$$

In the Schrödinger representation,

$$f = [\Omega_0, [\Omega_1, x]] + [\Omega_1, v_0] , \tag{4.9.23}$$

where $v_0 = [\Omega_0, x]$ is the velocity operator for the homogeneous medium.

Problem 4.9.2. Taking into account that the operator $i\partial/\partial k$ corresponds to the position operator x derive an explicit expression for $f(k, k')$.

The equation of motion for the coordinate of the packet center of gravity takes the form

$$\frac{d^2}{dt^2}\langle x\rangle = \varepsilon\langle f\rangle\,, \tag{4.9.24}$$

where in the approximation under consideration, the mean value in the right hand side must be taken with respect to the solution $\psi_0(x, t)$ of the unperturbed problem. If we consider this solution to be known, then $\langle f\rangle$ is a given function of time and $\langle x\rangle$ is found by integration. Thus, the motion of the packet center of gravity is found without solving the perturbed wave equation (the analog of the Ehrenfest theorem in quantum mechanics).

We can extend this result to the correlation functions of coordinate and velocity. To this end, it is necessary to find the right-hand sides of (4.5.14) or (4.5.25) which in the considered case will be of order ε. Hence, their averaging is carried on with respect to the solution of the unperturbed problem, and the correlation functions are found by simple integration.

Problem 4.9.3. Write the expression for the right-hand side of the equation

$$\frac{d^3}{dt^3}K(x, x) = \varepsilon F(t)\,, \tag{4.9.25}$$

which defines the change of the width of the packet in time.

Let us study qualitatively the propagation of a packet through a local inhomogenity. During this, it is obvious that the packet must be partially reflected. It is easy to see, however, that the reflected wave is of order ε and hence the energy of the reflected packet is of order ε^2. Thus, in the first approximation with respect to ε, the energy characteristics of the reflected packet are of order ε^2.

In order to find the asymptotic change of the velocity of the center of gravity of the packet, it is necessary to integrate (4.9.24) with respect to time from minus to plus infinity. Taking into account that the mean value of the right-hand side of (4.9.24) is taken in the unperturbed state, we have

$$\begin{aligned}\langle f\rangle &= \langle\psi_0(k, t)\,|\,f(k, k')\,|\,\psi_0(k', t)\rangle \\ &= \langle\psi_0^0(k)\,|\,e^{t|\Omega_0(k)-\Omega_0(k')|}f(k, k')\,|\,\psi_0^0(k')\rangle\,.\end{aligned} \tag{4.9.26}$$

When integrating over time the factor

$$\delta[\Omega_0(k) - \Omega_0(k')] \sim \delta(k - k')\,,$$

appears in the kernel of the operator, so that only $f(k, k)$ will contribute to the integration over time.

Problem 4.9.4. Verify that $f(k, k) = 0$.

We see that the center of gravity of the packet behaves as a material point, which moves on a smooth plane with small local unevenness. In this analogy, the energy of the packet plays the role of mass and the average velocity $\langle v \rangle$ is proportional to the momentum.

4.10 Local Defects

Let us now study the inhomogenity of mass and elastic bonds, localized in a small region, but without any assumptions about the smallness of the perturbation.

As a first step, we introduce examples of fundamental types of local defects.

1) An elastic bond with a rigid foundation at the point $x = 0$. The corresponding kernel of the operator L_1 has the form

$$L_1(x, x') = \alpha\delta(x)\,\delta(x'), \tag{4.10.1}$$

$$L_1(k, k') = \frac{\alpha}{2\pi} = \text{const}. \tag{4.10.2}$$

Obviously, in this case, the condition of invariance with respect to translation is not fulfilled.

2) A defect of the elastic moduli of the medium at the point $x = 0$

$$L_1(x, x') = \alpha a^2\delta'(x)\,\delta'(x'), \tag{4.10.3}$$

$$L_1(k, k') = \frac{\alpha a^2}{2\pi} kk', \tag{4.10.4}$$

where a is a scaling parameter.

3) A defect of an elastic bond in the chain between masses at points $x = 0$ and $x = a$.

$$L_1(x, x') = \alpha[\delta(x + a) - \delta(x)]\,[\delta(x' + a) - \delta(x')], \tag{4.10.5}$$

$$L_1(k, k') = \frac{\dot{\alpha}}{2\pi}(\mathrm{e}^{-\mathrm{i}ak} - 1)\,(\mathrm{e}^{\mathrm{i}ak'} - 1). \tag{4.10.6}$$

4) A defect of mass or density at the point $x = 0$.

$$L_1(x, x', t) = \alpha\delta(x)\,\delta(x')\,\partial_t^2, \tag{4.10.7}$$

$$L_1(k, k', \omega) = -\frac{\alpha}{2\pi}\omega^2 . \tag{4.10.8}$$

In all these examples, the admissible range of variation of the parameter α, which characterizes the magnitude of the defect, must be detemined from the condition of stability. In some cases, this range is obvious from the physical considerations. Thus, for example 1, only the values $\alpha > 0$ are admissible since a negative elastic bond with a rigid foundation would lead to instability. But, for the Example 2, the admissible range of α depends on the dispersion law and is not obvious a priori. A simple method of determining the admissible boundaries of variation of α will be indicated below.

Generalizing these examples, we assume by definition, that a local defect is described by an operator L_1 with the kernel of the form.

$$L_1(k, k', \omega) = \frac{\alpha}{2\pi} b(\omega)\, \gamma(k)\, \overline{\gamma(k')} , \tag{4.10.9}$$

where $b(\omega)$ and $\gamma(k)$ characterize the type of the defect.

For a system of local defects

$$L_1(k, k', \omega) = \frac{1}{2\pi} \sum_i \alpha_i b_i(\omega)\, \gamma_i(k)\, \overline{\gamma_i(k')} . \tag{4.10.10}$$

Note that an arbitrary local inhomogenity can be approximated by a set of δ-functions and their derivatives (expansion in multipoles). In this case, the kernel of the operator L_1 is also written in the form (4.10.10), i.e. from the mathematical point of view, the expansion of a local inhomogenity in multipoles is equivalent to the system of local defects.

The complete information about the solution of the wave equation, as we already know, is contained in the retarded Green's function $G^{\mathrm{r}}(x, x', t)$. We now proceed to its construction and consider first the case of one local defect.

Let us transform the equation for Green's function

$$(L_0 + L_1)\, G^{\mathrm{r}} = I , \tag{4.10.11}$$

by applying to both parts the Green's function G_0^{r} for the homogeneous medium. We have

$$G^{\mathrm{r}} + G_0^{\mathrm{r}} L_1 G^{\mathrm{r}} = G_0^{\mathrm{r}} \tag{4.10.12}$$

or in a more detailed form in the (k, ω) representation

$$G^{\mathrm{r}}(k, k', \omega) + \frac{\alpha}{2\pi} b(\omega)\, G_0^{\mathrm{r}}(k, \omega)\, \gamma(k) \int \overline{\gamma(k'')}\, G^{\mathrm{r}}(k'', k', \omega)\, dk''$$

$$= G_0^{\mathrm{r}}(k, \omega)\, \delta(k - k') . \tag{4.10.13}$$

Multiplying by $\overline{\gamma(k)}$ and integrating, we find

$$[1 + \alpha g(\omega)] \int \overline{\gamma(k)}\, G^{\mathrm{r}}(k, k', \omega)\, dk = \gamma(k')\, G_0^{\mathrm{r}}(k', \omega)\,, \tag{4.10.14}$$

where

$$g(\omega) = \frac{1}{2\pi} b(\omega) \int |\gamma(k)|^2\, G_0^{\mathrm{r}}(k, \omega)\, dk\,. \tag{4.10.15}$$

If the condition of stability is fulfilled, then the retarded function $G^{\mathrm{r}}(\omega)$ must be analytic in the upper half ω-plane (Sect. 4.1 and Appendix B). Because $G_0^{\mathrm{r}}(\omega)$ is analytic there, this requirement is equivalent to the absence of roots of the function $1 + \alpha g(\omega)$ in the upper half plane. This condition, by itself, defines the admissible region of variation of the parameter α.

Let us investigate the behavior of $1 + \alpha g(\omega)$ in the upper half ω-plane ($\omega = \omega' + \mathrm{i}\omega''$). For $b(\omega) = 1$, we have from (4.1.35) for the Green's function of a homogeneous medium:

$$\operatorname{Im}\{1 + \alpha g(\omega)\} = \frac{\alpha\omega'\omega''}{\rho_0} \int \frac{|\gamma(k)|^2\, dk}{|\omega^2(k) - \omega^2|^2}\,. \tag{4.10.16}$$

This expression tends to zero as $\omega' \to 0$, because the integral is bounded. When $\omega'' \to 0$ the integral diverges and the expression has a finite limit.

Problem 4.10.1. Show that

$$\lim_{\omega'' \to 0} \operatorname{Im}\{1 + \alpha g(\omega)\} = \frac{\alpha\theta(\omega_{\max}^2 - \omega^2)}{2\omega\rho_0} k'(\omega)\, |\gamma(k(\omega))|^2\,, \tag{4.10.17}$$

where $k(\omega)$ is the function inverse to $\tilde{\omega}(k)$.

It follows from here, that on the real axis, zeros of $1 + \alpha g(\omega)$ are possible only for those values $|\omega| > \omega_{\max}$ for which $\operatorname{Re}\{1 + \alpha g(\omega)\} = 0$. The corresponding real poles of $G^{\mathrm{r}}(\omega)$ do not cause any difficulties due to the accepted rule for passing around such poles. We shall see below that these poles correspond to local oscillations.

Thus, it is sufficient to consider $\operatorname{Re}\{1 + \alpha g(\omega)\}$ on the imaginary axis. The condition of the absence of roots of the equation

$$\operatorname{Re}\{1 + \alpha g(\omega'')\} = 1 + \frac{\alpha}{2\pi\rho_0} \int \frac{|\gamma(k)|^2\, dk}{\omega^2(k) + (\omega'')^2} = 0\,, \tag{4.10.18}$$

obviously has the form

$$\alpha > \alpha_0 = -\frac{1}{\sup \operatorname{Re}\{g(\omega'')\}}\,, \tag{4.10.19}$$

where, in accordance with (4.10.18),

$$\sup \operatorname{Re}\{g(\omega'') = \frac{1}{2\pi\rho_0} \int \frac{|\gamma(k)|^2}{\omega^2(k)} dk \,. \tag{4.10.20}$$

These expressions determine the region of admissible values of α in the examples 1–3.

Problem 4.10.2. Show that for the above-given Examples 1 and 2, respectively,

$$\alpha_0 = 0, \quad \alpha_0 = -\frac{2\pi\rho_0}{a^4} \left(\int \frac{k^2}{\omega^2(k)} dk \right)^{-1}, \tag{4.10.21}$$

and for the above Example 3

$$\alpha_0 = -\frac{1}{4} a\rho_0 \omega_{\max}^2 \tag{4.10.22}$$

with the dispersion law

$$\omega^2 = \omega_{\max}^2 \sin^2 \frac{ak}{2}, \quad |k| \leqslant \frac{\pi}{a}, \tag{4.10.23}$$

which corresponds to the interaction of nearest neighbors in the chain.

Problem 4.10.3. Show that for the case $b(\omega) = -\omega^2$ the condition (4.10.19) is conserved, but

$$\sup \operatorname{Re}\{g(\omega'')\} = \frac{1}{2\pi\rho_0} \int |\gamma(k)|^2 dk, \tag{4.10.24}$$

and hence, for the continuous medium, $\sup \operatorname{Re}\{g(\omega'')\} = \infty$ and for the chain, $\sup \operatorname{Re}\{g(\omega'')\} = (a\rho_0)^{-1}$. Correspondingly for Example 4, $\alpha_0 = 0$ and $\alpha_0 = -a\rho_0$.

Thus, in the admissible region $\alpha > \alpha_0$ we found we can solve (4.10.14) and substitute the result in (4.10.13). The final expression for G^{r} is

$$G^{\mathrm{r}} = G_0^{\mathrm{r}} - G_0^{\mathrm{r}} R G_0^{\mathrm{r}}, \tag{4.10.25}$$

where the operator R has the kernel

$$R(k, k', \omega) = \frac{\alpha}{2\pi} \gamma(k) \overline{\gamma(k')} r(\omega), \tag{4.10.26}$$

$$r(\omega) = \frac{b(\omega)}{1 + \alpha g(\omega)}, \tag{4.10.27}$$

and the function $g(\omega)$ is determined by (4.10.15).

Let us summarize. The construction of the Green's function G^{r} for individual local defect is reduced to the calculation of $g(\omega)$ according to (4.10.15). Let us present the results for the models discussed above.

Example 1. a) For the law of dispersion (4.10.22) (a chain)

$$\rho_0 g(\omega) = -\frac{|\omega|\theta(\omega^2 - \omega_{\max}^2) + \omega\theta(\omega_{\max}^2 - \omega^2)}{a\omega^2\sqrt{\omega^2 - \omega_{\max}^2}}\,; \tag{4.10.28}$$

b) for the law of dispersion $\omega^2 = v_0^2 k^2$ (a string)

$$\rho_0 g(\omega) = \frac{\mathrm{i}}{2v_0(\omega + \mathrm{i}0)} = \frac{\pi}{2v_0}\,\delta(\omega) + \frac{\mathrm{i}}{2v_0\omega}\,; \tag{4.10.29}$$

c) for the law of dispersion $\omega^2 = d_0^2 k^4$ (transverse vibrations of a rod)

$$\rho_0 g(\omega) = \frac{-1 + \mathrm{i}\,\mathrm{sgn}\{\omega\}}{8\sqrt{d_0|\omega|^3}}\,. \tag{4.10.30}$$

Note that, as expected, the Example 1b is obtained from Example 1a, if it is assumed that $\omega \ll \omega_{\max} = 2v_0/a$

Example 2. For the law of dispersion $\omega^2 = v_0^2 k^2$, $|k| \leqslant \pi/a$ (the Debye quasi-continuum)

$$\rho_0 g(\omega) = \frac{a}{v_0^2} + \frac{a^2\omega}{2\pi v_0^3}\ln\left|\frac{\pi v_0 - a\omega}{\pi v_0 + a\omega}\right| + \frac{\mathrm{i}a^2\omega}{2v_0^3}\,. \tag{4.10.31}$$

Example 3. For the chain

$$\frac{a\rho_0\omega_{\max}^2}{4}\,g(\omega) = 1 - \frac{|\omega|\theta(\omega^2 - \omega_{\max}^2) + \omega\theta(\omega_{\max}^2 - \omega^2)}{\sqrt{\omega^2 - \omega_{\max}^2}}\,. \tag{4.10.32}$$

Example 4. a) for the chain

$$\rho_0 g(\omega) = \frac{|\omega|\theta(\omega^2 - \omega_{\max}^2) + \omega\theta(\omega_{\max}^2 - \omega^2)}{a\sqrt{\omega^2 - \omega_{\max}^2}}\,; \tag{4.10.33}$$

b) for the string

$$\rho_0 g(\omega) = -\frac{\mathrm{i}\omega}{2v_0}\,; \tag{4.10.34}$$

c) for the rod

$$\rho_0 g(\omega) = \frac{1}{8}\sqrt{\frac{|\omega|}{d_0}}\,(1 - \mathrm{i}\,\mathrm{sgn}\{\omega\})\,. \tag{4.10.35}$$

As in Example 1, the Case 4b is obtained from Example 4a if we assume $\omega \ll \omega_{max}$

Let us note interesting limiting cases. It is easy to see that if $\omega \ll \omega_{max} = 2v_0/a$ then the functions $g(\omega)$ in the examples 2 and 3 coincide. However, as distinct from Examples 1 and 4, the dependence of $g(\omega)$ on the parameter a does not disappear. The limiting transition to a string $a \to 0$ gives $R \to 0$, i.e. at such a limit, waves are unaffected by the defect.

If, in the Examples 1 and 4, we take $\alpha \to \infty$, then the expression (4.10.26) for the kernel of the operator R becomes considerably simpler and

$$R(k, k', \omega) = [\int G_0^r(k'', \omega)\, dk'']^{-1} \tag{4.10.36}$$

is independent of k and k'. This case, obviously, corresponds to the fixed point $x = 0$, i.e. to a semibounded chain and to a string with one end fixed. However, this does not correspond to the semibounded rod due to the possibility of rotating the cross section of the rod.

If, in Example 3, with the law of dispersion (4.10.23) we set $\alpha = \alpha_0$ from (4.10.22), then this corresponds to cutting the bond and we obtain the semibounded chain with a free end. For this case

$$R(k, k', \omega) = -\frac{a\rho_0\omega_{max}^2}{8\pi} \frac{(e^{-iak}-1)(e^{iak'}-1)\sqrt{\omega^2-\omega_{max}^2}}{|\omega|\theta(\omega^2-\omega_{max}^2)+\omega\theta(\omega_{max}^2-\omega^2)} \tag{4.10.37}$$

The additional limiting transition $\omega \ll \omega_{max}$ and $ak, ak' \ll 1$ leads to a semibounded string with a free end.

$$R(k, k', \omega) = \frac{v_0^3}{\pi} \frac{kk'}{\omega + i0}, \quad v_0 = \frac{a\omega_{max}}{2}. \tag{4.10.38}$$

Taking into account the previous remark concerning Examples 2 and 3 we see that for the case of a defect of an elastic bond the final result depends essentially on the sequence of the limiting transitions.

Let us dwell briefly on the system of defects. The calculation of the Green's function is carried out in the same manner as that for the single defect. Instead of (4.10.14), taking into account (4.10.10), we have

$$\sum_j [\delta_{ij} + \alpha_j g_{ij}(\omega)] \int \overline{\gamma_j(k)}\, G^r(k, k', \omega)\, dk = \overline{\gamma_i(k')}\, G_0^r(k', \omega), \tag{4.10.39}$$

where

$$g_{ij}(\omega) = \frac{1}{2\pi} b_j(\omega) \int \overline{\gamma_i(k)}\, \gamma_j(k)\, G_0^r(k, \omega)\, dk. \tag{4.10.40}$$

The admissible region of values of the parameters α_j, is now determined by investigating the roots of the determinant of the matrix $\delta_{ij} + \alpha_j g_{ij}(\omega)$.

The Green's function G^r is still represented in the form (4.10.25), but the kernal of the operator R has the form

$$R(k, k', \omega) = \frac{1}{2\pi} \sum_{ij} \alpha_i b_i(\omega)\, M_{ij}(\omega)\, \gamma_i(k)\, \overline{\gamma_j(k')}\,, \tag{4.10.41}$$

where

$$M_{ij}(\omega) = [\delta_{ij} + \alpha_j g_{ij}(\omega)]^{-1}\,. \tag{4.10.42}$$

We saw above that $R(\omega)$ can have poles on the real axis ω for $\omega > \omega_{\max}$ which are determined, for the case of a single defect (4.10.9) by the equation

$$\mathrm{Re}\{1 + \alpha g(\omega)\} = 0\,. \tag{4.10.43}$$

Let us consider the structure of singular solutions which correspond to the discrete roots ω_l of this equation. Let us write the wave equation in the form

$$L^0 u = -\, L_1 u \tag{4.10.44}$$

and apply to both sides the Green's function G_0^r. Taking into account (4.10.9), we write the result in the form

$$u(k, \omega) = -\,\frac{\alpha b(\omega)}{2\pi}\, G_0^r(k, \omega)\, \gamma(k) \int \overline{\gamma(k')}\, u(k', \omega)\, dk'\,. \tag{4.10.45}$$

If we repeat the calculations, analogous to those carried out when constructing G^r, then we find that the equation has a nontrival solution only for frequencies $\omega_l > \omega_{\max}$ which satisfy (4.10.43). Let us examine the behavior of the function $u(x, \omega_l)$, which is determined by the integral

$$J(x) = \int \frac{\gamma(k)\, e^{ikx}}{\omega^2(k) - \omega_l^2}\, dk\,. \tag{4.10.46}$$

The main contribution to the integral is given by points in the neighborhood of $\omega^2(k) \sim \omega_{\max}^2$. Therefore, let us put $\omega^2(k) = \omega_{\max}^2 - \lambda^2 k^2$. Then

$$J(x) \sim e^{-\beta|x|}, \quad \beta = \frac{1}{\lambda} \sqrt{\omega_l^2 - \omega_{\max}^2}\,. \tag{4.10.47}$$

Thus solutions, corresponding to $\omega = \omega_l$ are concentrated only in a bounded region; in connection with this, they are called local oscillations as distinct from propagating waves. They are of interest for a number of applications, particularly for the theory of defects in crystal lattice [4.5].

Problem 4.10.4. Show that for the Examples 1*a* and 4*a*

$$\omega_l^2 = \frac{\omega_{\max}^2}{2} \left(1 + \sqrt{1 + \frac{4\alpha^2}{a^2 \rho_0^2 \omega_{\max}^2}}\right), \quad \alpha > 0\,,$$

$$\omega_l^2 = \frac{a^2 \rho_0^2 \, \omega_{max}^2}{a^2 \rho_0^2 - \alpha^2}, \quad -a\rho_0 < \alpha < 0 , \tag{4.10.48}$$

respectively.

From the last expression, it is seen that local oscillations in a chain are possible only in cases for which the mass of the defective atom is less than that of a normal one.

The Green's functions expressions, obtained here, will be used in Sect 4.14 for the problem of scattering by defects.

4.11 The Structure of the Green's Function of an Inhomogeneous Medium

In the previous two sections we considered particular cases of inhomogeneous media. Now, let us treat the structure of the Green's function for the general case of an inhomogeneous medium.

As a first step, let us represent the equation for an inhomogeneous medium in the form

$$Lu = q, \quad L = L_0 + L_1, \tag{4.11.1}$$

where

$$L_0 = \rho_0 \partial_t^2 + \Phi_0 \tag{4.11.2}$$

is the operator of the homogeneous medium and

$$L_1 = \rho_1 \partial_t^2 + \Phi_1 \tag{4.11.3}$$

is a perturbation, caused by the inhomogenity of the mass density and of elastic properties of the medium. We emphasize here that we do not assume the perturbation to be small.

Let us assume that the operator L_1 is localized, i.e. the functions $\rho_1(x)$ and $\Phi_1(x, x')$ vanish sufficiently rapidly at infinity, for example the latter have finite supports. Moreover, the perturbation must satisfy the obvious conditions: $\rho(x) = \rho_0 + \rho_1(x) > 0$ and $\Phi = \Phi_0 + \Phi_1$ is a positive definite operator.

We know that the retarded Green's function G^r is determined by the relations (4.1.6). As in the case of local defects, let us represent G^r in the form

$$G^r = G_0^r - G_0^r R G_0^r , \tag{4.11.4}$$

where G_0^r is the Green's function of the operator L_0. Then constructing G^r is equivalent to finding the operator R. This is more convenient since R is more directly connected with the perturbation L_1 and, in particular, has essentially the same region of localization (cf. Eqs. (4.10.10) and (4.10.36)). In sect. 4. 13 we shall also see that the operator R gives an explicit solution of the scattering problem.

Substituting (4.11.4) in (4.1.6) and appling L_0 from the right, we find the equation for R:

$$R + L_1 G_0^r R = L_1 . \tag{4.11.5}$$

For what follows, the relations

$$R = L_1 - L_1 G^r L_1, \quad G_0^r R = G^r L_1 . \tag{4.11.6}$$

will also be useful.

Problem 4.11.1. Verify these relations.

Problem 4.11.2. Using the first of the relations (4.11.6), find the region of localization of the kernal $R(x, x', t)$ in the plane (x, x') for the system of point defects of the type (4.10.1) or (4.10.7).

We can simplify (4.11.5) for R and reduce it to canonical form. With this in mind, let us use the factorization of the operator L_0 introduced in Sect 4.3.:

$$L_0 = \mathscr{L}_0 \mathscr{L}_0^+, \quad \mathscr{L}_0 = \partial_t + j\Omega_0, \quad \mathscr{L}_0^+ = \partial_t - j\Omega_0 , \tag{4.11.7}$$

and correspondingly let us factorize the Green's function G_0^r

$$G_0^r = \mathscr{G}_0 \mathscr{G}_0^+, \quad \mathscr{L}_0 \mathscr{G}_0 = I . \tag{4.11.8}$$

In the (k, ω) representation

$$\sqrt{\rho_0}\mathscr{G}_0(k, \omega) = \frac{\mathrm{i}}{(\omega + \mathrm{i}0) - \tilde{\omega}(k)}, \quad \sqrt{\rho_0}\mathscr{G}_0^+(k, \omega) = \frac{\mathrm{i}}{(\omega + \mathrm{i}0) + \tilde{\omega}(k)} . \tag{4.11.9}$$

Let

$$L_1' = \mathscr{G}_0^+ L_1 \mathscr{G}_0, \quad R' = \mathscr{G}_0^+ R \mathscr{G}_0 . \tag{4.11.10}$$

Then after multiplying (4.11.5) from the left by $\mathscr{G}_0^+$ and from the right by $\mathscr{G}_0$, one obtains the canonical form (omitting the prime)

$$R + L_1 R = L_1 . \tag{4.11.11}$$

Let us consider some cases when it is possible to solve expicitly this equation.

1) If the kernel of the original operator L_1 is degenerate, i.e. has the form (4.10.10), then the kernel of the transformed operator L_1 possesses the same

property. The solution of the equation is reduced to algebraic operations as was shown in Sect. 4.10. This method is also applicable for the approximate solution of (4.11.11), when the approximation of the operator L_1 by an operator with a degenerate kernel is acceptable. In particular this is justified, if the inhomogenity is concentrated in a small region or, equivalenty if sufficiently long waves are investigated. Then the indicated approximation is equivalent to expansion of the orginal kernel $L_1(x, x', t)$ in multipoles.

2) The perturbation is small, i.e. $L_1 \sim \varepsilon$. Obviously,

$$R = L + O(\varepsilon). \tag{4.11.12}$$

3) Strong perturbation, i.e. $L_1 \sim \varepsilon^{-1}$. We have

$$R = I - L_1^{-1} + O(\varepsilon^2). \tag{4.11.13}$$

4) Let $L_1 = -2P$, where P is a projection operator, i.e. $P^2 = P$. Then $R = 2P$ is an exact solution of (4.11.11). If now

$$L_1 = -2P + \varepsilon A. \tag{4.11.14}$$

then, as it is easy to verify by the substitution

$$R = 2P + \varepsilon(1 - 2P)A(1 - 2P) + O(\varepsilon). \tag{4.11.15}$$

Let us proceed to the consideration of the analytical properties of the kernel of the operator R. We saw (Sect. 4.1) that the retarded Green's functions $G^r(\omega)$ and $G_0^r(\omega)$ are analytic in the upper half plane. However it then follows, in accordance with (4.11.4) and (4.11.10), that the original and transformed functions $R(\omega)$ possess the same properties.

Let us assume further that $L_1(x, x', \omega)$ decreases rapidly at infinity with respect to the spatial variables. The first of the relations (4.11.6) shows that this then takes place also for the kernel $R(x, x', \omega)$. Hence $R(k, k', \omega)$ is an entire analytical function of the variables k, k' and an analytic function of ω in the upper half plane. In the following section, we shall see that this determines analytical properties of the scattering matrix.

The asymptotic (as $t \to \infty$) behavior of the total energy ΔE of the field caused by the perturbation L_1 is directly expressed in terms of the operator R. Let the field $u(t)$ be induced by the action of external forces $q(t)$ and let $u_0(t)$ be the field that would have been induced in the homogeneous medium under the action of the same forces, i.e. it be a solution of the equation $L_0 u_0 = q$. Then integrating over time the expression (4.1.17) for the power of the external forces, in view of (4.1.8) and (4.11.14), we obtain

$$\Delta E = E - E_0 = (u_0 | \dot{R} | u_0), \tag{4.11.16}$$

where E is the energy of the field $u(t)$ as $t \to \infty$, E_0 is the same for $u_0(t)$ and the brackets indicate a functional of the form (4.8.7). Obviously, only the even component of $R(t)$ contributes to ΔE.

Observe that the Green's function $D(t)$ cannot be represented in a form similar to (4.11.4). Instead of this, taking into account (4.1.14) and (4.1.11), we have in the ω representation

$$\begin{aligned} D(\omega) &= [G^{\mathrm{r}}(\omega) - G^{\mathrm{r}}(-\omega)]\,\rho \\ &= D_0(\omega)\,\frac{\rho}{\rho_0} - 2\mathrm{i}\,\mathrm{Im}\{G_0^{\mathrm{r}}(\omega)\,R(\omega)\,G^{\mathrm{r}}(\omega)\,\rho\}\,. \end{aligned} \tag{4.11.17}$$

Problem 4.11.3. Show that (4.11.17) satisfies the conditions (4.1.13).

Let us now introduce the matrix Green's functions, which will be convenient, in particular, for the scattering problem. With this in view, let us proceed from scalar functions u, q to vector functions w and f, according to (4.1.37) and (4.1.39) and let us rewrite (4.11.1) in the matrix form

$$\hat{L}w = f, \quad \hat{L} = \hat{L}_0 + \hat{L}_1\,, \tag{4.11.18}$$

where, as is easily seen, the matrix operators $\hat{L}_0$ and $\hat{L}_1$ are defined by the expressions

$$\hat{L}_0 = \begin{pmatrix} \rho_0\partial_t & -\rho_0 \\ \Phi_0 & \rho_0\partial_t \end{pmatrix}, \quad \hat{L}_1 = \begin{pmatrix} 0 & 0 \\ L_1 & 0 \end{pmatrix}. \tag{4.11.19}$$

Let us represent the matrix Green function $\hat{G}^{\mathrm{r}}$ in the form, analogous to (4.11.4),

$$\hat{G}^{\mathrm{r}} = \hat{G}^{\mathrm{r}} - \hat{G}^{\mathrm{r}}\hat{R}\hat{G}^{\mathrm{r}}\,, \tag{4.11.20}$$

where $\hat{G}_0^{\mathrm{r}}$ is connected with the scalar Green's function D_0 by (4.1.42) and the matrix $\hat{R}$ has the form

$$\hat{R} = \begin{pmatrix} 0 & 0 \\ R & 0 \end{pmatrix}. \tag{4.11.21}$$

Problem 4.11.4. Verify the expression for $\hat{R}$ by substitution in the equation $\hat{L}\hat{G} = \hat{I}$.

Problem 4.11.5. Show that $\hat{R}$ satisfies the equation, (cf. (4.11.5)),

$$\hat{R} + \hat{L}_1\hat{G}_0^{\mathrm{r}}\hat{R} = \hat{L}_1 \tag{4.11.22}$$

and the relations analogous to (4.11.6):

$$\hat{R} = \hat{L}_1 - \hat{L}_1\hat{G}^{\mathrm{r}}\hat{L}_1, \quad \hat{G}_0^{\mathrm{r}}\hat{R} = \hat{G}^{\mathrm{r}}\hat{L}_1\,. \tag{4.11.23}$$

Instead of w it is convenient to introduce a new vector function ψ, whose components asymptotically (as $|t| \to \infty$) turn into the normalized wave packets investigated in Sects 4.3–6 if $q(t) \to 0$, when $|t| \to \infty$ For this, analogous to (4.6.2), let

$$\psi = \hat{A}w \quad w = \hat{A}^{-1}\psi, \tag{4.11.24}$$

where

$$\hat{A} = \frac{1}{2}\sqrt{\frac{\rho_0}{E}}\begin{pmatrix} -j\Omega_0 & 1 \\ j\Omega_0 & 0 \end{pmatrix}, \quad \hat{A}^{-1} = \sqrt{\frac{E}{\rho_0}}\begin{pmatrix} j\Omega_0^{-1} & -j\Omega_0^{-1} \\ 1 & 1 \end{pmatrix} \tag{4.11.25}$$

and E is the asymptotic (as $t \to \infty$) total energy of the solution of the initial equation (4.11.1). Here, it is assumed that $E \neq 0$.

Instead of the system (4.11.18), we now have

$$\hat{L}'\psi = f', \tag{4.11.26}$$

where

$$\hat{L}' = \hat{A}\hat{L}\hat{A}^{-1}, \quad f' = \hat{A}f. \tag{4.11.27}$$

The relations (4.11.24) can be regarded as a transition from the w-basis to the ψ-basis. In what follows, only the ψ-basis is considered and the primes are omitted.

Problem 4.11.6. Show that in the ψ-basis, the operator $\hat{L}_0$ and the Green's function $\hat{D}_0$ are diagonal and have the form

$$\hat{L}_0 = \begin{pmatrix} \partial_t + j\Omega & 0 \\ 0 & \partial_t - j\Omega \end{pmatrix}, \quad \hat{D} = \begin{pmatrix} e^{-j\Omega t} & 0 \\ 0 & e^{j\Omega t} \end{pmatrix}. \tag{4.11.28}$$

The Green's function $\hat{G}_0^r$ from (4.11.20) is connected with $\hat{D}_0$ by the usual relation $\rho_0\hat{G}_0^r(t) = \theta(t)\,\hat{D}_0(t)$, and for the operator $\hat{R}$ in the ψ-basis, we have

$$\hat{R} = \frac{1}{2}Rj\Omega_0^{-1}\begin{pmatrix} 1 & -1 \\ 1 & -1 \end{pmatrix} \tag{4.11.29}$$

Problem 4.11.7. Derive this expression from (4.11.21).

In the following sections we shall see how the formulae for the Green's functions presented here, are used in the scattering problem.

4.12 The Scattering Matrix

Let us proceed to the formulation of the nonstationary scattering problem. We consider the wave equation

$$Lu = 0, \quad L = L_0 + L_1, \qquad (4.12.1)$$

where the operators L_0 and L_1 are given by (4.11.12 and 13) and satisfy the condition mentioned there.

Let $u(t)$ be a solution of the wave equation for propagating waves, i.e. $u(t)$ does not contain local oscillations. Let us also assume that the energy E of the wave $u(t)$ is finite. Then, due to the localized character of the perturbation L_1, the asymptotic behavior of $u(t)$ as $t \to -\infty$ and $t \to \infty$ is given by wave packets, which are mostly located far away from the inhomogenity and hence approximately satisfy the equation $L_0 u = 0$. The scattering problem consists in finding the asymptotics of $u(t)$ as $t \to \infty$, the asymptotics of $u(t)$ as $t \to -\infty$ being given. This connection between the asymptotics is given a special form in the theory of scattering.

Let us introduce auxilary functions $u^{\text{in}}(t)$ and $u^{\text{out}}(t)$, satisfying the equations for the homogenous medium

$$L_0 u^{\text{in}} = 0, \quad L_0 u^{\text{out}} = 0 \qquad (4.12.2)$$

and the asymptotic conditions

$$\begin{aligned} &[u(t) - u^{\text{in}}(t)] \to 0 \quad \text{as} \quad t \to -\infty, \\ &[u(t) - u^{\text{out}}(t)] \to 0 \quad \text{as} \quad t \to \infty. \end{aligned} \qquad (4.12.3)$$

From the uniqueness of the Cauchy problem for the operator L follows the existence of one-to-one correspondences

$$u^{\text{in}}(t) \leftrightarrow u(t) \leftrightarrow u^{\text{out}}(t). \qquad (4.12.4)$$

The scattering problem is obviously equivalent to establishing the relation

$$u^{\text{out}} = S u^{\text{in}}. \qquad (4.12.5)$$

The operator S appearing here is called the scattering operator. The construction of S for a given L_1 (with L_0 fixed) is called the direct problem of scattering. Obtaining possible information about L_1, S being known, is called the inverse scattering problem. Methods of solving these problems and also the investigation of properties of the scattering operator form the subject of the theory of scattering.

It is convenient to pass to the matrix representation of the scattering problem. Using the ψ-basis introduced in the preceding section, we have instead of (4.12.5)

$$\psi^{\text{out}} = \hat{S}\psi^{\text{in}}, \tag{4.12.6}$$

where $\hat{S}$ is a matrix with operator components.

Note the essential difference between the relations (4.12.5) and (4.12.6). The vector functions $\psi^{\text{in}}(t)$ and $\psi^{\text{out}}(t)$ as solutions of equations which are of first order in time

$$\hat{L}_0\psi^{\text{in}} = 0, \quad \hat{L}_0\psi^{\text{int}} = 0 \tag{4.12.7}$$

are determined completely by their values at an arbitrary time $t = t_0$. Hence the matrix operator $\hat{S}$ can always be constructed in such a way that it connects values of $\psi^{\text{in}}(t)$ and $\psi^{\text{out}}(t)$ at one and the same time, i.e. (4.12.6) can be rewritten in the form

$$\psi^{\text{out}}(t) = \hat{S}\psi^{\text{in}}(t), \tag{4.12.8}$$

where $\hat{S}$ operates only on the spatial variable and t appears as a parameter. As distinct from $\hat{S}$, the operator S operates on both the spatial and time variables, since $u(t_0)$ does not yet determine the function $u(t)$.

Let us elucidate the basic properties of the matrix operator $\hat{S}$. From (4.12.4), it follows that $\hat{S}$ is nonsingular, i.e. the inverse operator $\hat{S}^{-1}$ exists. Furthermore due to energy conservation, we have

$$\langle\psi^{\text{out}}|\psi^{\text{out}}\rangle = \langle\hat{S}\psi^{\text{in}}|\hat{S}\psi^{\text{in}}\rangle = \langle\psi^{\text{in}}|\hat{S}^{+}\hat{S}\rangle = \langle\psi^{\text{in}}|\psi^{\text{in}}\rangle, \tag{4.12.9}$$

whence, due to the existence of $\hat{S}^{-1}$, it follows that $\hat{S}$ is a unitary operator

$$\hat{S}^{+}\hat{S} = \hat{S}\hat{S}^{+} = \hat{I}, \quad \text{or} \quad \hat{S}^{+} = \hat{S}^{-1}. \tag{4.12.10}$$

A question arises whether the operator $\hat{S}$ could be of convolution type, i.e. to reduce in the k representation to multiplication by a functional matrix. Let us show that as a rule this is impossible. In fact, in the ψ-basis under consideration, the components of any solution of the equation $L_0\psi = 0$, in accordance with (4.3.6), have the form

$$\psi_1(t) = \mathrm{e}^{-j\Omega_0 t}\varphi_1, \quad \psi_2(t) = \mathrm{e}^{j\Omega_0 t}\varphi_2, \tag{4.12.11}$$

where the φ_i are the values of $\psi_i(t)$, went $t = 0$. The operator $\hat{S}$ transforms one such solution into another, cf. (4.12.8), acting on the spatial variable only. But this means that nondiagonal components of the matrix $\hat{S}$ must contain in themselves the operation of spatial inversion which changes the sign of the operator $j\Omega_0$. It is obvious that the convolution-type operator cannot possess this property because in the k representation, it is nothing but multiplication by a function. Thus the operator $\hat{S}$ could have been a convolution type only if it were diagonal. However, since vanishing of the nondiagonal components means the absence of reflection of the waves from the inhomogenity $\hat{L}_1$, this situation is possible only in very special cases.

We can avoid this difficulty if we make some changes in the ψ-basis. Let P be the operator of spatial inversion $f(x) \to f(-x)$ or correspondingly $f(k) \to f(-k)$. Obviously, $\delta(x + x')$ and $\delta(k + k')$ are its kernels in the x and k representations, respectively. Let us take into account that the operator Ω_0 is invariant with respect to inversion, while the operator j changes its sign (Sect 4.3). Hence

$$Pe^{j\Omega_0 t} = e^{-j\Omega_0 t} . \tag{4.12.12}$$

Let us introduce the matrix operator

$$\hat{P} = \begin{pmatrix} 1 & 0 \\ 0 & P \end{pmatrix} . \tag{4.12.13}$$

where, as is easily seen, $\hat{P}^+ = \hat{P}$ and $\hat{P}\hat{P} = I$, i.e. $\hat{P}$ is a unitary operator. Let

$$\psi'(t) = \hat{P}\psi(t) . \tag{4.12.14}$$

In particular, in the (k, t) representation

$$\psi'(k, t) = e^{-i\tilde{\omega}(k)t}\varphi'(k), \quad \varphi'(k) = \begin{pmatrix} \varphi_1(k) \\ \bar{\varphi}_2(k) \end{pmatrix} . \tag{4.12.15}$$

Here $\tilde{\omega}(k) = \omega(k) \operatorname{sgn}\{k\}$ is the odd root of $\omega^2(k)$ introduced in Sect 4.3.

We can now rewrite (4.12.8) as

$$\psi'^{\text{out}}(t) = \hat{S}'\psi'^{\text{in}}(t) , \tag{4.12.16}$$

where the operator $\hat{S}'$ is connected with $\hat{S}$ by the relations

$$\hat{S}' = \hat{P}\hat{S}\hat{P}, \quad \hat{S} = \hat{P}\hat{S}'\hat{P} . \tag{4.12.17}$$

Transformation (4.12.14) to the new ψ'-basis can be presented most clearly in the (k, ω) representation. In accordance with (4.12.11), the components $\psi_1^{\text{in}}(k, \omega)$ and $\psi_2^{\text{in}}(k, \omega)$ are concentrated respectively on the dispersion curves $\omega = \tilde{\omega}(k)$ and $\omega = -\tilde{\omega}(k)$. In the ψ'-basis, both components are concentrated on one curve $\omega = \tilde{\omega}(k)$. The operator $\hat{S}'$ transforms them into $\psi_1'^{\text{out}}(k, \omega)$ and $\psi_2'^{\text{out}}(k, \omega)$, which are concentrated on the same curve. Thus, the operator $\hat{S}'$ possesses simpler properties and, as we shall see, for 'good' dispersion laws, it is a convolution type. Note that in the scattering theory, it is the operator $\hat{S}'$, which is usually called the scattering matrix, or S-matrix. We retain this terminology. In the following, primes in the notations will be omitted, since it will be always clear to which basis we refer.

The factors in (4.12.17) being unitary implies that the S-matrix is unitary also. In order to investigate other properties of the S-matrix, let us consider its action on the simplest harmonic wave ψ^{in}, which, in the ψ'-basis, has the form

$$\psi_1^{\text{in}}(x,t) = a_1(k)\, e^{i(kx-\omega t)}, \quad \psi_2^{\text{in}}(x,t) = a_2(k)\, e^{i(kx-\omega t)}, \tag{4.12.18}$$

where the fixed numbers ω and k belong to the dispersion curve $\omega = \tilde{\omega}(k)$. Obviously, the components $\psi_1^{\text{out}}(x,t)$ must have the same frequency ω, since the perturbation operator $\hat{L}_1$ does not change it. But, instead of the wave number k, after scattering, there must appear (generally speaking) a set of N wave numbers k_n which are solutions of the equation $\omega = \tilde{\omega}(k)$ for a given ω. Then, according to the initial assumptions that ψ^{in} is a propagating wave, we must exclude from the admissible spectrum of frequencies ω those frequencies which exceed the absolute maximum of $\tilde{\omega}(k)$, and also the extreme points of $\tilde{\omega}(k)$, since for these, the group velocity is zero. Thus, in the general case, to the function ψ^{in} in the form (4.12.18) correspond

$$\psi_1^{\text{out}}(x,t) = \sum_{n=1}^{N} b_{1n}(k)\, e^{i(k_n x-\omega t)},$$

$$\psi_2^{\text{out}}(x,t) = \sum_{n=1}^{N} b_{2n}(k)\, e^{i(k_n x-\omega t)}, \tag{4.12.19}$$

where $b_{1n}(k)$ and $b_{2n}(k)$ depend linearly on $a_1(k)$ and $a_2(k)$. It is clear that, when $N > 1$, this connection cannot be given by 2×2 functional matrix $\hat{S}(k)$, i.e., in the general case, the operator $\hat{S}$ cannot be a convolution type.

Let us show that when $N = 1$, i.e. for the increasing function $\tilde{\omega}(k)$, the operator $\hat{S}$ is a convolution type. In place of (4.12.19) we now have

$$\psi^{\text{out}}(x,t) = b(k)\, e^{i(kx-\omega t)}, \tag{4.12.20}$$

where the vector $b(k)$ is related linearly to the vector $a(k)$.

Let $\hat{T}_a$ be the displacement operator $\psi(x,t) \to \psi(x+a,t)$. Then for the vector $\psi^{\text{in}}(x,t)$ defined in (4.12.18),

$$\hat{T}_a \psi^{\text{in}}(x,t) = e^{ika} \psi^{\text{in}}(x,t). \tag{4.12.21}$$

Since the relation between $\psi^{\text{out}}(x,t)$ and $\psi^{\text{in}}(k,t)$ is linear, the same constant factor $\exp(ika)$ also multiplies $\psi^{\text{out}}(x,t)$. But due to (4.12.20) this is equivalent to a displacement, i.e.

$$\hat{T}_a \psi^{\text{out}} = \hat{S} \hat{T}_a \psi^{\text{in}}, \tag{4.12.22}$$

whence

$$\hat{S} = \hat{T}_{-a} \hat{S} T_a. \tag{4.12.23}$$

Thus we have shown that the operators $\hat{S}$ and $\hat{T}_a$ commute when acting on functions ψ^{in} of the type of the harmonic wave (4.12.18). From this it follows immediately that (4.12.22) is fulfilled for an arbitrary wave ψ^{in}, i.e. the operator S is a convolution one.

Problem 4.12.1. Let $\tilde{\omega}(k) = k$, i.e. L_0 is the usual wave operator without dispersion. By reasoning analogous to that carried out above, investigate the properties of S- matrix with respect to Lorentz transformations.

Problem 4.12.2. Let $\omega = \tilde{\omega}(k)$ be an arbitrary nonmonotonic curve, $N(\omega)$ be a number of roots of the equation $\omega = \tilde{\omega}(k)$ with a fixed ω and $N = \sup N(\omega)$. Show that if ψ^{in} and ψ^{out} are replaced by corresponding $2N$-dimensional vectors, then the $2N \times 2N$-matrix $\hat{S}$ is a convolution operator in the ψ'-basis.

In what follows, we assume for simplicity that $\omega = \tilde{\omega}(k)$ is a monotonic function. Moreover, since the points at which $\omega'(k) = 0$, are not included in the admissible set of values of wave numbers k, one can consider $\omega(k)$ to be a strictly increasing function and hence the inverse function $k(\omega)$ to exist. As we have shown, in this case, the action of the operator $\hat{S}$ in the k representation is reduced to multiplication by the functional matrix $\hat{S}(k)$. In the ψ'-basis

$$\begin{aligned}
\varphi_1^{\text{out}}(k) &= s_{11}(k)\, \varphi_1^{\text{in}}(k) + s_{12}(k)\, \varphi_2^{\text{in}}(k)\,, \\
\varphi_2^{\text{out}}(k) &= s_{21}(k)\, \varphi_1^{\text{in}}(k) + s_{22}(k)\, \varphi_2^{\text{in}}(k)\,,
\end{aligned} \tag{4.12.24}$$

where s_{ij} are the components of the S matrix, which must satisfy the unitarity condition (4.12.10). In the case under consideration, these conditions take the form

$$\begin{aligned}
&|s_{11}(k)|^2 + |s_{12}(k)|^2 = |s_{22}(k)|^2 + s_{21}(k)|^2 = 1\,, \\
&s_{11}(k)\, \bar{s}_{21}(k) = s_{12}(k)\, \bar{s}_{22}(k) = 0\,.
\end{aligned} \tag{4.12.25}$$

Note that from (4.12.15), it follows that the relations (4.12.24), when written in terms of the ψ-basis, have the form

$$\begin{aligned}
\varphi_1^{\text{out}}(k) &= s_{11}(k)\, \varphi_1^{\text{in}}(k) + s_{12}(k)\, \bar{\varphi}_2^{\text{in}}(k)\,, \\
\varphi_2^{\text{out}}(k) &= \bar{s}_{21}(k)\, \bar{\varphi}_1^{\text{in}}(k) + \bar{s}_{22}(k)\, \varphi_2^{\text{in}}(k)\,.
\end{aligned} \tag{4.12.26}$$

Let $\varphi_2^{\text{in}} = 0$, so that as $t \to \infty$ the vector $\psi^{\text{in}}(t)$ describes a wave packet approaching the inhomogenity from the left. As $t \to \infty$, the components $\psi_1^{\text{out}}(t)$ and $\psi_2^{\text{out}}(t)$ describe (in the ψ-basis), the transmitted and reflected wave packets, and their amplitudes are given by (4.12.26), when $\psi_2^{\text{in}}(k) = 0$. Therefore, the S matrix components $s_{11}(k)$ and $s_{21}(k)$ are called, respectively, the (left) transmission and reflection amplitudes, and the quantities $|s_{11}(k)|^2$ and $s_{21}(k)|^2$ are called transmission and reflection coefficients. The components $s_{22}(k)$ and $s_{12}(k)$ have a similar meanings. The first equalities in (4.12.25) give simple relations between the transmission and reflection coefficients.

Up to now we considered the S matrix as an operator which acts on the spatial variable. However in the ψ'-basis, any solution of the equation $\hat{L}_0\psi = 0$ in the (k, ω) representation, in accordance with (4.12.15), has the form

$$\psi(k, \omega) = 2\pi\varphi(k)\,\delta(\omega - \tilde{\omega}(k))\,. \tag{4.12.27}$$

According to the assumption made above, $\tilde{\omega}(k)$ is a strictly increasing function and the inverse function $k(\omega)$ exists. This allows us to rewrite the expression for $\psi(k, \omega)$ as

$$\psi(k, \omega) = 2\pi\varphi(\omega)\,[\omega - \tilde{\omega}(k)], \quad \varphi(\omega) \overset{\text{def}}{=} \varphi[k(\omega)]\,. \tag{4.12.28}$$

Correspondingly, instead of $\hat{S}(k)$ we now introduce

$$\hat{S}(\omega) \overset{\text{def}}{=} \hat{S}[k(\omega)]\,. \tag{4.12.29}$$

The formulae (4.12.24) retain their form, but with k replaced by ω. The S matrix in the t representation now becomes a convolution-type operator, which acts through the time variable. We shall see that this representation is convenient in some cases connected with thc analyticity of the S matrix. We emphasize that $\hat{S}(\omega)$ is defined only for the frequencies ω which belong to the dispersion function $\omega = \tilde{\omega}(k)$. This follows from the fact that $\hat{S}(\omega)$ connects two functions of the form (4.12.28). Therefore, if $|\tilde{\omega}(k)| < \omega_{\max}$, cf. (4.10.23), then frequencies $\omega > \omega_{\max}$ are not included into the region of definition of the S matrix. In particular, this concerns the frequencies of local vibrations.

Let us return to the general properties of the S matrix. The relation (4.12.23) is, although important, only one of the possible invariance properties of the S matrix. Let us investigate the properties of S matrix with respect to some other transformations.

4.12.1 Time Reversal.

Among the operators acting on the time variable, only the inversion operator $T\colon f(t) \to f(-t)$ is of interest. It is easily seen that it anticommutes with the operator ∂_t, which is contained in $\hat{L}$.

$$T\partial_t = -\,\partial_t T \quad \text{or} \quad \partial_t = -\,T\partial_t T\,. \tag{4.12.30}$$

Convolution-type operators (in particular, the operator for displacement in time) commute with all operators contained in $\hat{L}$, i.e. play the role of constant factors. Therefore, they cannot convey any information about the S matrix. Aside T, operators which are not of convolution type do not have good commutation relations with the operator ∂_t.

Let us write the initial equation (4.12.1) in the w-basis.

$$\hat{L}w = (\hat{L}_0 + \hat{L}_1)\, w = 0\,, \tag{4.12.31}$$

where the operators $\hat{L}_0$ and $\hat{L}_1$ are given by (4.11.19). It is easily verified that to the time reversal transformation $u \rightarrow Tu$ there corresponds the transformation

$$w \rightarrow \hat{T}w, \quad \hat{T} \stackrel{\text{def}}{=} T \begin{pmatrix} 1 & 0 \\ 0 & -1 \end{pmatrix}, \tag{4.12.32}$$

in the w-representation and the operator $\hat{L}$ commutes with this transformation, as it should:

$$\hat{L} = \hat{T}\hat{L}\hat{T}^{-1}\,. \tag{4.12.33}$$

The transformation (4.12.32) induces, in the w-basis, the transformation of the S matrix

$$\hat{S} \rightarrow \hat{T}\hat{S}\hat{T}^{-1}\,. \tag{4.12.34}$$

We are interested in the corresponding transformation of the S matrix in the ψ'-basis. Taking into account (4.11.24) and (4.12.14), we have

$$\hat{S} \rightarrow (\hat{P}\hat{A}\hat{T})\, \hat{S}\, (\hat{P}\hat{A}\hat{T})^{-1}\,, \tag{4.12.35}$$

where the matrices $\hat{A}$ and $\hat{P}$ are given by (4.11.25) and (4.12.13). When calculating the product of the matrices, it is necessary to take into account the rules of commutation: the operator T commutes with all spatial operators and for the operator P

$$Pj = -jP, \quad P\Omega_0 = \Omega_0 P\,. \tag{4.12.36}$$

Simple calculations yield

$$\hat{S}(k) \rightarrow \begin{pmatrix} \bar{s}_{22}(k) & \bar{s}_{21}(k) \\ \bar{s}_{12}(k) & \bar{s}_{11}(k) \end{pmatrix}. \tag{4.12.37}$$

On the other hand, the time reversal leads to the substitution $\psi^{\text{in}} \rightarrow \psi^{\text{out}}$ and vice-versa, and this corresponds to the substitution $\hat{S} \rightarrow \hat{S}^{-1}$ or taking into account unitarity, $\hat{S} \rightarrow \hat{S}^{+}$, i.e. we must equate $\hat{S}^{+}$ to the transformed matrix (4.12.37). Hence

$$\begin{pmatrix} \bar{s}_{11}(k) & \bar{s}_{21}(k) \\ \bar{s}_{12}(k) & \bar{s}_{22}(k) \end{pmatrix} = \begin{pmatrix} \bar{s}_{22}(k) & \bar{s}_{21}(k) \\ \bar{s}_{12}(k) & \bar{s}_{11}(k) \end{pmatrix}, \tag{4.12.38}$$

and we obtain the condition for the diagonal components of the S matrix

$$s_{11}(k) = s_{22}(k)\,. \tag{4.12.39}$$

Observe that if we trace back carefully the derivation of this condition, then we see that the following group properties of the self-adjoint operator L and their consequences were used in an essential way:

a) invariance of L with respect to time translation (energy conservation, unitarity of $\hat{S}$);

b) invariance of L_0 with respect to coordinate translation ($\hat{S}$ being a difference operator);

c) invariance of L with respect to time reversal.

Taking into account the unitarity properties (4.12.25) one can derive from (4.12.39) the equality of the left and right reflection coefficients (but not of their amplitudes!).

4.12.2. Linear Transformation on the Space Variable.

Let in the initial equation (4.12.1) the transformation $u \to Xu$ be performed, where X is a reversible linear operator, acting, on the space variable. The transformation can be interpreted as a transition to a new basis, for which

$$L \to XLX^{-1}\,. \tag{4.12.40}$$

Accordingly, the operator $\hat{L}$ in (4.12.31) is transformed such that

$$\hat{L} \to \hat{X}\hat{L}\hat{X}^{-1}, \quad X = \begin{pmatrix} X & 0 \\ 0 & X \end{pmatrix}. \tag{4.12.41}$$

We are interested in finding how the S matrix, initially given in the ψ'-basis, is transformed in this case.

Problem 4.12.3. Show that

$$\hat{S} \to \hat{X}'\hat{S}\hat{X}'^{-1}\,, \tag{4.12.42}$$

where

$$\hat{X}' = (\hat{P}\hat{A})\,\hat{X}\,(\hat{P}\hat{A})^{-1} = \frac{1}{2}\begin{pmatrix} X + \tilde{X} & (X - \tilde{X})\,P \\ P(X - \tilde{X}) & P(X + \tilde{X})\,P \end{pmatrix},$$

$$\tilde{X} \overset{\text{def}}{=} (j\Omega_0)\,X\,(j\Omega_0)^{-1}\,. \tag{4.12.43}$$

Let us note that if $X = I$, then $\hat{X}' = \hat{I}$, as one would expect.

The expression (4.12.42) allows one to obtain some new properties of the S-matrix through a special choice of operators X. Let $X = P$. Then

$$\hat{S}(k) \to \begin{pmatrix} s_{22}(k) & s_{21}(k) \\ s_{12}(k) & s_{11}(k) \end{pmatrix}. \tag{4.12.44}$$

Taking into account (4.12.39), we see that spatial reflection leaves the diagonal elements invariant. If L_1 is also invariant with respect to reflection,

$$L_1(x, x') = L_1(-x, -x'), \tag{4.12.45}$$

then the components of S matrix satisfy one more condition

$$s_{12}(k) = s_{21}(k). \tag{4.12.46}$$

Now let $X = T_a$ be the shift operator. Then the transformation of the S matrix can be considered as the result of shift of an inhomogenity:

$$L_1(x, x') \to L_1(x - a, x' - a). \tag{4.12.47}$$

A computation yields

$$\hat{S}(k) = \begin{pmatrix} s_{11}(k) & e^{-2ika}\, s_{12}(k) \\ e^{2ika}\, s_{21}(k) & s_{22}(k) \end{pmatrix}. \tag{4.12.48}$$

As in the preceding case, the diagonal elements remain invariant.

4.12.3. Invariance with Respect to Translation $u(x, t) \to u(x, t) + u^0$.

As distinct from the earlier ones, this transformation is not homogeneous (linear) and therefore requires an individual study. Let us assume that the initial operator L in (4.12.1) is invariant with respect to translation, i.e. u^0 = constant satisfies the wave equation. In the w-basis, $w^0(u^0, 0)$ corresponds to u^0, Obviously, in this case the condition

$$\hat{S}w^0 = w^0 \tag{4.12.49}$$

must be satisfied, i.e. in the x-representation

$$\int S_{11}(x, x')\, dx' = 1, \quad \int S_{21}(x, x')\, dx' = 0. \tag{4.12.50}$$

For finding out the conditions on the S-matrix in the ψ'-basis, it is necessary to carry out the inverse transition from the ψ'-basis to the w-basis and to substitute the elements of the S matrix in (4.12.50).

Problem 4.12.4. Verify that the first condition in (4.12.50) is equivalent to the equality (for a noncontinuous $\hat{S}(k)$, the values at $k = 0$ are understood in the sense of the left and right limits)

$$\mathrm{Re}\{s_{11}(0)\} + \frac{1}{2}[s_{12}(0) + \bar{s}_{21}(0)] = 1, \tag{4.12.51}$$

and the second condition is satisfied identically. (Hint: take into account that from the condition of invariance of L, it follows that $\tilde{\omega}(0) = 0$).

4.12.4 Invariance with Respect to the Transformation $u(x, t) \to u(x, t) + ax$

Note that this is a rotation for transverse oscillations of a rod (Sect. 3.9), and a homogeneous deformation for other cases. The conditions on the S-matrix are found analogously to the previous case. In the w-basis

$$\int S_{11}(x, x')\, x'\, dx' = x, \quad \int S_{21}(x, x')\, x'\, dx' = 0\,. \tag{4.12.52}$$

Problem 4.12.5. Verify that the first condition is equivalent to the relations

$$\begin{aligned} &f(0) = 1, \quad f'(0) = 0\,, \\ &f(k) \overset{\text{def}}{=} \mathrm{Re}\{s_{11}(k)\} - \frac{1}{2}\,[s_{12}(k) + \overline{s_{21}(k)})\,, \end{aligned} \tag{4.12.53}$$

and the second one to the relations

$$\begin{aligned} &F(0) = 0, \quad F'(0) = 0\,, \\ &F(k) \overset{\text{def}}{=} \tilde{\omega}(k)\,(\mathrm{Im}\{s_{11}(k)\} + \frac{\mathrm{i}}{2}\,[s_{12}(k) - \overline{s_{21}(k)}]\,; \end{aligned} \tag{4.12.54}$$

moreover from the invariance of L_0 the condition $\tilde{\omega}(0) = 0$ follows.

Let us now consider the number N of functional-degrees of freedom of the the S-matrix. Let us assume, for convenience, that the real function $f(k)$ has two degrees of freedom, since it is determined by two functions, defined over the half-axis. Then, an arbitrary 2×2 complex functional matrix has $N = 16$ degrees of freedom. But $\hat{S}(k) = \overline{S(-k)}$ is the Fourier transform of a real operator, which decreases N to 8. The unitarity condition (4.12.25) and the condition of symmetry (4.12.39), as easily verified, give a set of five real euations for the elements $s_{ij}(k)$, which gives $N = 3$. It is thus advisable to express the S matrix directly in terms of three functional degrees of freedom.

Problem 4.12.6. Show that when the previously listed symmetry conditions are satisfied, the S matrix can always be presented in the form

$$\hat{S}(k) = \mathrm{e}^{\mathrm{i}\zeta(k)} \begin{pmatrix} \cos \xi(k) & \mathrm{i}\mathrm{e}^{2\mathrm{i}\eta(k)} \sin \xi(k) \\ \mathrm{i}\mathrm{e}^{-2\mathrm{i}\eta(k)} \sin \xi(k) & \cos \xi(k) \end{pmatrix}. \tag{4.12.55}$$

where $\xi(k)$, $\eta(k)$ and $\zeta(k)$ are real odd functions of k.

Note that it is sometimes convenient to choose a representation which differs from the given one by the substitution $\cos[\xi(k)] \rightleftarrows \mathrm{i} \sin[\xi(k)]$. This happens when $s_{11}(k) = 0$ at $k = 0$, because in this case, the new representation ensures the continuity of $\xi(k)$ at the point $k = 0$.

The condition (4.12.46) of invariance with respect to reflection is now equivalent to $\eta(k) = 0$ and the condition (4.12.51) of invariance relative to translation is fulfilled identically in the representation (4.12.55) for continuous $\xi(k)$, $\eta(k)$ and $\zeta(k)$. In Sects. 4.13 and 4.14 other cases which decrease the degrees of freedom of the S matrix will be considered.

In summary, we see that using only the group properties of the operator L, without solving the equations, it is possible to obtain significant information concerning the structure of the scattering matrix and, as a corollary, about the scattering problem itself. In the next section, we shall investigate the structure of the S-matrix from a different point of view, i.e. through its connection with Green's functions.

4.13 Connection of the S-Matrix with Green's Functions

In order to construct an algorithm for the computation of the S-matrix and to investigate its analytical properties, it is convenient to have an explicit relation between the S-matrix and the Green's function G^r.

Let us replace the operator $\hat{L}_0$ by $\hat{L} - \hat{L}_1$ in the second equation (4.12.7) and rewrite the equation in the form

$$\hat{L}\psi^{\text{in}} = \hat{L}_1\psi^{\text{in}} . \tag{4.13.1}$$

For the moment, let us regard the right-hand side as a given external force. Since ψ^{in} is a wave with finite energy, and the perturbation $\hat{L}_1$ is localized, the 'force' is equal to zero as $|t| \to \infty$, and for finite values of t, decreases sufficiently rapidly as $|x| \to \infty$ This allows us to apply the retarded Green's function $\hat{G}^r$ to both sides of (4.13.1). Then the 'solution' has the form

$$\psi^{\text{in}} = \hat{G}^r\hat{L}_1\psi^{\text{in}} + \psi . \tag{4.13.2}$$

It is assumed here that the first term on the right-hand side tends to zero as $t \to -\infty$ and therefore we have added the solution of the homogeneous equation $\hat{L}\psi = 0$, which has the same asymptotic behavior as ψ^{in} when $t \to -\infty$

The next step will be to find the connection between ψ and ψ^{out}. With this in mind, let us rewrite the equation $\hat{L}\psi = 0$ as

$$\hat{L}_0\psi = -\hat{L}_1\psi \tag{4.13.3}$$

and apply to both sides the advanced Green's function $\hat{G}_0^a$. An argument, similar to that above, yields

$$\psi = -\hat{G}_0^a L_1\psi + \psi^{\text{out}} , \tag{4.13.4}$$

where it is assumed that $\psi(t) \to \psi^{\text{out}}(t)$ as $t \to \infty$.

Eliminating ψ from (4.13.2) and (4.13.4), we obtain

$$\psi^{\text{out}} = (\hat{I} - \hat{G}^{\text{r}}\hat{L}_1 + G_0^{\text{a}}L_1 - \hat{G}_0^{\text{a}}\hat{L}_1\hat{G}^{\text{r}}\hat{L}_1)\,\psi^{\text{in}} \tag{4.13.5}$$

or, using the identities (4.11.22,23) independent of the choice of a basis

$$\psi^{\text{out}} = [\hat{I} - (\hat{G}_0^{\text{r}} - \hat{G}_0^{\text{a}})\,R]\,\psi^{\text{in}}\,. \tag{4.13.6}$$

Let us take into account (4.1.44) and write the relation in more detail:

$$\psi^{\text{out}}(t) = \psi^{\text{in}}(t) - \frac{1}{\rho_0}\iint \hat{D}_0(t-t')\,\hat{R}(t'-t'')\,\psi^{\text{in}}(t'')\,dt'\,dt''\,. \tag{4.13.7}$$

It is desirable to find the connection between $\psi^{\text{out}}(t)$ and $\psi^{\text{in}}(t)$ at one and the same moment. For this, let us use the property (4.1.46) of the Green's function $\hat{D}_0(t-t')$ as a propagator

$$\psi^{\text{in}}(t'') = \hat{D}^0(t''-t)\,\psi^{\text{in}}(t)\,. \tag{4.13.8}$$

The substitution in (4.13.7) and comparison with (4.12.8) leads to the following result:

$$\hat{S} = \hat{I} - \frac{1}{\rho_0}\iint \hat{D}_0(-t')\,\hat{R}(t'-t'')\,\hat{D}_0(t'')\,dt'\,dt'' \tag{4.13.9}$$

or after a transformation of the integral

$$\hat{S} = \hat{I} - \frac{1}{2\pi\rho_0}\int \hat{D}_0(\omega)\,\hat{R}(\omega)\,\hat{D}_0(\omega)\,d\omega\,. \tag{4.13.10}$$

We have obtained the expression for S in the ψ-basis. However, in accordance with prior discussion the S matrix has a good representation in the ψ'-basis (4.12.14). The transition to this basis is carried out by the transformations

$$\hat{S} \to \hat{P}\hat{S}\hat{P}, \quad \hat{D}_0 \to \hat{P}\hat{D}_0\hat{P}, \quad \hat{R} \to \hat{P}\hat{R}\hat{P}\,. \tag{4.13.11}$$

For the transformed $\hat{D}_0$ with (4.11.28), (4.12.12) and (4.12.13) taken into account we find in the (k, ω) representation

$$\hat{D}_0(k, k', \omega) = 2\pi\delta(k-k')\,\delta(\omega - \tilde{\omega}(k))\,\hat{I}\,. \tag{4.13.12}$$

Similarly, from (4.11.29), we obtain the expression for the transformed matrix $\hat{R}$

$$\hat{R}(k, k', \omega) = \frac{\mathrm{i}}{2\tilde{\omega}(k')} \begin{pmatrix} R(k, k', \omega) & R(k, -k', \omega) \\ R(-k, k', \omega) & R(-k, -k', \omega) \end{pmatrix}. \tag{4.13.13}$$

Substitution of (4.13.12) and (4.13.13) into (4.13.10) yields

$$\hat{S}(k, k') = \hat{I} - \frac{2\pi}{\rho_0} \delta[\tilde{\omega}(k) - \tilde{\omega}(k')]\, \hat{R}[k, k', \tilde{\omega}(k)]\,. \tag{4.13.14}$$

Let us now take into account the assumption made in the preceding section that $\tilde{\omega}(k)$ is a strictly increasing function. Using the known properties of δ-function, we have

$$\delta[\tilde{\omega}(k) - \tilde{\omega}(k')] = \frac{1}{\tilde{\omega}'(k)} \delta(k - k')\,. \tag{4.13.15}$$

We see that the operator $\hat{S}$ is reduced as it should be to a multiplication by the functional matrix $\hat{S}(k)$ with the elements

$$\begin{aligned} s_{11}(k) &= s_{22}(k) = 1 - \frac{2\pi \mathrm{i} R[k, k, \tilde{\omega}(k)]}{[\omega^2(k)]'\, \rho_0}\,, \\ s_{12}(k) &= -\frac{2\pi \mathrm{i} R[k, -k, \tilde{\omega}(k)]}{[\omega^2(k)]'\, \rho_0}\,, \quad s_{21}(k) = -\frac{2\pi \mathrm{i} R[-k, k, \tilde{\omega}(k)]}{[\omega^2(k)]'\, \rho_0}\,. \end{aligned} \tag{4.13.16}$$

Here we have used the symmetry properties of the kernel $R(k, k', \omega)$, cf. (2.4.9) and (2.4.12), which lead to the equality of the diagonal elements. The group origin of this equality was explained earlier. By analogy to (4.12.29) we can rewrite (4.13.16) in the ω representation

$$\begin{aligned} s_{11}(\omega) &= s_{22}(\omega) = 1 - \frac{\pi \mathrm{i} k'(\omega)}{\omega \rho_0} R[k(\omega), k(\omega), \omega]\,, \\ s_{12}(\omega) &= -\frac{\pi \mathrm{i} k'(\omega)}{\omega \rho_0} R[k(\omega), -k(\omega), \omega]\,, \\ s_{21}(\omega) &= -\frac{\pi \mathrm{i} k'(\omega)}{\omega \rho_0} R[-k(\omega), k(\omega), \omega]\,. \end{aligned} \tag{4.13.17}$$

The formulae (4.13.16) and (4.13.17) give the desired explicit relation between the S matrix and the Green's function. We now list some immediate consequences of the representation of the S-matrix which we have just obtained.

First of all, we see that in the absence of inhomogenity, i.e. as $R \to 0$, the S-matrix coincides with the unit matrix, as it should. Further, in all those cases studied in Sect 4.11, when R could be obtained exactly or approximately, the S-matrix is constructed immediately. The corresponding expressions of the S-matrix for local defects will be studied in sect 4.14.

Problem 4.13.1. Show that if at $k = 0$, the reflection coefficient is 1, then $s_{12}(0) = s_{21}(0) = -1$ [Hint: Use (4.13.16) with $k = 0$ and the unitarity property (4.12.25)].

In the general case, a part of the information about inhomogenity is lost in the S-matrix. In fact, as it follows from (4.11.5), the function $R(k, k', \omega)$ is in a one-to-one correspondence with $L_1(k, k', \omega)$. At the same time, the elements of the S matrix are defined by the values of $R(k, k', \omega)$ on the 'diagonal' which, in general, does not permit one to reconstruct L_1. However, in a number of cases, for example for the usual local wave equation the operator L_1 is defined by a function of one variable and under some additional restrictions the inverse scattering problem is well posed [4.6].

The formulae (4.13.17) also allow one to draw some conclusions concerning the analytical properties of the S matrix. By construction, the function $R(k, k', \omega)$ is analytic in the upper half ω-plane. If the perturbation L_1 decreases in the x representation sufficiently rapidly, the function $R(k, k, \omega)$ is analytic in k and k'. Then in accordance with (4.13.17), the properties of the S-matrix are determined by the analytical continuation $k(\omega)$. The S matrix, for sufficiently well-behaved $k(\omega)$, is analytic in the upper half ω-plane. Note that the condition of analyticity leads to a decrease in the number of functional degrees of freedom of the S matrix.

Earlier (Sect. 4.4–6) we saw that, from the energy characteristics point of view, each physical quantity y can be considered to correspond to a Hermitian operator $\hat{y}$. For the mean value $\langle \hat{y} \rangle$ in the state ψ^{out}, we have, in accordance with (4.6.7),

$$\langle \hat{y} \rangle = \langle \psi^{\text{out}} | \hat{y} | \psi^{\text{out}} \rangle . \tag{4.13.18}$$

Using (4.12.8) we find

$$\langle \hat{y} \rangle = \langle \psi^{\text{in}} | \hat{S}^{+} \hat{y} \hat{S} | \psi^{\text{in}} \rangle . \tag{4.13.19}$$

Problem 4.13.2. Let $\psi_2^{\text{in}} = 0$. Show that the energy E and velocity $\langle v \rangle$ of the transmitted wave are given by the expressions

$$E = \frac{1}{2\pi} |s_{11}(k)\, \varphi_1^{\text{in}}(k)|^2\, dk ,$$

$$\langle v \rangle = \frac{1}{2\pi E} \int \tilde{\omega}'(k) |s_{11}(k)\, \varphi_1^{\text{in}}(k)|^2\, dk . \tag{4.13.20}$$

Finally, let us consider the asymptotics behavior of $\psi^{\text{out}}(t)$. In the ψ'-basis

$$\psi^{\text{out}}(x, t) = \int e^{i[kx - \tilde{\omega}(k)t]}\, \hat{S}(k)\, \varphi^{\text{in}}(k)\, dk . \tag{4.13.21}$$

For sufficiently smooth $\hat{S}(k)$ and $\varphi^{in}(k)$ the asymptotic behavior, as $t \to \infty$, of integrals of such a form was studied in Sect. 4.2, where it was shown that it is of the order of $t^{-\lambda}$. The quantity λ depends on the stationary points of the argument of the exponent and on the real roots of the factor in the exponent. In the general case, singularities of the S-matrix can give additional contributions to the asymptotic behavior.

4.14 Scattering on Local Defects

Let us now use the results of Sect. 4.10 and 4.13 to solve problems concerning scattering on defects.

We start with the case of a single defect of the type (4.10.9). Taking into account (4.10.26) and (4.13.16), we find for the elements of S-matrix

$$s_{11}(k) = s_{22}(k) = 1 - i\alpha |\gamma(k)|^2 \frac{r[\tilde{\omega}(k)]}{[\omega^2(k)]' \rho_0},$$

$$s_{ij}(k) = - i\alpha\gamma^2(\pm k) \frac{r[\tilde{\omega}(k)]}{[\omega^2(k)]' \rho_0} \quad (i \neq j), \tag{4.14.1}$$

where the upper (lower) sign corresponds to $s_{12}(s_{21})$.

Accordingly, in the ω representation, compare with (4.13.17),

$$s_{11}(\omega) = s_{22}(\omega) = 1 - \frac{i\alpha k'(\omega)}{2\omega\rho_0} |\gamma(\omega)|^2 r(\omega),$$

$$s_{ij}(\omega) = - \frac{i\alpha k'(\omega)}{2\omega\rho_0} \gamma^2(\pm \omega) r(\omega), \tag{4.14.2}$$

where $\gamma(\omega)$ denotes $\gamma[k(\omega)]$

Recall that by construction $r(\omega)$ is analytic in the upper half ω-plane. If $k(\omega)$ and $\gamma(\omega)$ are analytic functions, then $\hat{S}(\omega)$ is analytic in the upper half-plane.

Let us transform (4.14.1) and (4.14.2) into the form (4.12.55). Substituting (4.1.35) in (4.10.15), we find an explicit expression for $\mathrm{Im}\{g(\omega)\}$

$$\mathrm{Im}\{g(\omega)\} = \frac{k'(\omega)}{2\omega\rho_0} |\gamma(\omega)|^2 . \tag{4.14.3}$$

On the other hand, from (4.10.27) and (4.10.15) and considering (4.14.3), follows the relation

$$\frac{\mathrm{Im}\{r(\omega)\}}{|r(\omega)|^2} = - \frac{\alpha k'(\omega)}{2\omega\rho_0} |\gamma(\omega)|^2 . \tag{4.14.4}$$

Transforming this to the k representation and substituting in (4.14.1), we finally obtain

$$s_{11}(k) = s_{22}(k) = e^{i\xi(k)} \cos \xi(k)\,,$$
$$s_{ij}(k) = ie^{i[\xi(k) \pm 2\eta(k)]} \sin \xi(k)\,, \tag{4.14.5}$$

where

$$\xi(k) = \arg\{r(\tilde{\omega}(k))\}, \quad \eta(k) = \arg\{\gamma(k)\}\,. \tag{4.14.6}$$

Comparison with (4.12.55) shows that in the S-matrix the number of functional degrees of freedom was decreased to two due to the relation $\zeta(k) = \xi(k)$.

Problem 4.14.1. Verify that this relation is equivalent to the condition

$$|s_{11}(k)|^2 = \mathrm{Re}\{s_{11}(k)\}\,. \tag{4.14.7}$$

We see that in this case the number of degrees of freedom of the S-matrix coincides with the number of degrees of freedom of a single defect of the type (4.10.9), which is mainly determined by prescribing $\gamma(k)$ (recall that $b(\omega) = -\omega^2$ or $b(\omega) = 1$). This in principle permits us to reconstruct $\gamma(k)$ in terms of $\xi(k)$ and $\eta(k)$. Let us outline the corresponding algorithm.

The second relation in (4.14.6) defines arg $\{\gamma(k)\}$ immediately. For finding $|\gamma(k)|^2$, let us equate $\mathrm{Im}\{s_{11}^{-1}(k)\}$ from (4.14.1) and (4.14.5). Taking into account (4.10.15) and (4.10.27), we obtain a singular integral equation

$$\frac{1}{\pi} \int \frac{\Gamma(\omega')\, d\omega'}{\omega - \omega'} - \mathrm{ctg}\, \xi(\omega) \cdot \Gamma(\omega) = \frac{2\omega}{b(\omega)}\,, \tag{4.14.8}$$

where

$$\Gamma(\omega) = \alpha k'(\omega) |\gamma[k(\omega)]|^2, \quad \xi(\omega) = \xi[k(\omega)]\,. \tag{4.14.9}$$

Integral equations of such a type are well-studied and their solution can be constructed in closed form [4.7]. For finding the constant α in terms of $\Gamma(\omega)$ it is sufficient to normalize $\gamma(k)$, for example, in the form $\gamma(0) = 1$.

Thus for the S-matrix which satisfies the condition $\zeta(k) = \xi(k)$ or, equivalently, (4.14.7), the inverse scattering problem is, in principle, well-posed. Let us emphasize that here the structure of the defect is fixed in the form (4.10.9). An inhomogenity L_1 of another structure, equivalent to the defect (4.10.9) in the sense of the scattering problem, can correspond to the same S-matrix.

The elements of the S-matrix for the models of defects considered in Sect. 4.10 are given below in Table 4.1.

First of all, note that in all the cases except 1c and 4c, the S-matrices are analytic in the upper half k-or ω-plane for the admissible values $\alpha > \alpha_0$. The S-matrix for the rod is an exception due to the non-analyticity of the function $\tilde{\omega}(k) = d_0 k^2 \,\mathrm{sgn}\{k\}$. In this case $\hat{S}(k)$ is discontinuous on the real axis at the point $k = 0$.

Table 4.1. Components of S-matrix

	$s_{11}^{-1}(k) - 1$	$s_{ij}^{-1}(k)$
1a	$\frac{2i\alpha}{\alpha\rho_0\omega_{max}^2 \sin ak}$	$\frac{ia\rho_0\omega_{max}^2}{2\alpha}\sin ak - 1$
1b	$\frac{i\alpha}{2\rho_0 v_0^2 k}$	$\frac{2i\rho_0 v_0^2 k}{\alpha} - 1$
1c	$\frac{i\alpha}{8\rho_0 d_0^2 k^3 - \alpha \operatorname{sgn} k}$	$\frac{8i\rho_0 v_0^2 k}{\alpha} - i \operatorname{sgn} k - 1$
2	$\frac{i}{2}\left(\frac{1}{av_0 k} + \frac{\rho_0 v_0}{a^2\alpha k} + \frac{1}{\pi v_0}\ln\left\|\frac{\pi - ak}{\pi + ak}\right\|\right)^{-1}$	$1 - 2i\left(\frac{1}{av_0 k} + \frac{\rho_0 v_0}{a^2\alpha k} + \frac{1}{\pi v_0}\ln\left\|\frac{\pi - ak}{\pi + ak}\right\|\right)$
3	$\frac{4\alpha i}{4\alpha + a\rho_0\omega_{max}^2}\operatorname{tg}\frac{ak}{2}$	$e^{\mp iak} \times \left(1 - i\frac{4\alpha + a\rho_0\omega_{max}^2}{4\alpha}\operatorname{ctg}\frac{ak}{2}\right)$
4a	$-\frac{i\alpha}{a\rho_0}\operatorname{tg}\frac{ak}{2}$	$-1 - \frac{ia\rho_0}{\alpha}\operatorname{ctg}\frac{ak}{2}$
4b	$-\frac{i\alpha k}{2\rho_0}$	$-1 - \frac{2\rho_0}{\alpha k}$
4c	$-\frac{ik}{8\rho_0 + \alpha\|k\|}$	$-1 - \frac{8\rho_0 + \alpha\|k\|}{\alpha k}$

Problem 4.14.2. Verify that for all the limiting transitions to semibounded media indicated in Sect. 4.10 the condition $s_{11}(k) = 0$ is satisfied, but it is not satisfied for the rod with a fixed point.
Problem 4.14.3. Verify that for the string with fixed (free) end, reflection with change (without change) of sign, in accordance with known results, takes place.

The summary of the group properties of the S-matrix for the examples studied in given in Table 4.2. The upper signs correspond to the model and the lower ones correspond to the S-matrix. The upper sign + means the fulfillment of the corresponding property of group invariance for the model, the lower sign + points out the fulfillment of the additional relations for the S-matrix enumerated in Sect 4.12. It is obvious that the combination of $\pm$ signs is imposible. On the other hand, the presence of the combination $\mp$ in the examples 2 and 3, shows that the additional relations for the S-matrix do not give sufficient invariance conditions for the model

Construction and investigation of the S-matrix for a system of local defects can be carried out by a similar method by using explicit expressions for $R(k, k', \omega)$ from Sect. 4.10. Let us present, as an illustration, expressions of the ele-

Table 4.2. Symmetry properties of S-matrix

Group	1			2	3	4		
	a	b	c			a	b	c
Space reflections	+ +	+ +	+ +	+ +	− −	+ +	+ +	+ +
Translations	− −	− −	− −	+ +	+ +	+ +	+ +	+ +
Rotations (homogeneous deformation)	+ +	+ +	+ +	− +	− +	+ +	+ +	+ +

ments of $\hat{S}(k)$ for two similar defects, which are obtained by displacing the defect (4.10.9) to the points $x = x_0$ and $x = -x_0$.

$$\begin{aligned}\Delta(k)\, s_{11}(k) &= [1 + \alpha \mathrm{Re}\{g(k)\}]^2 + \alpha^2[\mathrm{Im}\{g(k)\}]^2 \sin^2 2x_0k - \alpha^2 f^2(k)\,,\\ \Delta(k)\, s_{ij}(k) &= -2\mathrm{i}\alpha e^{\pm 2\mathrm{i}\eta(k)}\, \mathrm{Im}\{g(k)\}[(1 + \alpha\, \mathrm{Re}\{g(k)\}) \cos 2x_0k - \alpha f(k)]\,.\end{aligned} \tag{4.14.10}$$

Here $g(k) \overset{\mathrm{def}}{=} g(\tilde{\omega}(k))$, $\eta(k)$ are defined by (4.10.15) and (4.14.6) and

$$f(k) = \frac{b[\tilde{\omega}(k)]}{2\pi\rho_0} \int \frac{|\gamma(k')|^2 \cos(2x_0k')\, dk'}{\omega^2(k') - \omega^2(k)}\,.$$

$$\begin{aligned}\Delta(k) = [1 + \alpha\, \mathrm{Re}\{g(k)\}]^2 - \alpha^2[\mathrm{Im}\{g(k)\}]^2 \sin^2(2x_0k)\\ - \alpha^2 f^2(k) + 2\mathrm{i}\alpha\, \mathrm{Im}\{g(k)\}\, [1 + \alpha\, \mathrm{Re}\{g(k)\} - \alpha f(k) \cos(2x_0k)]\,.\end{aligned} \tag{4.14.11}$$

Problem 4.14.4. Verify that the condition (4.14.7) is fulfilled only in the limiting case $x_0 = 0$, which corresponds to the transition to one defect.
Problem 4.14.5. Verify that, for $\gamma(k) = \gamma(-k)$, the condition (4.12.46) of invariance with respect to spatial inversion is sulfilled.

In conclusion, let us make a remark concerning the inverse scattering problem. It follows from the above, that, for, in order that this problem be well posed, it is necessary for the perturbation L_1 to have three functional degrees of freedom. The inverse problem can be, in principle, well posed, for example, in the case of two local defects, one of which is invariant with respect to the spatial inversion.

4.15 Notes*

About time-dependent Green's functions see, e.g., [B8.24]. Investigations of the asymptotic behavior of waves in a medium with dispersion have been carried

*The citations refer to the bibliography to be given in [1.12]

out by many authors; see, e.g. [B5.8]. The energy method of analysis of wave processes is based on [B3.12]. A generalization to the three-dimensional case is given in [B8.16].

Extensive mathematical and physical literature is devoted to the scattering problem, see, e.g., [B8.18,24] which contain detailed references. The contents of Sect. 4.10–14 are based on [B3.11].

5. Nonlinear Waves

This chapter is an introduction to nonlinear waves in dispersive media. We show how the interaction of nonlinear and nonlocal effects leads to such basic phenomenon as the soliton.

5.1 Korteweg-de Vries Model

In the preceding chapters, we investigated wave processes in linear media with dispersion. We will now study, using simple models as an example, new effects which arise due to taking into account nonlinearity. In this chapter, we confine ourselves to elementary methods of investigation. An analytical method of solution of some classes of nonlinear equation, which are particularly inferesting, will be investigated in the next chapter.

Let us first state our conventions regarding notation. In this and in next chapters it will be convenient to denote derivatives by appropriate subscripts, for example, ξ_t, ξ_x. The derivatives of higher order N will be written in the form ξ_{Nx}. However, in a number of cases, we will be forced to use explicitly the operators ∂_t and ∂_x.

Let us start with the consideration of nonlinearity in its pure form, i.e. for sufficiently long wavelengths, when the dispersion may be ignored. Let us write down the factorized wave equation, which describes a wave $\xi(x, t)$, which propagates with a velocity v_0 in the positive direction along the x-axis

$$\xi_t + v_0 \xi_x = 0 . \tag{5.1.1}$$

The simplest nonlinear model can be obtained, by adding to the wave equation a term which is of the second order with respect to the field variable ξ. From the natural condition that ξ = constant satisfies the equation, follows the representability of this term in the form of $\xi\xi_x$. Since long wavelengths are being investigated, it is of no use to introduce higher derivatives.

Thus, under the appropriate choice of variables, the equation

$$\xi_t + v_0 \xi_x + \xi\xi_x = 0 \tag{5.1.2}$$

approximately describes the long waves, provided the amplitude is not very large.

Obviously, in terms of the variables chosen, ξ has the dimension of velocity. The change of variables $x \to x - v_0 t$ or $\xi \to \xi - v_0$, constituting a transition to a moving system of coordinates, reduces the equation to the form

$$\xi_t + \xi\xi_x = 0 . \tag{5.1.3}$$

This equation admits a simple mechanical interpretation, namely it describes approximately the flow noninteracting particles moving with constant velocities.

Let the velocities of the particles have a sinusoidal distribution at the initial moment (dotted curve in Fig. 5.1). The sinuisoid deforms in the course of time. Roughly speaking, the velocities of points of the curve are proportional to their distance from the x-axis. The points of the curve with $\xi > 0$ are displaced to the right and those with $\xi < 0$ to the left. The front of the wave becomes steeper and as a consequence, there is an increase of concentration of the particles $\rho(\xi)$ in the region of the maximum slope of the front. During further growth of the slope, the density tends to infinity and a moment comes when the wave turns over. These features of slope growth and turn over are well studied in hydrodynamics. So long as the slope of the front increases, the contribution of the short wave part of the spectrum also increases, and when the process approaches the point of turn over, (5.1.3) becomes, in general, invalid and an additional consideration of dispersion and dissipation are necessary. We shall ignore the influence of the latter and will be interested only in the interaction of the effects of nonlinearity and dispersion.

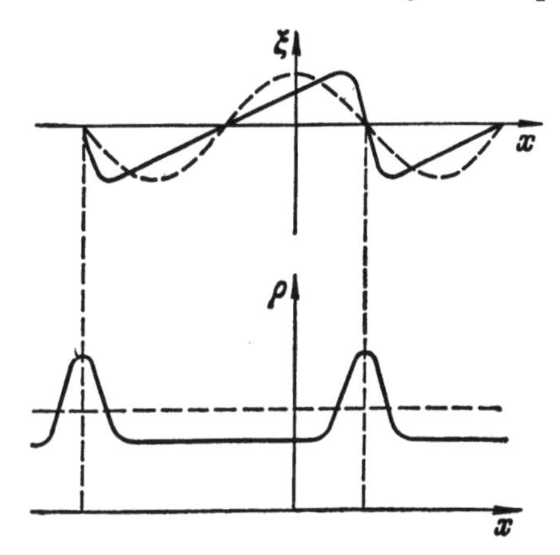

Fig.5.1. Evolution of velocity ξ and density ρ

In order to take into account the dispersion, it is necessary to replace the operator $v_0\partial$ in (5.1.2) by the operator $j\Omega$, which was studied in the previous chapter. Assuming the dispersion to be weak we restrict ourselves to the first approximation in $j\Omega$, i.e. in the expansion of $j\Omega(k)$, we preserve only the linear and cubic terms. Then instead of (5.1.2) we obtain the equation

$$\xi_t + v_0\xi_x + b\xi_{xxx} + \xi\xi_x = 0 . \tag{5.1.4}$$

The parameter $b = \pm v_0 l^2$ (l is a scaling parameter) characterizes the deviation of the law of dispersion from a linear one. When $b > 0$, the group velocity $v(k)$ decreases with increasing k and the medium is called negatively dispersing. We saw earlier that such a law of dispersion is typical for a chain. When $b < 0$, the velocity increases with k and the medium is called positively dispersing.

If, as above, we pass to a moving coordinate system, then instead of (5.1.3) we obtain

$$\xi_t + b\xi_{xxx} + \xi\xi_x = 0 \,. \tag{5.1.5}$$

Observe that the above equation is invariant with respect to the transformation

$$x \to -x, \quad \xi \to -\xi, \quad b \to -b \,, \tag{5.1.6}$$

i.e. in the moving coordinate system, the wave process possesses a mirror symmetry for media with negative and positive dispersions. Therefore, in the following without loss of generality it can be assumed that $b > 0$.

Equation (5.1.4) was investigated for the first time by Korteweg and de Vries in 1895 in connection with the problem of waves on shallow water [5.1]. Thus it is now called the Korteweg-de Vries, or KdV-equation.

Since the KdV-equation takes into account nonlinearity and dispersion in the lowest approximation, it must appear wherever these effects are important. In fact, it found wide applications not only in hydrodynamics but also for the description of wave processes in a plasma, elastic medium, etc. This universal significance of KdV-equation allows to speak about the Korteweg-de Vries model, which itself is of interest.

Many works are devoted to the KdV-equation and its different generalizations; it has been investigated analytically as well as with the help of computer "experiments". Such attention to the KdV-model is due to a number of reasons: its practical significance, interesting phenomena which it describes and, last but certainly not least, its nontrival mathematical content.

Let us now proceed to a qualitative investigation of the behavior of the solutions of the KdV-equation which are called simple waves. We saw that the dispersion and nonlinearity lead to, in a sense, opposite effects: spreading out and slope increase. A situation is possible when these effects compensate each other and the solution has the form of a stationary wave $\xi(x - vt)$. Substituting this expression in (5.1.4), we find the equation for $\xi(x)$

$$(v_0 - v)\,\xi_x + b\xi_{xxx} + \xi\xi_x = 0 \,, \tag{5.1.7}$$

or after integration

$$b\xi_{xx} = (v - v_0)\,\xi - \frac{1}{2}\xi^2 + \text{const.} \tag{5.1.8}$$

By displacement of the origin of reference for ξ, the constant of integration can always be set equal to zero. Let us assume $c = v - v_0$ and introduce the function

$$W(\xi) = -\frac{c\xi^2}{2} + \frac{\xi^3}{6}. \tag{5.1.9}$$

Then (5.1.8) can be written in the form

$$b\xi_{xx} = -\frac{\partial}{\partial\xi} W(\xi) \tag{5.1.10}$$

and interpreted as an equation of motion of a material point in a potential well $W(\xi)$. Here the x-coordinate plays the role of the time t.

The function $W(\xi)$ is shown in Fig. 5.2 and the integral curves in the phase plane with the coordinates ξ and ξ_x are shown in Fig. 5.3.

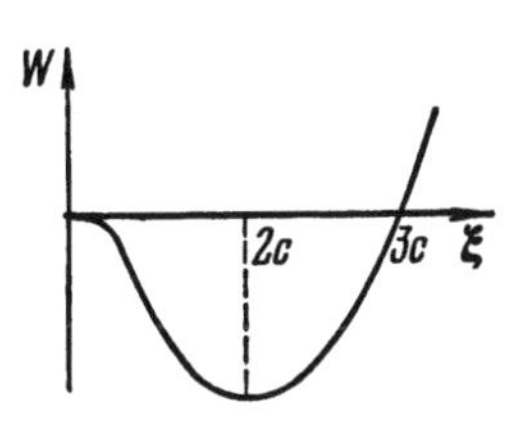

Fig.5.2. Potential $W(\xi)$

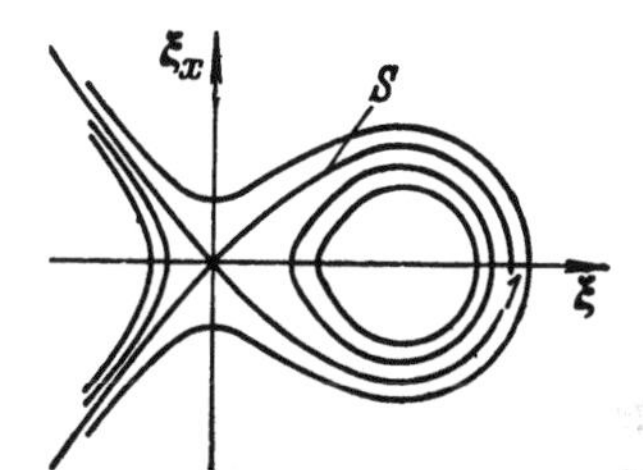

Fig.5.3. Phase diagram

The loop of the separatrix S bounds the region of periodic motions which correspond to closed trajectories in the phase plane. For small amplitudes the point undergoes harmonic oscillations in the neighborhood of the equilibrium state $\xi = 2c$, i.e. the wave in the moving coordinate system has the form

$$\xi \simeq 2c + \xi_0 \exp\left[i\sqrt{\frac{c}{b}}(x - ct)\right]. \tag{5.1.11}$$

With increasing amplitude, the sinusoidal behavior is distorted. As shown in Fig. 5.4, the particle spends most of its "time" in the region of small ξ's, where the restoring force is smaller and quickly jumps through the region of larger ξ's. The exact solution is expressed in terms of the Jacobi elliptical function.

If amplitude increases still further, the distance between the peaks tends to infinity and in the limit, one peak is left, which corresponds to the separatrix S in Fig. 5.3. The corresponding solution is known as a solitary wave or a soliton.

In the considered case, the form of the soliton can be found by an elementary integration of (5.1.8). Taking into account the boundary conditions at infinity, we have

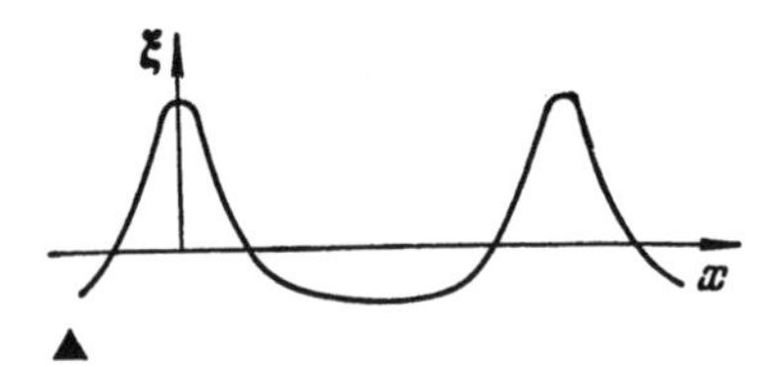

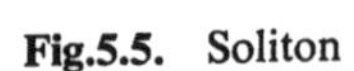

Fig.5.4. Velocity as a function of coordinate

Fig.5.5. Soliton

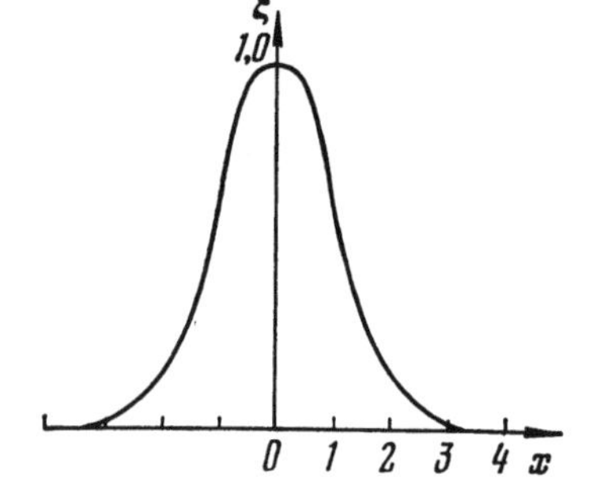

$$\xi = \xi_0 \operatorname{sech}^2\left(\frac{x - ct}{\lambda}\right). \tag{5.1.12}$$

Here the amplitude ξ_0 and the width λ of the soliton are connected with its velocity c according to

$$\xi_0 = 3c, \quad \lambda = 2\sqrt{\frac{b}{c}} = 2l\sqrt{\frac{v_0}{c}}, \tag{5.1.13}$$

where the relation $b = v_0 l^2$ was used.

We see that for a fixed value of b, solitons form a one-parameter family, where the parameter can be taken to be the amplitude or the relative velocity of the soliton. The latter is always positive and hence the absolute velocity v is greater than v_0. In the appropriate dimensionless variables, the soliton is written in the universal form

$$\xi = \operatorname{sech}^2(x - t), \tag{5.1.14}$$

which does not contain any parameters. The graph of this function is shown in Fig. 5.5.

When investigating nonstationary solutions of the KdV-equation (5.1.5), dimensionless variables are usually introduced by setting

$$\xi' = \frac{\xi}{\xi_0}, \quad x' = \frac{x}{\lambda}, \quad t' = \frac{\xi_0}{\lambda}t. \tag{5.1.15}$$

Here ξ_0 and λ are the characteristic amplitude and width of the initial disturbance. In terms of these variables, (5.1.5) takes the form (we omit the primes)

$$\xi_t + \xi\xi_x + \frac{1}{\sigma^2}\xi_{xxx} = 0, \tag{5.1.16}$$

where the dimensionless parameter

$$\sigma = \lambda\sqrt{\frac{\xi_0}{b}} = \frac{\lambda}{l}\sqrt{\frac{\xi_0}{v_0}} \tag{5.1.17}$$

characterizes the degree of nonlinearity, the dispersion being fixed. In particular, it has one and the same value σ_s for all solitons. From (5.1.13) we find

$$\sigma_s = \sqrt{12}\,. \tag{5.1.18}$$

For $\sigma \ll \sigma_s$ and not too large time a simple wave, localized at the initial moment, behaves similarly to the packets considered in the previous chapter.

When $\sigma \gg \sigma_s$ the behavior of essentially nonlinear, simple waves, as has been shown by explicit numerical computations [5.2], can be divided into three stages.

At the initial stage, the dispersion does not play any role and the process is described by (5.1.3). The characteristic times for this stage must be smaller than the time t_* for turn over.

At times of the order of t_*, short-wavelength oscillations arise at the trailing edge of the wave and the effects of dispersion begin to be noticeable. This stage is called transitional.

As the amplitudes of the oscillations increase, the latter break down into separate groups, each of which can approximately be characterized by the local value of the parameter σ. If these values are smaller than σ_s, then the group behaves as a weakly nonlinear packet. For larger σ, the process of breaking down continues, until for $\sigma \sim \sigma_s$, solitons apear, as shown in Fig. 5.6.

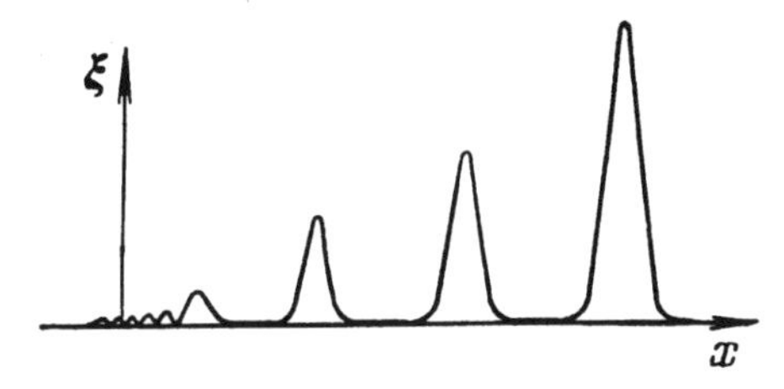

Fig.5.6. Asymptotic solitions and the wave packet

Solitons move with velocities $v > v_0$, their apexes lying asymptotically on one and the same straight line, since their velocities are proportional to their amplitudes. At the same time, the weakly nonlinear short-wavelength packet, as we know from the results of the previous chapter, moves with an average velocity which is smaller than v_0 (as $b > 0$) and lags behind the solitons. Since the packet is spreading out at the same time and its amplitude tends to zero, the asymptotic behavior of the solution is nothing but a system of solitons.

In the general case of initial disturbance, the picture can be more complex. In particular, it is necessary to take into account that, in a medium with negative dispersion, solitons can appear only for a disturbance having positive amplitude. For negative amplitudes, the disturbance evolves into a nonlinear wave "tail."

5.2 Connection Between the KdV-Model and Nonlinear Wave Equation

Let us take, as the original linear model, the model of an elastic medium with dispersion, which we have studied well,

$$\rho u_{tt} + \Phi u = 0\,, \tag{5.2.1}$$

where Φ is a linear operator of elastic energy with the kernel $\Phi(x - x')$.

To this model corresponds the Lagrangian (2.4.1), the elastic energy being a quadratic functional of the displacement. From the condition of translational invariance of the energy, it follows, according to (2.5.5), that the elastic energy is a quadratic functional of the strain u_x.

If we want to take into account a nonlinearity in the first approximation, then to the expression for elastic energy must be added a term, which is cubic with respect to the displacement–the so-called approximation of cubic anharmonicity. Obviously, this term must be a cubic functional of the strain u_x due to the condition of translational invariance.

Let us consider sufficiently long wave lengths. Then, the additional term will depend only on the strain but not on its derivatives, and the operator Φ can be written in the lowest approximation, which takes into account the dispersion. The corresponding wave equation takes the form

$$u_{tt} - v_0^2(u_{xx} + l^2 u_{4x} - \chi u_x u_{xx}) = 0\,. \tag{5.2.2}$$

Here v_0 is the velocity of sound in the linear model, l is a scaling parameter which must be considered to be small. The choice of positive sign in front of l^2 corresponds to the assumption that the medium has negative dispersion, i.e. the group velocity decreases with increasing wave number. The parameter of nonlinearity χ must be also considered to be small.

Problem 5.2.1. Verify that when $\chi > 0$, the elastic force increases more rapidly under compression from the state of equilibrium than it does under tension.

Note that the equation is invariat with respect to replacing of x by $-x$, as it should be, since this causes the change of the sign of the displacement u. After differentiating the equation with respect to x and denoting $\xi = u_x$, we get

$$\xi_{tt} - v_0^2[\xi_{xx} + l^2\xi_{4x} - \chi(\xi\xi_{xx} + \xi_x^2)] = 0\,. \tag{5.2.3}$$

Let us write this equation in a dimensionless form, by setting

$$x' = \frac{x}{\lambda}, \quad t' = \frac{v_0}{\lambda}t, \quad \xi' = \frac{\xi}{\xi_0}, \tag{5.2.4}$$

where ξ_0 is the amplitude of a wave, λ is a characteristic length, for example, the width of a soliton or the wavelength for a periodic process. Then the equation takes the form (primes are omitted)

$$\xi_{tt} - \xi_{xx} - 2\beta\xi_{4x} + 2\alpha(\xi\xi_x)_x = 0\,,$$
$$\beta = \frac{t^2}{2\lambda^2}, \quad \alpha = \frac{\xi_0\chi}{2}\,. \qquad (5.2.5)$$

The equation contains two independent small parameters α and β, which characterize the degree of nonlinearity and dispersion. Three cases are possible: $\beta \gg \alpha$, $\beta \ll \alpha$, $\beta \sim \alpha$. In the last case, the effects of dispersion and nonlinearity are of the same order and can counterblance each other.

In the next approximation, it is necessary to add to the wave equation terms quadratic in α and β. Among these the term of order $\alpha\beta$ is of fundamental significance, since under independent variation of α and β, it can be assigned a "one and a half" order. Naturally, it is necessary to take the additional terms in such a combination that the equation of motion remains the Euler equation of the corresponding variational problem.

Along with (5.2.5), let us study the KdV-equation

$$\xi_t + \xi_x + \beta\xi_{xxx} - \alpha\xi\xi_x = 0 \qquad (5.2.6)$$

(it differs from (5.1.4), by the substitution $\xi \to -\xi$).
Differentiating this equation with respect to time and replacing, ξ_t by its expression from (5.2.6), we obtain

$$\xi_{tt} - \xi_{xx} - 2\beta\xi_{4x} + 2\alpha(\xi\xi_x)_x$$
$$- \beta^2\xi_{6x} - \alpha^2(\xi^2\xi_x)_x + \alpha\beta\left(2\xi\xi_{xx} + \frac{1}{2}\xi_x^2\right)_{xx} = 0\,. \qquad (5.2.7)$$

Comparison with (5.2.5) shows that the two equations coincide to within terms quadratic in α and β. Thus, the solutions of the KdV-equation (5.2.6) exactly satisfy the wave equation (5.2.7) and approximately satisfy the original wave equation (5.2.5).

Equation (5.2.7) is not a unique wave equation, whose set of solutions contains all the solutions of the KdV-equation. Let us construct one more example of such an equation. To this end, let us introduce the linear operator

$$D(\xi) = \partial + \beta\partial^3 - \frac{1}{3}\alpha(\partial\xi + \xi\partial)\,. \qquad (5.2.8)$$

which depends on $\xi(x, t)$ (here $\partial\xi$ is the product of the operator ∂ and the operator of multiplication by ξ). Its adjoint $D^+(\xi)$ is defined in the usual way

$$\langle\varphi|D(\xi)|\psi\rangle = \langle D^+(\xi)\,\varphi|\psi\rangle\,, \qquad (5.2.9)$$

where φ, ψ are arbitrary smooth finite functions.

Problem 5.2.2. Show that

$$D^{+}(\xi) = -D(\xi)\,. \tag{5.2.10}$$

It is easily seen that the KdV-equation (5.2.6) can be written in the form

$$[\partial_t + D(\xi)]\,\xi = 0\,. \tag{5.2.11}$$

But, it follows from here that all the solutions of KdV-equation exactly satisfy the wave equation

$$[\partial_t - D(\xi)]\,[\partial_t + D(\xi)]\,\xi = 0\,. \tag{5.2.12}$$

Problem 5.2.3. Verify that this equation coincides with (5.2.5) to within quadratic terms and differs from (5.2.7) only by the term $\alpha\beta$.

The approximate correspondence between (5.2.5) and (5.2.12) can be considered as an analog of the exact factorization of the linear wave equation (4.3.7).

Let us forget temporarily the approximate character of KdV-equation and attempt to consider the corresponding wave equations (5.2.7) or (5.2.12) as exact models of certain elastic media. Unfortunately, in doing this, we encounter a serious difficulty, namely the wave equations (5.2.7) and (5.2.12) cannot be obtained from any variational principle, i.e. they are not potential.[1]

For the convenience of interpreting the results in terms of the elastic medium, let us pass from the strain $\xi = u_x$ to the displacement u. Then the KdV-equation (5.2.6) takes the form (after taking into account the boundary conditions at infinity)

$$u_t + u_x + \beta u_{xxx} - \frac{1}{2}\alpha u_x^2 = 0\,. \tag{5.2.13}$$

We shall call this the KdV-equation in displacements.

Problem 5.2.4. Show that, in the corresponding wave equation, the most general term of order $\alpha\beta$, which can be obtained from a Lagrangian, is given by the expression

$$\frac{\partial}{\partial x}\,(2u_x u_{xxx} + u_{xx}^2)$$

and differs from the analogous terms in (5.2.7) and (5.2.12).

---

[1]The operator is called potential if it is the Euler operator for some Lagranigan. About conditions of potentiality see for example [5.3].

Problem 5.2.5. Verify that, if we add to the KdV-equation (2.13) a new term $-\alpha\beta[(4a-1)u_x u_{xxx} + 2au_{xx}^2]/2$, where a is an arbitrary parameter, then the corresponding wave equation will be potential, to within terms cubic in α, β.

However, a new difficulty arises here: the modified KdV-equation will not possess good properties. To elucidate this question, let us turn to the original KdV-equation (5.2.6), and introduce the functional

$$H[\xi] = \frac{1}{2}\int\left(-\xi^2 + \frac{\alpha}{3}\xi^3 + \beta\xi_x^2\right)dx\,. \tag{5.2.14}$$

Then (5.2.6) can be written as

$$\partial_t\xi = \frac{\partial}{\partial x}\frac{\delta H[\xi]}{\delta\xi}\,, \tag{5.2.15}$$

where $\delta/\delta\xi$ is the functional derivative [5.3]. In the next chapter, we shall see that the representation of the equation in such a form means that the (infinite dimensional) system, can be derived from Hamiltonian functional $H[\xi]$ where the operator $\partial/\partial x$ plays the role of the symplectic form. The existence of the Hamiltonian of a system is a necessary condition for a number of important properties, in particular for the existence of conservation laws.

Problem 5.2.6. Verify that (5.2.6), with a term of the order of $\alpha\beta$ added, will be Hamiltonian, if and only if it has the structure $\partial_x(2\xi\xi_{xx} + \xi_x^2)$ and the corresponding term for the KdV-equation in displacements (5.2.13) has the form $2u_x u_{xxx} + u_{xx}^2$.

Comparison with Problem 5.2.5 shows that this condition cannot be fulfilled for any value of the parameter a. It follows from here that the correspondence between the Hamiltonian equation of the KdV type and the wave equation is violated already in the approximation of the "one and a half" order.

Taking into account the above reasoning, let us attempt to adopt another point of view. As the original equation, let us take the potential wave equation obtained in the required approximation, which has a clear physical interpretation in terms of the elastic medium. The equation of first order in time, which has the same stationary solutions, but with a fixed velocity sign, will be considered as the corresponding (modified) KdV-equation. As regards nonstationary solutions, they will coincide only in the first order in α and β.

In the next section, we shall investigate an example of such a correspondence and shall learn that the modified KdV-equation turns out to be Hamiltonian.

5.3 Deformed Soliton

Let us return to the wave equation (5.2.2) or, equivalently to (5.2.5). It contains terms linear in α and β. Regarding the parameters α and β as independent, let us add to the equation a cross term of the order of $\alpha\beta$, which does not destroy the potential nature of the equation. The equation obtained will be of order "one and a half" with respect to α and β. Investigating the stationary solution of this equation, we shall see that the retention of the term of the order of $\alpha\beta$ is fully justified.

Let us write the equation in which we are interested in terms of the dimensional variables, cf. (5.2.2),

$$u_{tt} - v_0^2\left[u_{xx} + l^2 u_{4x} - \chi u_x u_{xx} - \sigma\left(u_x u_{xxx} + \frac{1}{2}u_{xx}^2\right)_x\right] = 0 \qquad (5.3.1)$$

or in terms of the strain $\xi = u_x$:

$$\xi_{tt} - v_0^2\left[\xi_{xx} + l^2\xi_{4x} - \chi(\xi\xi_{xx} + \xi_x^2) - \sigma\left(\xi\xi_{xx} + \frac{1}{2}\xi_x^2\right)_{xx}\right] = 0\,. \qquad (5.3.2)$$

It is easily verified that this equation is obtained from the Lagrangian

$$L = \frac{\rho}{2}u_t^2 - \frac{\rho v_0^2}{2}\left(\xi^2 - l^2\xi_x^2 - \frac{\chi}{3}\xi^3 + \frac{\sigma}{2}\xi^2\xi_{xx}\right). \qquad (5.3.3)$$

Comparison of (5.3.1) and (5.3.2) with (5.2.2) and (5.2.3) shows that they differ only by the term with the coefficient σ; it is obviously of the order of $\alpha\beta$.

Let us search for a solution of (5.3.2) in the form of a stationary wave $\xi(x - vt)$. The function $\xi(x)$ must satisfy the equation

$$(w-1)\,\xi - l^2\xi_{xx} + \frac{1}{2}\chi\xi^2 + \sigma\left(\xi\xi_{xx} + \frac{1}{2}\xi_x^2\right) = 0\,,$$

$$w \stackrel{\text{def}}{=} \frac{v^2}{v_0^2} \geqslant 0. \qquad (5.3.4)$$

The constants of integration here are set equal to zero, since $\xi(x)$ is assumed to be bounded as $|x| \to \infty$ and $\xi = 0$ must satisfy the equation.

We see that (5.3.4) can be related to the modified KdV-equation

$$\xi_t + \xi_x + l^2\xi_{xxx} - \chi\xi\xi_x - \sigma(\xi\xi_{xxx} + 2\xi_x\xi_{xx}) = 0\,, \qquad (5.3.5)$$

which, for the stationary wave $\xi(x - wt)$, can be also reduced to (5.3.4). When $\sigma = 0$ (and after the additional substitution $\xi \to -\xi$) Eq. (5.3.5) coincides with the usual KdV-equation (5.1.4). However in this case, a solution in the form of a soliton exists, which in the corresponding dimensionless variables, can be written in the universal form (5.1.4). We recall that this is possible only when

$w > 1$, i.e. when $v > v_0$. It is convenient to pass to such dimensionless variables also in the general case $\sigma \neq 0$. Let

$$x = \frac{2l}{\sqrt{w-1}}\,x', \quad \xi = \frac{3(w-1)}{\chi}\,\xi'. \tag{5.3.6}$$

Then (5.3.4) takes the form (primes are omitted)

$$(1-\gamma\xi)\,\xi_{xx} - \frac{1}{2}\gamma\xi_x^2 - 4\xi - 6\xi^2 = 0\,, \tag{5.3.7}$$

where the only dimensionless parameter γ is defined by the expression

$$\gamma = \frac{3(w-1)\,\sigma}{\chi l^2}\,. \tag{5.3.8}$$

We remark that if, in addition, we introduce a dimensionless time and pass to a moving coordinate system, then the modified KdV-equation can be written in the form

$$\xi_t + \partial\left[(1-\gamma\xi)\,\xi_{xx} - \frac{1}{2}\gamma\xi_x^2 - 3\xi - 6\xi^2\right] = 0. \tag{3.5.9}$$

For the stationary wave $\xi(x-t)$, it is reduced to (5.3.7). Equation (5.3.7) is easily integrated. To this end we multiply it by ξ_x and we find the first integral

$$\xi_x = 2\sqrt{\frac{\xi^2+\xi^3}{1-\gamma\xi} + c}\,. \tag{5.3.10}$$

The integral curves in the phase plane for $\gamma > 0$ and $-1 < \gamma < 0$ are shown in Fig. 5.7. Let us compare these with the curves in Fig. 5.3, which (after replacing ξ by $-\xi$) corresponds to the case $\gamma = 0$. We see that when $-1 < \gamma < \infty$,

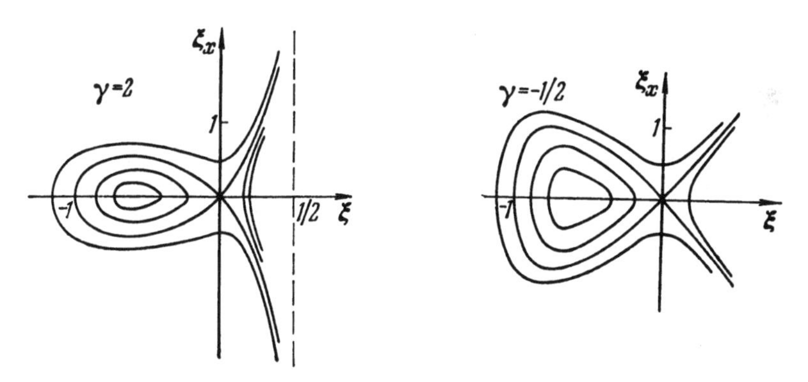

Fig.5.7. Phase diagrams

the loop of the separatrix bounds the region of closed integral curves, which correspond to periodic solutions $\xi(x)$, of the types shown in Fig. 5.4. It is obvious that the loop itself is obtained as $c = 0$ and corresponds to the soliton-like solution. It is essential that the amplitude of the soliton does not depend on γ and is equal to unity. In terms of the original variables, in accordance with (5.3.6)

$$\xi_0 = \frac{3(w-1)}{\chi}, \tag{5.3.11}$$

i.e. the usual relation between amplitude and velocity is preserved. Changing the parameter γ leads merely to a deformation of the soliton.

Setting $c = 0$ in (5.3.10) and integrating, we find the implicit expression for the soliton when $\gamma \geqslant 0$

$$x = \tanh^{-1}\theta + \sqrt{\gamma}\cot^{-1}\frac{1}{\sqrt{\gamma}}\theta \tag{5.3.12}$$

and when $-1 < \gamma \leqslant 0$

$$x = \tanh^{-1}\theta - \sqrt{-\gamma}\tanh^{-1}\frac{1}{\sqrt{-\gamma}}\theta\,. \tag{5.3.13}$$

Here

$$\theta = \sqrt{\frac{1-\gamma\xi}{1+\xi}}, \tag{5.3.14}$$

and, for $x \geqslant 0$, the principal value of the arcctg is to be taken. On the negative x-axis, $\xi(x)$ is continued such that the result is an even function.

Let us proceed to the investigation of the solution obtained. When $\gamma = 0$, we have the usual soliton

$$\xi(x) = -\operatorname{sech}^2 x \tag{5.3.15}$$

with the asymptotic behavior

$$\xi(x) \sim e^{-2x} \quad \text{as} \quad x \to \infty. \tag{5.3.16}$$

Using the representations

$$\tanh^{-1}\theta = \frac{1}{2}\ln\frac{\theta+1}{\theta-1}, \quad \tan^{-1}\theta = -\frac{i}{2}\ln\frac{i\theta-1}{i\theta+1}, \tag{5.3.17}$$

we find that $\xi(x)$ has the same asymptotic behavior for any γ in the admissible region $-1 < \gamma < \infty$.

The curvature κ of the function $\xi(x)$ at the point of minimum $x = 0$, i.e. when $\xi = -1$, in accordance with (5.3.7) is equal to

$$\kappa = \frac{2}{1+\gamma}. \tag{5.3.18}$$

This gives a geometrical interpretation to the parameter γ. When $\gamma \to -1$ the curvature tends to infinity and hence $\xi(x)$ has an angular point. It is not difficult to find the limit of the function $\xi(x)$ when $\gamma \to -1$

$$\xi(x) = -e^{-2|x|}. \tag{5.3.19}$$

The degeneracy of the loop of separatrix into straight segments with a discontinuity at the point $\xi = -1$ corresponds to this limit.

Strictly speaking, the initial assumptions about sufficient smoothness of $\xi(x)$ are not fulfilled for this solution, but it is of interest, provided the equations (5.3.2) or (5.3.5) are considered as model ones.

In the other limiting case as $\gamma \to \infty$ or $\kappa \to 0$, the soliton diffuses. From (5.3.12), neglecting the first small term, we obtain

$$\xi(x) = -\cos^2\frac{x}{\sqrt{\gamma}} \quad \left(\frac{x}{\sqrt{\gamma}} < \frac{\pi}{2}\right), \tag{5.3.20}$$

from which it is seen that the width of the soliton is of the order of $\sqrt{\gamma}$. The loop of the separatrix, as it follows from (5.3.10), coincides with the narrow ellipse

$$\frac{\gamma}{4}\xi_x^2 + \xi^2 + \xi = 0 \tag{5.3.21}$$

everywhere, excluding a small neighborhood of the origin.

In Fig. 5.8, the deformation of the soliton as γ is varied is shown. The dotted curve shows the usual soliton. The corresponding curves for the displacement $u(x)$ are shown in Fig. 5.9. We see that in terms of the displacement, the solution

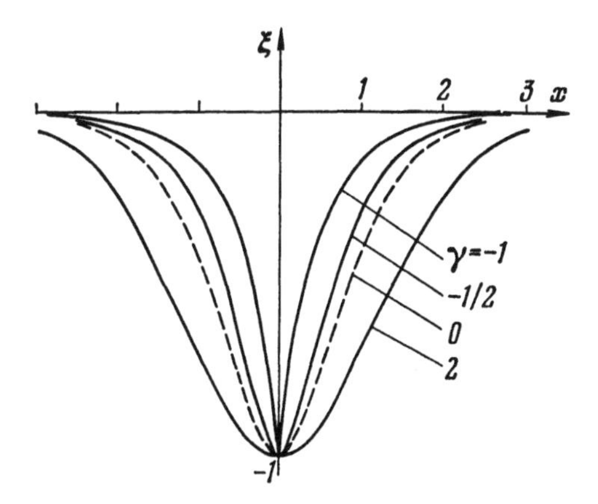

Fig.5.8. A family of solitons

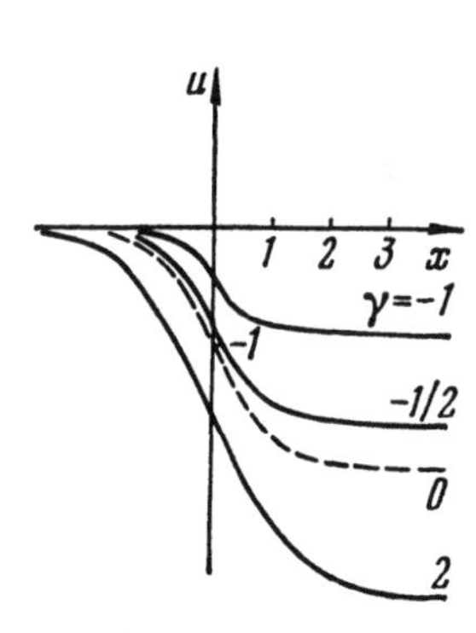

Fig.5.9. A family of shock waves

has the form of a shock wave. With decreasing γ, the discontinuity Δu decreases but the slope of the front increases.

We can also interpret the results obtained in terms of the original modified KdV-equation (5.3.5). The parameters of the equation being fixed, we have a family of solitons, whose amplitude and form depend on the velocity $w - 1$ (in the moving coordinate system). Because γ varies between the limits $-1 < \gamma < \infty$, it follows from (5.3.8) that the velocity of a soliton can be negative, as distinct from the usual KdV-equation.

Observe also that the limiting case $\gamma \to -1$ corresponds to the well known Stokes waves on the surface of a liquid. The modified KdV-equation is the simplest model describing such a phenomenon.

5.4 The Nonlinear Chain

In this section we shall investigate wave processes in an unbounded chain of particles with mass m, connected by nonlinear springs. For simplicity, the interaction of nearest neighbors is only assumed.

Let

$$\zeta(n) \overset{\text{def}}{=} \nabla u(n) = \frac{u(n) - u(n-1)}{a} \tag{5.4.1}$$

be the relative change of length of the spring under limiting displacements of particles, and $\varphi(\zeta)$ be the elastic energy of the spring. Then the Lagrangian of the chain has the form (the dependence on time is not shown explicitly)

$$L = \frac{m}{2} \sum_n u_t^2(n) - \frac{1}{2} \sum_n \varphi \left[\frac{u(n) - u(n-1)}{a} \right], \tag{5.4.2}$$

To the Lagrangian corresponds the equation of motion

$$m u_{tt}(n) + a^{-1} \{\varphi'[\zeta(n)] - \varphi'[\zeta(n+1)]\} = 0\,. \tag{5.4.3}$$

By obvious transformation this equation can be represented in the form

$$m\zeta_{tt} - \Delta\varphi'(\zeta) = 0\,, \tag{5.4.4}$$

where Δ is the second-difference operator

$$\Delta f(n) \overset{\text{def}}{=} a^{-2} [f(n+1) + f(n-1) - 2f(n)]\,. \tag{5.4.5}$$

Problem 5.4.1. Write the equations of motion of a chain with interaction of an arbitrary number of neighbors in a form analogous to (5.4.4).

In the anharmonic approximation

$$\varphi(\zeta) = \frac{1}{2!}\varphi_0''\zeta^2 + \frac{1}{3!}\varphi_0'''\zeta^3\,, \tag{5.4.6}$$

whence it is seen that $c_0 = a^{-1}\varphi_0''$ is the elastic constant of the harmonic model in the zeroth long-wavelength approximation. The equation of motion (5.4.4) takes the form

$$\zeta_{tt} - v_0^2\Delta\left(\zeta - \frac{1}{2}\chi\zeta^2\right) = 0\,, \tag{5.4.7}$$

where

$$v_0^2 = \frac{\varphi_0''}{m} = \frac{c_0}{\rho}\,,\quad \chi = -\frac{\varphi_0'''}{\varphi_0''}\,. \tag{5.4.8}$$

In order to write (5.4.7) in the long-wavelength approximation, first let us pass to the quasicontinuum, associating an analytic function $u(x)$ with a function of discrete argument $u(n)$. Then as is easily seen, $\exp(a\partial)\,u(x)$ corresponds to the function $u(n+1)$, and for the second difference operator we have

$$\Delta = \frac{2}{a^2}(\cosh a\partial - 1) = \partial^2\left(1 + \frac{a^2}{12}\partial^2 + \dots\right). \tag{5.4.9}$$

In the lowest approximation, which takes into account dispersion, (5.4.7) takes the form

$$\zeta_{tt} - v_0^2\partial^2\left(1 + \frac{a^2}{12}\partial^2\right)\left(\zeta - \frac{1}{2}\chi\zeta^2\right) = 0\,. \tag{5.4.10}$$

Let us compare this with (5.3.2), written in the same approximation. We see that there is a discrepancy in the last term, since

$$\frac{1}{2}(\zeta^2)_{xx} = \zeta\zeta_{xx} + \zeta_x^2\,,$$

and in accordance with (5.3.2), the coefficient 1/2 must be in front of ζ_x^2. But we have seen that (5.3.2) is the most general equation which could be obtained in the given approximation. At the same time, (5.4.10), also obtained from variational principle, does not conform to this.

In order to resolve this paradox let us turn to the original equation (5.4.3), which is Euler's equation for the Lagrangian (5.4.2) and write it in the form

$$\rho u_{tt} + a^{-2}\left[\varphi'(\zeta) - \varphi'(e^{a\partial}\zeta)\right] = 0 \tag{5.4.11}$$

or in the anharmonic approximation (5.4.6)

$$u_{tt} + \frac{v_0^2}{a}\left\{(1 - e^{a\partial})\,\zeta - \frac{1}{2}\chi[\zeta^2 - (e^{a\partial}\zeta)^2]\right\} = 0\,. \tag{5.4.12}$$

Let us take into account that, in accordance with (5.4.1),

$$\zeta = \nabla u = \frac{1}{a}(1 - e^{-a\partial})\,u\,, \tag{5.4.13}$$

and hence

$$\frac{1}{a}(1 - e^{a\partial})\,\zeta = \frac{2}{a^2}(1 - \cosh a\partial)\,u \simeq \partial^2\left(1 + \frac{a^2}{12}\partial^2\right)u\,,$$

$$\frac{1}{a}[\zeta^2 - (e^{a\partial}\xi)^2] = \frac{4}{a^3}(1 - \cosh a\partial)\,u\cdot(\sinh a\partial)\,u$$

$$\simeq -\partial\left[u_x^2 + \frac{a^2}{6}\left(u_x u_{xxx} + \frac{1}{2}u_{xx}^2\right)\right]. \tag{5.4.14}$$

Substitution into (5.4.12) leads to (5.3.1) if we set

$$l^2 = \frac{a^2}{12},\quad \delta = \frac{\chi a^2}{12} = \chi l^2\,, \tag{5.4.15}$$

i.e., Eq. (5.4.3) coincides with (5.3.1) in the corresponding approximation as it should.

We now observe that the transition from (5.4.3) to (5.4.4) is carried out by applying the operator

$$\nabla = \frac{1}{a}(1 - e^{-a\partial})\,, \tag{5.4.16}$$

and from (5.3.1) to (5.3.2) by simple differentiation. Therefore, (5.4.10) and (5.3.2) are equivalent in the given approximation, though formally different.[2]

Problem 5.4.2. Prove directly the equivalence of (5.4.10) and (5.3.2). (Hint: Take into account, that the operator $\nabla\partial^{-1}$ has an inverse over the class of functions under consideration, and use the connection between ξ and ζ).

Note that if we write (5.4.10) in the form (5.3.2) and then transform to the variables (5.3.6), then, according to (5.4.15) and (5.3.8) the only dimensionless parameter γ is given by the expression

$$\gamma = 3(w - 1)\,. \tag{5.4.17}$$

[2] We have already met a similar nonuniqueness of representation of approximate equations in weak nonlocal models (Sect, 5.2.9) when investigating the Timoshenko's equations.

The connection between the approximate equations and the Korteweg-de Vries equation and their stationary solutions was considered above and now we again turn to the exact equation for the chain (5.4.4).

In the works of Toda [5.5] a special law of interaction $\varphi(\xi)$, for which a whole class of particular solutions of (5.4.4) can be constructed, was proposed and investigated. Namely, in the Toda model, an exponential law of interaction of masses is assumed

$$\varphi(\zeta) = \frac{ac_0}{\chi^2}\left(e^{-\chi\zeta} - 1 + \chi\zeta\right). \tag{5.4.18}$$

This is shown in Fig. 5.10.

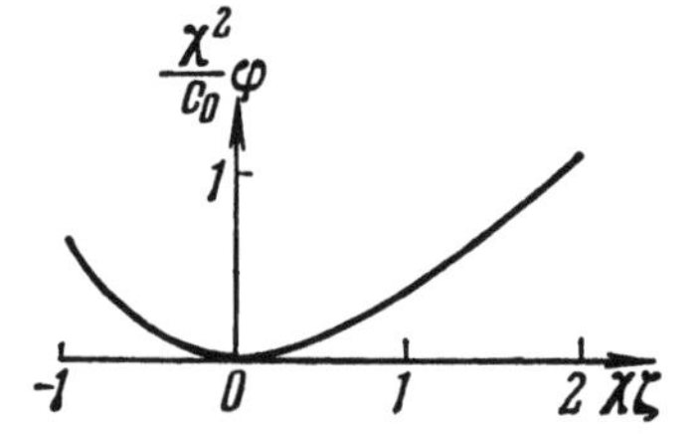

Fig.5.10. Law of interaction for the Toda model

For small values of ζ

$$\varphi(\zeta) = ac_0\left(\frac{1}{2!}\xi^2 - \frac{1}{3!}\chi\zeta^3 + \dots\right). \tag{5.4.19}$$

and the equations of motion in the anharmonic approximation coincide with (5.4.7).

Let us introduce new variables

$$x' = \frac{x}{a}, \quad t' = \frac{v_0 t}{a}, \quad \zeta' = \chi\zeta. \tag{5.4.20}$$

Then, for the Toda model, (5.4.4) takes the form (primes are omitted)

$$\zeta_{tt} - \Delta(1 - e^{-\zeta}) = 0. \tag{5.4.21}$$

It is convenient also to introduce the variable s, which plays the role of a potential with respect to ζ and u,

$$\zeta = -\Delta s, \quad u = -\partial s \tag{5.4.22}$$

and satisfies the equation

$$\ln(1 + s_{tt}) - \Delta s = 0. \tag{5.4.23}$$

Let us consider some partial solutions of the equations (5.4.21) and (5.4.23).

Periodic sequence of impulses: As in the case of KdV-equation, there exist solutions in the form of periodic propagating waves $\zeta(\kappa n \mp \nu t)$ like those shown in Fig. 5.4. These solutions are expressed in terms of elliptic functions and integrals.

Soliton: If the period of a wave tends to infinity, the wave degenerates into a soliton or, equivalently into a shock wave in terms of displacement. One verifies directly that (5.4.21) and (5.4.23) are satisfied by solution of the form

$$s = \ln \cosh(\kappa n \mp \nu t) + \text{const}\,, \tag{5.4.24}$$

$$1 - \partial s = \ln\left[\cosh \kappa - \nu \tanh(\kappa n \mp \nu t)\right], \tag{5.4.25}$$

$$1 - e^{-\zeta} = -\nu^2 \operatorname{sech}^2(\kappa n \mp \nu t)\,, \tag{5.4.26}$$

where the parameters ν and κ are connected by the relation

$$\nu = \sinh \kappa\,. \tag{5.4.27}$$

It is easily seen that the parameter κ, in turn, is connected with the amplitude ζ_0 of the soliton

$$\zeta_0 = -2\ln \cosh \kappa\,. \tag{5.4.28}$$

For small values of the parameter κ, the amplitude $\zeta \simeq -\kappa^2$. The soliton velocity ν/κ grows with increasing amplitude, and tends to unity as $\kappa \to 0$.

Collision of two solitons: The following solution is more general than (5.4.24):

$$s = \ln\left[\cosh(\mu_1 n - \lambda_1 t) + b\cosh(\mu_2 n - \lambda_2 t + \delta)\right] + \text{const}\,. \tag{5.4.29}$$

Here δ is an arbitrary constant and λ_1, λ_2 and b are functions of μ_1, μ_2. If $\mu_1 = \mu_2$, then this solution does in fact coincide with (5.4.24), and if $\mu_1 \neq \mu_2$, then it represents two solitons.

Two cases are to be distinguished: solitons moving in the opposite and in the same directions.

In the first case

$$\lambda_1 = 2\sinh\frac{\mu_2}{2}\cosh\frac{\mu_1}{2}, \quad \lambda_2 = 2\sinh\frac{\mu_1}{2}\cosh\frac{\mu^2}{2}, \quad b = \cosh\frac{\mu_1}{2} \Big/ \cosh\frac{\mu_2}{2}\,. \tag{5.4.30}$$

Asymptotically, as $t \to -\infty$ $(i = 1, 2)$

$$1 - e^{-\zeta} = -\nu_i^2 \operatorname{sech}^2(\kappa_i - \nu_i t + \theta)\,, \tag{5.4.31}$$

and, as $t \to \infty$,

$$1 - e^{-\zeta} = -\nu_i^2 \operatorname{sech}^2(\kappa_i \nu - v_i t - \theta)\,, \tag{5.4.32}$$

where the parameters of the solitons are given by the expressions

$$\kappa_t = \frac{\mu_1 \pm \mu_2}{2}, \quad \nu_t = \frac{\lambda_1 \pm \lambda_2}{2}, \quad \theta = \frac{\delta - \ln b}{2}. \tag{5.4.33}$$

For solitons moving in the same direction

$$\lambda_1 = 2\sinh\frac{\mu_1}{2}\cosh\frac{\mu_2}{2}, \quad \lambda_2 = 2\sinh\frac{\mu_2}{2}\cosh\frac{\mu_1}{2}, \quad b = \sinh\frac{\mu_1}{2} \Big/ \sinh\frac{\mu_2}{2}. \tag{5.4.34}$$

The asymptotics as $t \to \pm\infty$, is the same as above.

We see that the asymptotic forms of solitons and their velocities are preserved, but during the interaction, they are accelerated since after a collision, they overtake the phases they had before the collision. It is interesting to note that such a phenomenon was first discovered by a numerical solution of the KdV-equation.

Let us now present the results of numerical experiments which were carried out by Oyama and Saito [5.6] with the Toda model. In Fig. 5.11 the collision of three solitons is shown. The horizontal arrows show the velocities of the solitons, and the vertical one shows the point of collision. The picture is qualitatively the same as during the collision of two solitons.

Some peculiar phenomena arise if one changes the form of a normal soliton. By the latter we understand the exact solution of (5.4.24) shown by the dotted line in Fig. 5.12. The calculations were carried out for a finite chain. If the initial

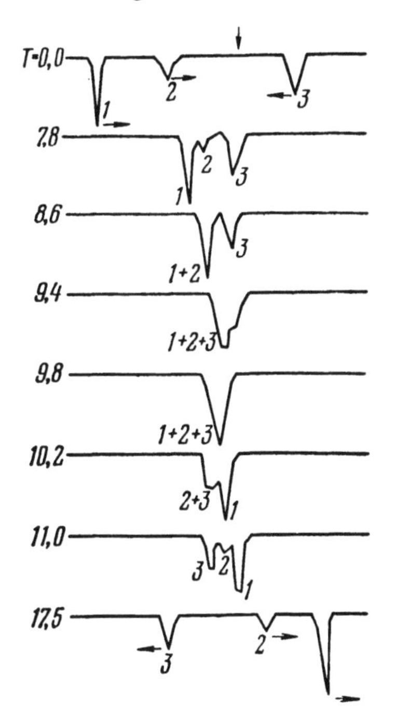

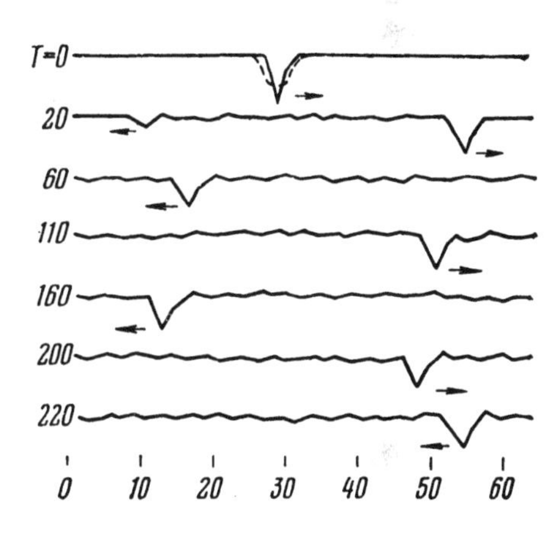

Fig.5.12. Evolution of the soliton form

Fig.5.11. Collision of three solitions

deformation is not very strong, the soliton in the course of time breaks down into a normal soliton and a short wave ripple, which lags behind the soliton. The soliton is reflected from both ends of the chain and the ripple is propagated throughout the entire chain. Subsequently, the soliton moves through the ripple without a change of its velocity. The form and the amplitude of the soliton can fluctuate, but on the average, they coincide with those of a normal soliton.

With increasing initial deformation, the soliton breaks up into two normal ones accompanied by a short wave ripple. Subsequently, the picture remains stable.

Having obtained these results, the authors of the numerical experiment became interested in the behavior of a soliton against a background of stronger perturbations. It was shown that, if the energy of the random background exceeds some threshold value of the order of 2—3%, then the soliton is decelerated as if it were subjected to viscous friction, and its energy is gradually transferred to the background. Thus, as distinct from the case of weak ripple, if the energy of the background is higher than the threshold value, the soliton becomes unstable.

We see that a far reaching analogy between the Toda chain, the nonlinear wave equation and the KdV-equation exists. In the next chapter, deep reasons for this analogy will be elucidated.

5.5 Conservation Laws

In the investigation of nonstationary solutions of equations of Korteweg-de Vries type, the conservation laws are important. Thus, (5.2.6) can be written in divergence form (for simplicity, let us set $\alpha = 1$)

$$\partial_t \xi + \partial \left(\beta \xi_{xx} - \frac{1}{2} \xi^2\right) = 0 , \tag{5.5.1}$$

from which, the conservation of the integral

$$I_1 = \int \xi(x, t)\, dx \tag{5.5.2}$$

follows for solutions $\xi(x, t)$, which satisfy the conditions $\xi \to 0$ as $|x| \to \infty$.

Analogously, multiplying (5.1.16) by ξ and ξ^2, we find

$$\frac{1}{2} \partial_t(\xi^2) + \partial\left[-\frac{1}{3} \xi^3 + \beta \left(\xi\xi_{xx} - \frac{1}{2} \xi_x^2\right)\right] = 0 , \tag{5.5.3}$$

$$\frac{1}{3} \partial_t(\xi^3 + 3\beta\xi_x^2) + \partial\left[-\frac{1}{4} \xi^4 + \beta(\xi^2\xi_{xx} + 2\xi_t\xi_x) - \beta^2\xi_{xx}^2\right] = 0 , \tag{5.5.4}$$

and we have two more invariants

$$I_2 = \frac{1}{2} \int \xi^2 dx, \quad I_3 = \frac{1}{3} \int (\xi^3 + 3\beta\xi_x^2)\, dx\,. \tag{5.5.5}$$

Problem 5.5.1. Write down analogous invariants for the modified KdV-equation (5.3.9).

An infinite sequence of invariants I_m can be constructed, all I_m having the form of polynomials of increasing degree in β. The existence of an infinite number of invariants for a nonlinear equation is an indication of its exceptional nature. Later on it will become clear, with what this special structure of the KdV-equation is connected.

The invariants I_1 and I_2 are usually interpreted as "momentum" and "energy." From the results of Sect. 5.2, it follows that I_3 is distinguished from the Hamiltonian only by a numerical multiplier. As regards the other invariants, a physical meaning can hardly be attributed to them, if one considers the KdV-equation as an approximate one, but they are essential for an exact KdV-model.

If the invariant I_m is written in the form

$$I_m = \int Q_m(\xi)\, dx\,, \tag{5.5.6}$$

then a general expression for the density $Q_m(\xi)$, can be constructed in the form of a polynomial in β. The first two terms have the form

$$Q_m(\xi) = \frac{\xi^m}{m} - \beta \frac{(m-1)(m-2)}{2} \xi_x^2 \xi^{m-2}\,. \tag{5.5.7}$$

This expression is useful in cases when β can be considered as a small parameter, but it is unsuitable for a soliton. In fact, it can be shown that, in that case, all the terms in $Q_m(\xi)$ have the same order $\sqrt{\beta}$. For the exact value of the invariant, in the case of the soliton (5.1.12), when $\alpha = 1$, the expression [5.2]

$$I_m = 2\sqrt{\beta}\, \frac{6^m[(m-1)!]^2}{(2m-1)!}\, v^{m-1/2}\,, \tag{5.5.8}$$

is valid, and it will be used in the next section.

We saw earlier that the correspondence between the nonlinear wave equation and KdV-equation appears only to linear approximation in α and β. An analysis shows that, in this approximation, the energy characteristics of the KdV-equation possess a number of "nice" properties; in particular, the law of conservation of velocity of the centre of gravity of the packet is fulfilled. However, taking the term of order $\alpha\beta$ into account, violates this law.

Problem 5.5.2. Show that the kinetic energy T and the potential energy Φ of the solution of the wave equation (5.2.2) in the form of a soliton, which moves with a velocity v, are equal to

$$T = \frac{12\rho v_0^2 l}{\chi^2} w(w-1)^{3/2},$$

$$\Phi = \frac{12\rho v_0^2 l}{\chi^2}\left[1 - \frac{13}{5}(w-1)\right](w-1)^{3/2}, \tag{5.5.9}$$

where $w = v^2/v_0^2$,

We see that, as distinct from the case of the packet, which satisfies a linear equation, the kinetic and potential energies are not equal to one another (though in the considered case, they are certainly conserved). As $w \to 1$, the amplitude of the soliton decreases and an approximate transition to a linear model should occur. Hence, to within quantities of the next order of smallness relative to $w - 1$, in this case $T \simeq \Phi$.

5.6 Decay of the Initial Perturbation and the Distribution Function of Solitons

Let us now consider in more detail the behavior of a localized initial disturbance in the KdV-model with strong nonlinearity i.e., when $\sigma \gg \sigma_*$ or, equivalently, when β is sufficiently small. As shown in Sect. 5.1, the initial perturbation quickly decays into solitons and a short-wave packet which lags behind them. The latter spreads out in the course of time and its amplitude tends to zero. Moreover, the soliton component of the solution is due mainly to the positive part of the initial perturbation, and the packet is due to the negative part.

In this and the next sections we shall consider only the soliton part of the solution, assuming that the number N of solitons generated is sufficiently large so that it is possible to introduce a distribution function $F(v)$ of the solitons with respect to velocities: $F(v)\Delta v$ is the number of solitons in the velocity interval $(v, v + \Delta v)$.

By definition, the number of solitons N and the distribution function $F(v)$ are connected with the asymptotic behavior of the solution and cannot depend on time because the solitons do not interact asymptotically. However, N and $F(v)$ are uniquely determined by the initial data $\xi(x, 0)$ and hence can be considered as functionals of the solution $\xi(x, t)$ of the KdV-equation, which do not depend on time. Naturally a question arises about the connection of these invariants with the invaiants I_m indicated above. Such a connection is proved to exist and it permits one to express N and $F(v)$ directly in terms of the initial disturbance. Below we shall see that these results follow from the general theory of the KdV-equation, based on the method of the inverse scattering problem. However, it is also expedient to construct the desired connection in an elementary way.

It is not difficult to prove that the moments of the distribution function $F(v)$ are proportional to the corresponding invariants of the KdV-equation. In fact, the invariants I_m, as is easily seen, are additive for non-interacting packets. But the solitons are asympotically uncoupled and hence the value of the invariant I_m for the system of solitons is equal to the sum of the values of the invariant for each of the solitons. If the number of solitons is large, the summation can be replaced by integration and in view of (5.5.8), we have

$$I_m = g(m) \int_0^\infty F(v)\, v^{m-1/2}\, dv\,, \tag{5.6.1}$$

where

$$g(m) = 2\sqrt{\beta}\, \frac{6^m[(m-1)!]^2}{(2m-1)!}\,. \tag{5.6.2}$$

Let us write the initial disturbance in the form

$$\xi(x, 0) = \xi_0 \varphi\left(\frac{x}{l}\right) \geqslant 0\,, \tag{5.6.3}$$

where ξ_0 and l are a characteristic amplitude and width. Substituting in (5.5.6), we find

$$I_m = \frac{\xi_0^m}{m} \int \varphi^m\left(\frac{x}{l}\right) dx + O(\beta)\,. \tag{5.6.4}$$

Comparison with (5.6.1) allows to express the moments of the function in terms of the initial disturbance

$$\int_0^\infty F(v)\, v^{m-1/2}\, dv = \frac{\xi_0^m}{mg(m)} \int \varphi^m\left(\frac{x}{l}\right) dx\,. \tag{5.6.5}$$

Passing to a dimensionless variable $y = x/l$ and assuming for simplicity $\varphi(y)$ to be an even function, we obtain

$$\int_0^\infty F(v)\, v^{m-1/2}\, dv = \frac{2l\xi_0^m}{mg(m)} \int_0^\infty \varphi^m(y)\, dy\,. \tag{5.6.6}$$

We shall assume that $\varphi(y)$ decreases monotonically for $y \geqslant 0$ from the value $\varphi(0) = 1$ to $\varphi(\infty) = 0$. Then the change of the variables

$$z = \varphi(y), \quad y = -\psi(z) \quad (0 \leqslant z \leqslant 1) \tag{5.6.7}$$

transforms (5.6.6) to the form

$$\int_0^\infty F(v)\, v^{m-1/2}\, dv = \frac{2l\xi_0^m}{mg(m)} \int_0^1 z^m \psi'(z)\, dz\,. \tag{5.6.8}$$

Let

$$f(k) = \int F(\eta^2) e^{-ik\eta}\, d\eta\,. \tag{5.6.9}$$

It is easy to verify that the moments of the function $F(v)$ are expressed in terms of the derivatives of $f(k)$ at the point $k = 0$:

$$\int_0^\infty F(v)\, v^{m-1/2}\, dv = (-1)^m f^{(2m)}(0)\,. \tag{5.6.10}$$

Substitution of this expression in (5.6.8) yields

$$f^{(2m)}(0) = -2l \frac{(-1)^m \xi_0^m}{mg(m)} \int_0^1 z^m \psi'(z)\, dz\,. \tag{5.6.11}$$

This defines an even analytic function $f(k)$. It is directly verified that it can be represented as a Bessel integral transform of the function $\psi'(z)$:

$$f(k) = -\frac{l}{2\sqrt{\beta}} \int_0^1 Y_0\left(k\sqrt{\frac{2}{3}\xi_0 z}\right) \psi'(z)\, dz\,, \tag{5.6.12}$$

which is equivalent to (5.6.11).

Taking into account properties of the Bessel transform, we see that $f(k)$ and hence also $F(v)$ are connected one-to-one with $\psi'(z)$ or ultimately with $\xi(x, 0)$. Transforming from $f(k)$ to $F(v)$ in (5.6.12), and in view of (5.6.9), we obtain the Karpman formula [5.2]

$$F(v) = -\frac{\sqrt{3}}{2\pi} \frac{l}{\sqrt{\beta}} \int_{\frac{3v}{2\xi_0}}^1 \frac{\psi'(z)\, dz}{\sqrt{2\xi_0 z - 3v}}\,, \quad v < \frac{2}{3}\xi_0 \tag{5.6.13}$$

and $F(v) = 0$ when $v > 2\xi_0/3$. This can also be written in the form

$$F(v) = \frac{\sqrt{3}}{4\pi} \int \frac{dy}{\sqrt{2\xi_0 \varphi(y) - 3v}}\,, \tag{5.6.14}$$

where the region of integration is determined by the condition that the argument of the square root be positive.

By integration over v, we find the total number of the solitons

$$N = \frac{l}{\pi} \sqrt{\frac{\xi_0}{6\beta}} \int_{\varphi(y)>0} dy \sqrt{\varphi(y)}\,. \tag{5.6.15}$$

Let us investigate the given formulae by an example for which the exact distribution of solitons is known. Let the initial disturbance be

$$\xi(x, 0) = \varphi(x) = \operatorname{sech}^2 x\,, \tag{5.6.16}$$

i.e. $\xi_0 = 1$, $l = 1$. Substitution in (5.6.14), after simple calculation, yields

$$F(v) = \frac{1}{4\sqrt{\beta v}}, \quad v < \frac{2}{3}\,, \tag{5.6.17}$$

and $F(v) = 0$ when $v > 2/3$. Hence

$$N = \int_0^\infty F(v)\, dv = \frac{1}{\sqrt{6\beta}}\,. \tag{5.6.18}$$

On the other hand, if we insert

$$\beta = \frac{1}{6p(p+1)}\,, \tag{5.6.19}$$

where p is a positive integer, into the KdV-equation, then it can be shown [5.7] that the initial disturbance (5.6.16) exactly splits up into p solitons with velocities

$$v_n = \frac{2n^2}{3p(p+1)}\,, \quad n = 1, 2, \ldots, p\,, \tag{5.6.20}$$

whence it is seen, in particular, that the maximum velocity $v_{max} = v_p$, does not exceed 2/3, in complete agreement with (5.6.17).

Let $p \gg 1$ and let us consider large values of n. In that case, it can be assumed that the velocities v_n change continuously and the distribution function $F(v)$ can be introduced.

Problem 5.6.1. Show that it follows from (5.6.20) when $n \gg 1$ that

$$F(v) = p\sqrt{\frac{3}{8v}}, \quad v < \frac{2}{3}\,. \tag{5.6.21}$$

Comparison with (5.6.16), with (5.6.18) taken into account, shows that both distribution functions and hence also the total numbers of solitons coincide, as one would expect.

In conclusion, we note an apparent paradox, connected with the above-indicated correspondence $F(v) \leftrightarrow \xi(x, 0)$. It seems that one can inter-change $\xi(x, 0)$ with the solution $\xi(x, t)$ of the KdV-equation with the initial condition $\xi(x, 0)$, but this leads to a contradiction because $F(v)$ is an invariant, independent of time. The paradox is removed, if we observe that, when deriving (5.6.14), $\xi(x, 0)$ was assumed to be even and monotonic, and this condition is not preserved in time due to the oscillating nature of the solution.

5.7 The Soliton Gas

Let us imagine a situation in which the initial disturbance consists of several peaks, each of which corresponds to a large local value σ or, aquivalently, a small value of β. Then the peaks must decay into solitons and we can write the further evolution of the initial disturbance in terms of a soliton gas.

Let us preliminarily consider the interaction of two solitons. Typical cases of such interaction are shown in Figs. 5.13, 14, 15, taken from [5.8]. The solitons in Figs. 5.13 and 14 have velocities in the same direction but different relative amplitudes, and Fig. 5.15 presents a picture of a frontal collision of two solitons having the same amplitude. In all these cases, the final result of the interaction consists in a change of the phases of the colliding solitons. Let $\alpha(v_i, v_j)$ be the change of phase of the i-th soliton with the velocity v_i before the collision with the j-th soliton with the velocity v_j. In the next chapter the N-soliton solution of the KdV equation will be analyzed, and it will be shown that

$$\alpha(v_i, v_j) = \frac{4 \operatorname{sgn}\{v_i - v_j\}}{\sqrt{v_i}} \ln \left| \frac{\sqrt{v_i} + \sqrt{v_j}}{\sqrt{v_i} - \sqrt{v_j}} \right| . \tag{5.7.1}$$

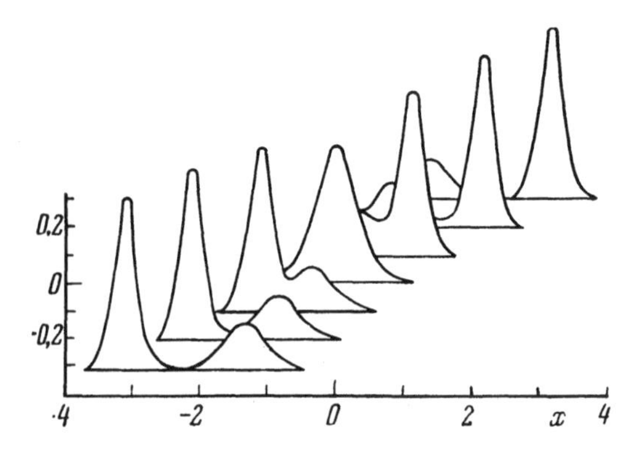

Fig.5.13.

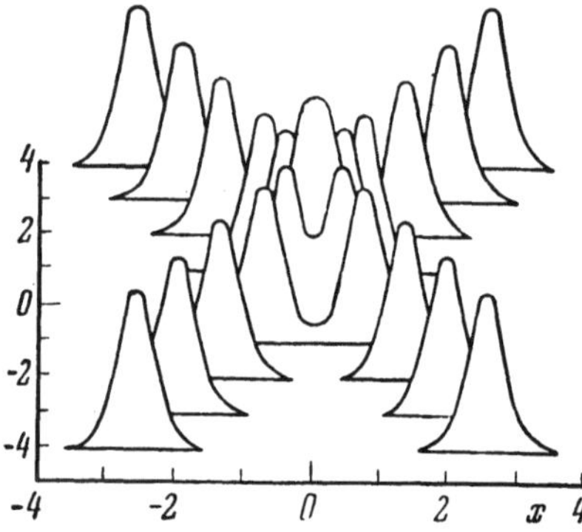

Fig.5.15.

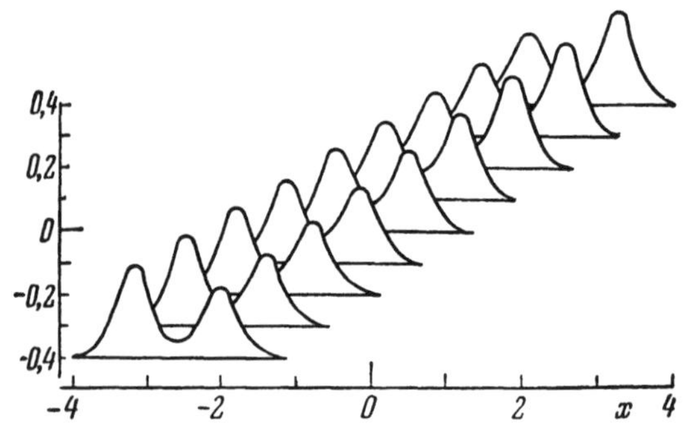

Fig.5.14.

Fig.5.13–15. Collison of solitons

The formulae for the wave equation, Toda chain and other equations, which have N-soliton solutions, have a similar form [5.4].

Let us consider the soliton gas as an assembly of N particles, each of them being fully characterized by prescribing two kinematic variables: the coordinate x_i and the velocity v_i. As distinct from the usual gas, $\dot{x}_i \neq v_i$, since the velocities of solitons during collision are conserved and the coordinates are changed. We shall assume that the soliton gas is sufficiently dilute, i.e. the changes in phases are small and occur discontinuously. Then the equations for x_i and v_i are obviously

$$\dot{x}_i = v_i + \sum_j \alpha(v_i, v_j)\, \delta(t - t_{ij}) ,$$
$$\dot{v}_i = 0 , \tag{5.7.2}$$

where t_{ij} is the time of collision. Taking into account the relative smallness of the changes of phase, we find

$$t - t_{ij} = \frac{x_i - x_j}{v_i - v_j} + O(\alpha) , \tag{5.7.3}$$

i.e. the moments of collision in (5.7.2) can be determined in the approximation of unperturbed motion of solitons.

Let $z_i^1 = x_i$, $z_i^2 = v_i$ and write (5.7.2) in the form

$$\dot{z}_i^\lambda = w_i^\lambda(z), \quad \lambda = 1, 2, \tag{5.7.4}$$

where, (5.7.2) and (5.7.3) being taken into account, $w_i^\lambda(z)$ has the form

$$w_i' = z_i^2 + \sum_j \alpha(z_i^2, z_j^2)\, \delta \left(\frac{z_i^1 - z_j^1}{z_i^2 - z_j^2}\right) .$$
$$w_i^2 = 0 . \tag{5.7.5}$$

It is known [5.9] that the complete statistical information about a system of N particles is contained in the N-particle distribution function $f_N(\dots , z_i, \dots .)$, which satisfies the Liouville equation

$$\frac{\partial f_N}{\partial t} + \sum_{i\lambda} \frac{\partial}{\partial z_i^\lambda} (w_i^\lambda f_N) = 0 . \tag{5.7.6}$$

In practice, it is sufficient to know, for most of the problems, the single particle distribution function $f(z)$, which is obtained from f_N by integrating over all z_i except $z_1 = z$. The Boltzmann equation for $f(z)$ is, as usual, obtained by the integration of the Liouville equation over all $z_i \neq z$.

Problem 5.7.1. Show that the equation for $f(x)$ has the form (compare [5.10], where it was obtained by a different approach)

$$\frac{\partial f}{\partial t} + \frac{\partial}{\partial x}(\tilde{v}f) = 0\,,$$

$$\tilde{v}(x, v, t) = v + \int |v - v'|\,\alpha(v, v')\,f(x, v', t)\,dv'\,. \tag{5.7.7}$$

It is essential that, in view of the nature of the problem, explicit expressions for f_N and f, can be found with the same accuracy to which the Liouville equation (5.7.4) and hence also (5.7.7) are valid. In fact, the original equations (5.7.2) are integrated and we have

$$x_i = v_i t + x_i^0 + \sum_j \alpha(v_i, v_j)\,\theta(t - t_{ij})\,,$$

$$v_i = \text{const.} \tag{5.7.8}$$

These expressions can be interpreted as a pointwise transformation from the variables $z_i(x, v)$ to the variables $Z_i(X, V)$ where

$$X_i = V_i t + X_i^0, \quad V_i = v_i\,. \tag{5.7.9}$$

With the accepted degree of accuracy, one can set

$$t - t_{ij} = \frac{X_i - X_j}{V_i - V_j}, \tag{5.7.10}$$

and the transformation takes the form

$$z_i^\lambda = Z_i^\lambda + \xi_i^\lambda(Z)\,, \tag{5.7.11}$$

where $\xi_i^\lambda(Z)$ is a small addition, defined by the expression (5.7.8).

Obviously, the inverse formula has the form

$$Z_i^\lambda = z_i^\lambda - \xi_i^\lambda(z)\,. \tag{5.7.12}$$

Let us associate with $f_N(z)$ the function $F_N(Z)$ with the help of the relation

$$\int_\omega f_N(z)\,dz = \int_\Omega F_N(Z)\,dZ\,, \tag{5.7.13}$$

where ω is an arbitrary region in $2N$-dimensional phase space of z's, Ω is its image, and z and Z denote the points of the spaces. Then

$$f_N(z) = F_N[Z(z)]\,\Delta(z)\,, \tag{5.7.14}$$

where

$$\Delta(z) = 1 - \operatorname{div}\xi(z) \tag{5.7.15}$$

is the Jacobian of the transformation.

Taking into account (5.7.11), we have

$$f_N(z) = F_N(z) - \operatorname{div}[\xi(z) F_N(z)]\,. \tag{5.7.16}$$

But $F_N(Z)$ is the distribution function of noninteracting particles and hence it can be represented in the form of single-particle distribution function

$$F(Z_t, t) = F_0(X_t - V_t t, V_t)\,. \tag{5.7.17}$$

This allows one to obtain, by integration, the explicit expression for a single-particle distribution function of interacting solitons.

Problem 5.7.2. Show that

$$f(x, v, t) = f_0(x - vt, v) + \frac{\partial}{\partial x}\{f_0(x - vt, v) \iint I(x - x', v, v', t) f_0(x' - v't, v')\, dx'\, dv'\}\,. \tag{5.7.18}$$

where

$$I(x - x', v, v', t) = \alpha(v, v')\left[\theta\left(\frac{x - x'}{v - v'} - t\right) - \theta\left(\frac{x - x'}{v - v'}\right)\right]. \tag{5.7.19}$$

The second term in (5.7.18) accounts for the contribution of collisions. Since it vanishes at $t = 0$, the function $f_0(x, v)$ is the initial value of $f(x, v, t)$. We leave it to the reader to verify that the collision term, as it should be, is equal to zero, if, when $t > 0$, the solitons with velocities v and v' cannot collide.

The distribution function of non-interacting solitons, as easily seen, has the assymptotic behavior

$$f_0(x - vt, v) = t^{-1}F_0(v)\,\delta\left(v - \frac{x}{t}\right) - t^{-2}F_1(v)\,\delta'\left(v - \frac{x}{t}\right) + \cdots \tag{5.7.20}$$

Here the $F_m(v)$ are the moments of the function $f_0(x, v)$, $F_0(v)$ coinciding with the invariant found in the preceding section. If we substitute this asymptotic decomposition in (5.7.18), we obtain the asymptotic behavior for $f(x, v, t)$ while the collision term makes a contribution only in the terms of the order of t^{-2} and higher.

5.8 Notes*

There exists extensive literature devoted to nonlinear waves in media with dispersion. For details we refer the reader to an excellent survey by *Kadomtsev* and *Karpman* [D.27], as well as to the book by *Karpman* [B4.28] and to review articles [B4.1].

*The citations refer to the bibliography to be given in [1.12]

6. Inverse Scattering Method

The method of inverse scattering is a new method of mathematical physics, which is applicable to a class of nonlinear wave equations. This chapter was written by V.E. Zakharov who made a significant contribution to the development of the method.

6.1 Basic Idea of the Method

The Korteweg-de Vries equation considered in the previous chapter plays a special role in the theory of nonlinear waves. During its investigation, the new method of the inverse scattering was discovered [6.1]. This method, which at the beginning looks like an artificial trick for solving a particular problem, gradually became an effective mathematical tool, which permits the exact solution of a number of difficult and physically interesting nonlinear problems, which did not allow an analytical approach earlier. The field of application of the method of the inverse scattering is constantly expanding. Following a historical tradition, we start the account of this method with the KdV-equation, which, for convenience, we write in the form

$$u_t - 6uu_x + u_{xxx} = 0\,. \tag{6.1.1}$$

Let us consider two linear differential operators L and A, defined on complex-valued functions $\psi(x)$, $-\infty < x < \infty$,

$$L = -\frac{d^2}{dx^2} + u(x,t)\,, \tag{6.1.2}$$

$$A = -4\mathrm{i}\frac{d^3}{dx^3} + 3\mathrm{i}\left(u\frac{d}{dx} + \frac{d}{dx}u\right), \tag{6.1.3}$$

and calculate the commutator of these operators

$$[L, A] = LA - AL = -\mathrm{i}(6uu_x - u_{xxx})\,. \tag{6.1.4}$$

In (6.1.2, 3) it is assumed that $u(x, t)$ is a solution of the KdV-equation (6.1.1). The operator L depends on time t as a parameter, and

$$\frac{\partial L}{\partial t} = \frac{\partial u}{\partial t}\,. \tag{6.1.5}$$

Comparing (6.1.4, 5) with (6.1.1), we see that the KdV-equation (6.1.1) is identical to the operator relation

$$\frac{\partial L}{\partial t} = \mathrm{i}\,[L, A]\,. \tag{6.1.6}$$

Let us assume further that the function $u(x)$ (using the quantum mechanical analogy, we shall call it the potential) satisfies the condition ($\alpha > 0$)

$$|u(x)| < \frac{c}{|x|^{2+\alpha}} \quad \text{as} \quad |x| \to \infty\,, \tag{6.1.7}$$

and let us consider the eigenvalue problem for the operator L

$$L\psi = -\frac{d^2\psi}{dx^2} + u(x, t)\,\psi = \lambda\psi\,. \tag{6.1.8}$$

The operator L has no more than a finite number of discrete eigenvalues $\lambda = -\kappa_n^2$ which are all negative and nondegenerate. We define the corresponding eigenfunctions ψ_n by normalization

$$\psi_n \to \mathrm{e}^{-\kappa_n x}, \quad x \to +\infty\,. \tag{6.1.9}$$

From this definition, it follows that the $\psi_n(x)$'s are real functions and

$$\psi_n(x) \to c_n \mathrm{e}^{\kappa_n x} \quad \text{as} \quad x \to -\infty\,. \tag{6.1.10}$$

The operator L also has a continuous spectrum $\lambda = k^2$ occupying the half axis $0 < \lambda < \infty$. We define eigenfunctions of the continuous spectrum with the normalization

$$\psi(x, k) \to \mathrm{e}^{\mathrm{i}kx}, \quad x \to +\infty\,. \tag{6.1.11}$$

These eigenvalues are twofold degenerate: apart from $\psi(x, k)$, to the same $\lambda = k^2$ there corresponds the eigenfunction $\psi^*(x, k)$.

The solutions of (6.1.8) determined by the condition (6.1.11) are called Jost functions. When $x \to -\infty$, they have the asymptotic behavior

$$\psi(x, k) \to a(k)\,\mathrm{e}^{\mathrm{i}kx} + b(k)\,\mathrm{e}^{-\mathrm{i}kx}\,. \tag{6.1.12}$$

We shall call the set of values λ_n, c_n, $a(k)$, $b(k)$ (again using quantum mechanical language) the scattering data for the operator L.

A Jost function $\psi(x, k)$ obviously satisfies the relation

$$\psi(x, -k) = \psi^*(x, k)\,.$$

It follows from here that

$$a^*(-k) = a(k), \quad b^*(-k) = b(k)\,. \tag{6.1.13}$$

Let us denote the Wronskian, for arbitrary functions f and g by $\{f, g\} = fg_x - gf_x$. If f and g are solutions of (6.1.8), then $(d/dx)\{f, g\} = 0$. Calculating $\psi(x, k)$, $\psi^*(x, k)$ as $x \to \pm\infty$, we obtain

$$|a(k)|^2 - |b(k)|^2 = 1\,. \tag{6.1.14}$$

Let us set $\lambda = \lambda_n$ in (6.1.8) and differentiate with respect to time. We obtain

$$\frac{\partial L}{\partial t}\psi_n + L\frac{\partial \psi_n}{\partial t} = \frac{\partial \lambda_n}{\partial t}\psi_n + \lambda_n\frac{\partial \psi_n}{\partial t}\,. \tag{6.1.15}$$

Substituting $\partial L/\partial t$ from (6.1.6) in (6.1.15) we find

$$(L - \lambda_n)\left(\frac{\partial \psi_n}{\partial t} + \mathrm{i}A\psi_n\right) = \frac{\partial \lambda_n}{\partial t}\psi_n\,. \tag{6.1.16}$$

Multiplying by ψ_n and integrating, we have

$$\int_{-\infty}^{\infty} \psi_n(L - \lambda_n)\left(\frac{\partial \psi_n}{\partial t} + \mathrm{i}A\psi_n\right)dx = \frac{\partial \lambda_n}{\partial t}\int_{-\infty}^{\infty} \psi_n^2\,dx\,. \tag{6.1.17}$$

As a consequence of the self-adjointness of the operator L, the left-hand side of (6.1.10) vanishes. Hence

$$\frac{\partial \lambda_n}{\partial t} = 0\,, \tag{6.1.18}$$

i.e. all the eigenvalues of the operator L are integrals of the equation (6.1.1).

Taking (6.1.18) into account, the relation (6.1.16) takes the form

$$(L - \lambda_n)\left(\frac{\partial \psi_n}{\partial t} + \mathrm{i}A\psi_n\right) = 0; \tag{6.1.19}$$

from here it follows that the quantity $\partial\psi_n/\partial t + \mathrm{i}A\psi_n$ at any moment is an eigenfunction of the operator L

$$\frac{\partial \psi_n}{\partial t} + \mathrm{i}A\psi_n = r(t)\,\psi_n\,,$$

where $r(t)$ is an arbitrary function of time.

Let us choose $r(t) = -4\kappa_n^3$. Then we have for ψ_n

$$\frac{\partial \psi_n}{\partial t} + 4\psi_{nxxx} - 6u\psi_{nx} - 3u_x\psi_n + 4\kappa_n^3\psi_n = 0\,. \tag{6.1.20}$$

Through this choice of $r(t)$, the function ψ_n, defined by the normalization (6.1.9) is a solution of (6.1.20). From (6.1.20), we obtain as $x \to -\infty$

$$\frac{\partial \psi_n}{\partial t} + 4\psi_{nxxx} = -4\kappa_n^3 \psi_n . \tag{6.1.21}$$

Substituting the asymptotic form (6.1.10) in (6.1.21), we see that

$$\frac{\partial c_n}{\partial t} + 8\kappa_n^3 c_n = 0; \quad c_n(t) = c_n(0)\, \mathrm{e}^{-8\kappa_n^3 t} . \tag{6.1.22}$$

Analogously, differentiating (6.1.8) with $\lambda = k^2$ being constant, we find that $\psi(x, k)$ satisfies the equation

$$(L - k^2)\left(\frac{\partial \psi}{\partial t} + \mathrm{i}A\psi\right) = 0 ,$$

whence

$$\frac{\partial \psi}{\partial t} + \mathrm{i}A\psi = \theta(t)\, \psi + \xi(t)\, \psi^* .$$

Assuming that the Jost function defined by the asymptotic form (6.1.11), satisfies this equation, we obtain

$$\theta(t) = -4\mathrm{i}k^3; \quad \xi(t) = 0 .$$

We have for ψ, as $x \to -\infty$

$$\frac{\partial \psi}{\partial t} + 4\frac{\partial^3 \psi}{\partial x^3} = -4\mathrm{i}k^3 \psi . \tag{6.1.23}$$

Substituting the asymptotic form (6.1.12) in (6.1.23), we have

$$\frac{\partial a(k)}{\partial t} = 0, \quad \frac{\partial b(k)}{\partial t} = -8\mathrm{i}k^3 b(k) ,$$

$$b(k, t) = b(k, 0)\, \mathrm{e}^{-8\mathrm{i}k^3 t} . \tag{6.1.24}$$

The formulae (6.1.18, 22 and 24) show that if $u(x, t)$ satisfies the KdV-equation, then the scattering data for the operator L are changed in time according to a very simple law. This gives the possibility of solving the Cauchy problem for the KdV-equation, following the scheme:

$$u(x, 0) \xrightarrow{I} \lambda_n, c_n(0), a(k), b(k, 0) \xrightarrow{II} \lambda_n , c_n(t), a(k), b(k, t) \xrightarrow{III} u(x, t) . \tag{6.1.25}$$

In this scheme, the first and the third stages are nontrivial. In the first stage, the scattering data must be found from the given potential $u(x, 0)$, i.e. a direct scattering problem is to be solved. In the last stage, it is necessary to reconstruct the potential $u(x, t)$ from the given scattering data. The solution of this problem is the subject of the inverse scattering problem for the operator L; it has given its name to the method.

Although neither the solution of the direct nor, moreover, the solution of the inverse problem for the operator L is a simple task, the possibility of application of the scheme (6.1.25) to the KdV-equation contains in itself great simplification, because the scheme (6.1.25) reduces the solution of a nonlinear partial differential equation to sufficiently well studied operations on ordinary linear differential equations.

It is clear from what has been stated above that the method of the inverse scattering is applicable for those equations, for which there exists a pair of operators L and A [(L, A)-pair] such that the operational relation (6.1.6) is identical to the equation under investigation. It is necessary that the direct and inverse problems of scattering for the operator L be uniquely solvable. The operators L and A need not be differential, they can be difference and even (at least the operator A) integral as well. In these cases the equation under consideration is differential, difference or integro-differential. If one defines the operators L and A over vector functions, one can extend the method to a system of equations.

It is clear that only exceptional equations possess an (L, A)-pair and can be solved by the inverse scattering method. Fortunately, however, among the solvable equations are those which present basic interest from the point of view of their physical and mechanical applications. Examples of such equations will be given in Sect. 6.9.

6.2 Inverse Scattering Problem for the Operator $L = -d^2/dx^2 + u(x)$

In the present section, we shall investigate the problem of the reconstruction of the potential $u(x)$ from the scattering data $a(k)$, $b(k)$, λ_n, c_n. To achieve this let us consider along with the Jost function $\psi(x, k)$, the function $\varphi(x, k)$ defined as the solution of (6.1.8) with the boundary condition

$$\varphi(x, k) \to e^{ikx} \quad \text{as} \quad x \to -\infty\,, \tag{6.2.1}$$

and also the Jost functions

$$\bar{\psi}(x, k) \to e^{-ikx} \quad \text{as} \quad x \to +\infty\,,$$

$$\bar{\varphi}(x, k) \to e^{-ikx} \quad \text{as} \quad x \to -\infty\,.$$

All four Jost functions can be defined for complex k's; obviously, on the real k- axis

$$\tilde{\psi}(x, k) = \psi^*(x, k), \quad \tilde{\varphi}(x, t) = \varphi^*(x, k).$$

The pairs of functions $(\psi, \tilde{\psi})$ and $(\varphi, \tilde{\varphi})$ are pairs of linearly independent solutions of (6.1.8). On the real axis, from the asymptotic relations (6.1.12), the exact relations

$$\begin{aligned} \psi(x, k) &= a(k)\, \varphi(x, k) + b(k)\, \tilde{\varphi}(x, k), \\ \tilde{\psi}(x, k) &= b^*(x, k)\, \varphi(x, k) + a^*(k)\, \tilde{\varphi}(x, k) \end{aligned} \tag{6.2.2}$$

follow. Solving these equations with respect to $\tilde{\varphi}(x, k)$ we obtain

$$\tilde{\varphi}(x, k) = a(k)\, \tilde{\psi}(x, k) - b^*(k)\, \psi(x, k). \tag{6.2.3}$$

Let us ascertain that the Jost function $\psi(x, k)$ can be represented in the form

$$\psi(x, k) = e^{ikx} + \int_x^\infty K(x, y)\, e^{iky}\, dk, \tag{6.2.4}$$

where $K(x, y)$ is some kernel. The representation (6.2.4) is called a triangular representation. Substituting (6.2.4) in (6.1.8), we obtain after simple, though long, calculations

$$\int_x^\infty \left[\frac{\partial^2 K(x, y)}{\partial x^2} - \frac{\partial^2 K(x, y)}{\partial y^2} - u(x)\, K(x, y)\right] e^{iky}\, dy \\ - e^{ikx}\left[u(x) + 2\frac{d}{dx} K(x, x)\right] = 0. \tag{6.2.5}$$

Since this equality is satisfied identically in k, the kernel $K(x, y)$ must satisfy the hyperbolic equation

$$\frac{\partial^2 K(x, y)}{\partial x^2} - \frac{\partial^2 K(x, y)}{\partial y^2} = u(x)\, K(x, y) \tag{6.2.6}$$

and the condition on the characteristic

$$u(x) + 2\frac{d}{dx} K(x, x) = 0. \tag{6.2.7}$$

If we also impose the condition of decreasing as $x \to +\infty$ along the set of characteristics $y = x + c$

$$K(x, x + z) \to 0 \quad \text{as} \quad x \to +\infty, \tag{6.2.8}$$

then we obtain the Cauchy-Goursat problem for the equation (6.2.6). This problem (see e.g. [6.2]) is uniquely solvable.

As $x \to +\infty$, (6.2.6) degenerates into the wave equation and for its arbitrary solution $K(x, y)$, we have

$$K(x, y) \to K_1(x + y) + K_2(x - y)\,. \tag{6.2.9}$$

The condition (6.2.6) means that

$$K_2(\xi) \equiv 0, \quad K_1(\xi) \to 0 \quad \text{as} \quad \xi \to +\infty\,. \tag{6.2.10}$$

It follows from here that the solution of the Cauchy-Goursat problem (6.2.6–8) determines the triangular representation (6.2.4) of some solution $\psi(x, k)$ of (6.1.8) (the integral in (6.2.4) converges!). Since as $x \to +\infty$, we have $K(x, y) \to K_1(x + y)$, where $K_1(\xi)$ is an asymptotically vanishing function, then this solution has the asymptotic form

$$\psi(x, k) \simeq \mathrm{e}^{ikx} + \mathrm{e}^{ikx}\, O\left(\frac{1}{x}\right) \quad \text{as} \quad x \to +\infty\,, \tag{6.2.11}$$

i.e. it is a Jost function.

From the existence of the representation (6.2.4) for a Jost function $\psi(x, k)$, it follows that this function is analytic in the upper half k-plane and has the asymptotic behavior

$$\psi(x, k) \simeq e^{ikx}\left[1 + \mathrm{O}\left(\frac{1}{k}\right)\right], \quad |k| \to \infty, \quad \mathrm{Im}\{k\} > 0\,. \tag{6.2.12}$$

From the relation $\tilde{\psi} = \psi^*$ for $\mathrm{Im}\{k\} = 0$, it follows that the function $\tilde{\psi}(x, k)$ has the triangular representation

$$\tilde{\psi}(x, k) = \mathrm{e}^{-ikx} + \int_x^\infty K(x, y)\,\mathrm{e}^{-iky}\,dy \tag{6.2.13}$$

with the same kernel $K(x, y)$ and is analytic in the lower halfplane.

Quite analogously it can be proved that the functions $\varphi(x, k)$ and $\tilde{\varphi}(x, k)$ have a triangular representation with a kernel $L(x, y)$

$$\varphi(x, k) = \mathrm{e}^{ikx} + \int_{-\infty}^{x} L(x, y)\,\mathrm{e}^{iky}\,dy\,, \tag{6.2.13a}$$

$$\tilde{\varphi}(x, k) = \mathrm{e}^{-ikx} + \int_{-\infty}^{x} L(x, y)\,\mathrm{e}^{-iky}\,dy\,. \tag{6.2.13b}$$

It follows from the last formula that $\tilde{\varphi}(x, k)$ is analytic in the upper halfplane and has there the asymptotic form

$$\bar{\varphi}(x, k) \simeq e^{-ikx}\left[1 + O\left(\frac{1}{k}\right)\right]. \qquad (6.2.14)$$

For the coefficient $a(k)$, there is an easily verifiable formula

$$a(k) = -\frac{\{\psi . \bar{\varphi}\}}{2ik}, \qquad (6.2.15)$$

from which it follows that $a(k)$ is also analytic in the upper halfplane.

Substituting the asymptotic formulae (6.2.12–14) for $\bar{\varphi}$ and ψ into (6.2.15), we can convince ourselves that the function $a(k)$ has the asymptotic form $a(k) \simeq 1 + O(1/k)$.

It follows from the analyticity of $a(k)$ that the principal term of the asymptotic expansion of the function $\psi(x, k)$ in the upper halfplane ($k = k_1 + i\kappa$), as $x \to -\infty$, is exponentially increasing:

$$\psi(x, k) \to a(k_1 + i\kappa)\, e^{(ik_1 - \kappa)x}, \quad x \to -\infty, \quad \kappa > 0. \qquad (6.2.16)$$

From the analyticity of $\psi(x, k)$ and the definition (6.1.9) of the eigenfunction ψ_n, it follows that $\psi_n = \psi(x, i\kappa_n)$ and that ψ_n has the triangular representation

$$\psi_n = e^{-\kappa_n x} + \int_x^\infty K(x, y)\, e^{-\kappa_n y}\, dy. \qquad (6.2.17)$$

Since $\psi_n \to 0$, as $x \to -\infty$, the function $a(k)$ vanishes at the points $k = i\kappa_n$, and since the discrete eigenvalues of the operator L are non-degenerate, these zeros are simple.

From the asymptotic formula (6.1.10), it follows that

$$\psi_n = \psi(x, i\kappa_n) = c_n \bar{\varphi}(x, i\kappa_n).$$

Let us consider two solutions g_1 and g_2 of the two equations (6.1.8) with one and the same $u(x)$, but with different eigenvalues λ_1 and λ_2. Multiplying the first equation by g_2 and subtracting from it the second one multiplied by g_1, we obtain

$$\frac{d}{dx}\{g_2, g_1\} = (\lambda_2 - \lambda_1)\, g_2 g_1. \qquad (6.2.18)$$

Differentiating g_1 with respect to λ_1 and letting $\lambda_2 \to \lambda_1 = \lambda$ we find for two arbitrary solutions of (6.1.8)

$$\frac{d}{dx}\left\{g_2, \frac{\partial g_1}{\partial \lambda}\right\} = -g_2 g_1. \qquad (6.2.19)$$

Let $g_1 = \psi(x, i\kappa)$, $g_2 = \bar{\varphi}(x, i\kappa)$, and $\partial g_1/\partial\lambda = \partial\psi/\partial\lambda = -(2\kappa)^{-1}\, \partial\psi(x, i\kappa)/\partial\kappa$. Further, let $\kappa = \kappa_n$. Then $g_2 = 1/c_n^{-1}\, \psi(x, i\kappa_n)$ and $\{g_2, \partial g_1/\partial\lambda\} \to 0$ as $x \to +\infty$.

Integrating the relation (6.2.19), we obtain

$$\left\{\tilde{\varphi}, \frac{\partial \psi}{\partial \kappa}\right\}_{\substack{\kappa=\kappa_n \\ x\to-\infty}} = -\frac{2k_n}{c_n} \int_{-\infty}^{\infty} \psi_n^2(x)\, dx\,. \tag{6.2.19a}$$

As $x \to -\infty$, $\kappa = \kappa_n$, see (6.2.15, and $\partial\psi/\partial x \to (\partial a/\partial\kappa) \exp(-\kappa_n x)$. Substituting this expression into (6.2.19a), we obtain

$$\left.\frac{\partial a}{\partial \kappa}\right|_{\kappa_n} = \mathrm{i}\left.\frac{\partial a}{\partial k}\right|_{k=\mathrm{i}\kappa_n} = \frac{2\kappa_n}{c_n} \int_{-\infty}^{\infty} \psi_n^2(x)\, dx\,. \tag{6.2.20}$$

Let us divide (6.2.3) by $a(k)$, subtract $\exp(-\mathrm{i}kx)$ from both sides of the equality obtained, multiply the result by $(1/2\pi) \exp(\mathrm{i}ky)$ and integrate from $k = -\infty$ to $k = \infty$:

$$\frac{1}{2\pi}\int_{-\infty}^{\infty}\left(\frac{\tilde{\varphi}(x,k)}{a(k)} - \mathrm{e}^{-\mathrm{i}kx}\right)\mathrm{e}^{\mathrm{i}ky}\,dk = \frac{1}{2\pi}\int_{-\infty}^{\infty}[\tilde{\varphi}(x,k) - \mathrm{e}^{-\mathrm{i}kx}]\,\mathrm{e}^{\mathrm{i}ky}\,dk$$
$$- \frac{1}{2\pi}\int_{-\infty}^{\infty}\frac{b^*(k)}{a(k)}\psi(x,k)\,\mathrm{e}^{\mathrm{i}ky}\,dy\,. \tag{6.2.21}$$

When $y > x$, in the left-hand side, we are dealing with an integral of a function which is analytic in the upper halfplane, decreases exponentially along any complex direction, and is of the order of $O(1/k)$, as $k \to \pm\infty$. This integral is equal to the sum of the residues.

$$\frac{1}{2\pi}\int_{-\infty}^{\infty}\left(\frac{\tilde{\varphi}(x,k)}{a(k)} - \mathrm{e}^{-\mathrm{i}kx}\right)\mathrm{e}^{\mathrm{i}ky}\,dk = \mathrm{i}\sum_n \frac{\tilde{\varphi}(x,\mathrm{i}\kappa_n)}{a_k'(\mathrm{i}\kappa_n)}$$
$$= \mathrm{i}\sum_n \frac{\psi(x,\mathrm{i}\kappa_n)}{c_n a_k'(\mathrm{i}\kappa_n)} = -\sum M_n^2\psi(x,\mathrm{i}\kappa_n)\,, \tag{6.2.22}$$

where $M_n^2 = \left(2\kappa \int_{-\infty}^{\infty} \psi_n^2(x)\,dx\right)^{-1}$. Substituting in (6.2.21, 22) the triangular representations (6.2.4, 13), one can find after simple transformations, that the kernel $K(x, y)$ satisfies the integral equation

$$K(x,y) = F(x+y) + \int_x^{\infty} K(x,s)F(s+y)\,ds\,, \tag{6.2.23}$$

where

$$F(\xi) = -\sum M_n^2 \mathrm{e}^{-\kappa_n\xi} + \frac{1}{2\pi}\int_{-\infty}^{\infty} s(k)\,\mathrm{e}^{\mathrm{i}k\xi}\,dk\,,$$
$$s(k) = \frac{b^*(k)}{a(k)}\,. \tag{6.2.24}$$

Equation (6.2.23) (the so-called Marchenko equation) allows the determination of the kernel of the triangular representation $K(x, y)$ and, along with this, the potential $u(x) = -2\, d/dxK(x, x)$ in terms of the scattering data κ_n, c_n, $a(k)$, $b(k)$. This equation can also be deduced by another simpler method.

To this end, it is necessary to solve (6.2.6) by the Fourier method, expanding $K(x, y)$ in eigenfunctions of the operator L. Choosing ψ_n, $\psi(x, k)$, $\tilde{\psi}(x, k)$ to be a complete set, we have

$$K(x, y) = \sum a_n\psi_n(x)\, e^{-\kappa_n y} + \int_{-\infty}^{\infty} c(k)\, \psi(x, k)\, e^{iky}\, dk$$

$$+ \int_{-\infty}^{\infty} s_1(k)\, \tilde{\psi}(x, k)\, e^{iky}\, dy\,.$$

From the condition (6.2.10) and asymptotic form $\tilde{\psi}(x, k) \sim \exp(-ikx)$, it follows that $s_1(k) = 0$. Finally, we obtain

$$K(x, y) = \sum a_n\psi_n(x)\, e^{-\kappa_n y} + \int_{-\infty}^{\infty} c(k)\, \psi(x, k)\, e^{iky}\, dk\,. \tag{6.2.25}$$

Substituting the triangular representation (2.4) in (6.2.25) we arrive at (6.2.23) where

$$F(\xi) = \sum_n a_n e^{-\kappa_n \xi} + \int_{-\infty}^{\infty} c(k)\, e^{ik\xi}\, d\xi\,. \tag{6.2.26}$$

With such a deduction of the Marchenko equation establishing, the connection between the function F and the scattering data is nontrivial. Let us perform this for the simplest case, when the operator L has no discrete eigenvalues. For this, let us note that as $x \rightarrow -\infty$, (6.2.6) degenerates into the wave equation and its solution takes the form (6.2.9) where

$$K_1(\xi) = \int c(k)\, a(k)\, e^{ik\xi}\, dk\,,$$
$$K_2(\xi) = \int c(k)\, b(k)\, e^{ik\xi}\, dk\,. \tag{6.2.27}$$

Substituting $K(x, y)$ in the form (6.2.9) in (6.2.23) and taking $y \simeq x$, we note that the integration is carried out over $s \simeq -y \gg x$. Therefore, the lower limit of the integration can be replaced by $-\infty$. We obtain

$$K_2(\xi) = \int_{-\infty}^{\infty} K_1(\xi + \xi')\, F(\xi')\, d\xi'\,. \tag{6.2.28}$$

This equation can easily be solved by the Fourier method. Using (6.2.26, 27), we find

$$c(k) = \frac{1}{2\pi}\, s(k) = \frac{1}{2\pi}\frac{b^*(k)}{a(k)}\,,$$

which agrees with the result (6.2.24) obtained earlier. We leave it to the reader, as an exercise, to prove the formula $a_n = c_n^2/a'_{\kappa_n} = -M_n^2$, which ends the solution of the inverse scattering problem by a more elementary method.

The Marchenko equation (6.2.23) is a Fredholm equation of the first kind at each x, with kernel equal to $-F(s + y)$. The discrete spectrum of the operator L makes a positive contribution to the kernel and this equation is uniquely solvable, if the values of $|s(k)|$ are sufficiently small. A more accurate estimate shows that it is in fact the condition $|s(k)| < 1$ that leads to this being valid [due to the equality (6.1.17)] for arbitrary potentials which decrease as $|x| \to \infty$.

6.3 *N*-Soliton Solutions of the KdV-Equation

From the formulae (6.1.9), (6.1.24), (6.2.22) and (6.2.24), it follows that

$$M_n^2(t) = M_8^2(0)\,\mathrm{e}^{8\kappa_n^3 t}, \quad s(k,t) = s(k,0)\,\mathrm{e}^{8ik^3 t}\,. \tag{6.3.1}$$

and that the function $F(\xi, t)$ satisfies the equation

$$\frac{\partial F}{\partial t} + 8\frac{\partial^3 F}{\partial \xi^3} = 0\,. \tag{6.3.2}$$

Instead of direct application of the scheme (6.1.24), the inverse scattering method can be used as a method of finding exact solutions of the KdV-equation. In fact, choosing arbitrary $M_n^2(0) > 0$, κ_n, $s(k,0)| < 1/2\pi$ and constructing with their help, by (6.2.24), the function $F(\xi)$ or simply choosing as $F(\xi, t)$, a solution of (6.3.2), we shall find an exact solution of the KdV-equation, determining $K(x, y)$ at all moments t through the solutions of the Marchenko equation.

As the first such example, we obtain the soliton, already considered earlier (Sect. 5.1). To see this let us set $s(k,0) = s(k,t) \equiv 0$ and

$$F(\xi) = -\,M_n^2(0)\,\mathrm{e}^{-\kappa\xi + 8\kappa^3 t}\,. \tag{6.3.3}$$

The Marchenko equation is solved by the method of separation of variables. Substituting (6.3.3) into (6.2.24) and setting

$$K(x,y) = \varphi(x,t)\,\mathrm{e}^{-\kappa y}\,,$$

we obtain

$$K(x,x) = \varphi(x,t)\,\mathrm{e}^{-\kappa x} = -\frac{2\kappa}{1 + \mathrm{e}^{2\kappa(x - vt - x_0)}}\,,$$

$$v = 4\kappa^2, \quad x_0 = -\frac{1}{2\kappa}\ln\frac{M_n^2(0)}{2\kappa}\,,$$

$$u(x,t) = -\,2\frac{d}{dx}K(x,x) = -\frac{2\kappa^2}{\mathrm{ch}^2 2\kappa(x - vt - x_0)}\,. \tag{6.3.4}$$

The soliton has the form of a "well," which propagates to the right with the velocity $v = 4\kappa^2$. The negativity of the soliton is connected with the negativity of the nonlinear term in (6.1.1). Such a choice of sign is made in order for the soliton to correspond to a potential well, when the equation $-d^2\psi/dx^2 + u(x)\psi = \lambda\psi$ is interpreted as the quantum-mechanical Schrodinger equation.

In the quantum-mechanical language the condition $s(k) = 0$ means that there is no scattering from the well at any k; thus a soliton is a "non-reflecting" potential.

In the case of a non-reflecting potential of a more general form, the operator L has N discrete eigenvalues and the potential is characterized by a set of $2N$ constants $-\kappa_n$ and $M_n^2(0)$. For such potentials, the function $F(\xi)$ has the form

$$F(\xi) = -\sum_{n=1}^{N} M_n^2(t)\, \mathrm{e}^{-\kappa_n \xi}\,. \tag{6.3.5}$$

Let us represent $K(x, y)$ in the form

$$K(x, y) = \sum_{n=1}^{N} K_n(x)\, \mathrm{e}^{-\kappa_n y}\,. \tag{6.3.6}$$

After substitution in (6.2.23) we obtain a system of linear algebraic equations

$$K_n(x, t) + M_n^2 \sum_{m=1}^{N} \frac{K_m(x, t)\, \mathrm{e}^{-(\kappa_n+\kappa_m)x}}{\kappa_n + \kappa_m} = -M_n^2\, \mathrm{e}^{-\kappa_n x}\,. \tag{6.3.7}$$

The matrix of the system is positive definite, and hence the system is uniquely solvable. Let Δ be the determinant of this system

$$\Delta = \det \left\| \delta_{nm} + M_n^2 \frac{\mathrm{e}^{-(\kappa_n+\kappa_m)x}}{\kappa_n + \kappa_m} \right\|. \tag{6.3.8}$$

From (6.3.6), we have

$$K(x, x) = \sum_{n=1}^{N} K_n(x)\, \mathrm{e}^{-\kappa_n x}\,. \tag{6.3.9}$$

Differentiating the logarithm of the determinant (6.3.8) with respect to x, we have

$$\frac{\partial \ln \Delta}{\partial x} = \sum_{n=1}^{N} \frac{\Delta_n}{\Delta} = \sum_{n=1}^{N} \mathrm{e}^{-\kappa_n x} \frac{\tilde{\Delta}_n}{\Delta}\,.$$

Here the Δ_n are determinants, which are obtained from Δ by differentiating its n-th column. Performing this differentiation explicitly, we have for the m-th element of the column,

$$-M_n^2\, \mathrm{e}^{-(\kappa_n+\kappa_m)x}\,.$$

The corresponding element of the column of the determinant $\tilde{\Delta}_n$ has the form $-M_m^2 \exp(-\kappa_m x)$ and coincides with the free term of the n-th equation of the system (6.3.7). Thus, according to the Kramer's formula, $\tilde{\Delta}_n/\Delta = K_n(x, t)$ and finally

$$K(x, x) = \frac{\partial}{\partial x} \ln \Delta ,$$

$$u(x) = -2 \frac{\partial^2}{\partial x^2} \ln \Delta . \tag{6.3.10}$$

The formulae (6.3.10), first obtained by Key and Moses [6.3], give an explicit expression for the non-reflecting potentials with N discrete eigenvalues. The analysis of these formulae shows that they describe only negative potentials $[u(x) < 0]$. However, it is obvious from the quantum-mechanical reasoning that the absence of scattering is possible only in the absence of "potential peaks", i.e. regions where $u(x) > 0$.

Let us study the asymptotic form of a "non-reflecting" initial condition for the KdV-equation, as $t \to \pm\infty$. Let us introduce the notations

$$A_n(x, t) = K_n(x, t)\, e^{-\kappa_n x} ,$$

$$\xi_n = x - 4\kappa_n^2 t - \frac{1}{2\kappa} \ln |M_n(0)|^2 . \tag{6.3.11}$$

The system (6.3.7) is rewritten in the form

$$A_n(x, t)\, e^{2\kappa_n \xi_n} + \sum_{m=1}^{N} \frac{A_m(x, t)}{\kappa_n + \kappa_m} + 1 = 0 , \tag{6.3.12}$$

$$K(x, x) = \sum_{n=1}^{N} A_n(x, t) . \tag{6.3.13}$$

Let us arrange the numbers κ_n in increasing order, $\kappa_{n+1} > \kappa_n$, and investigate the asymptotic form of a solution of the system (6.3.12) along straight lines $x - vt = \xi$ in the x, t-plane. First, assume that the velocity satisfies the inequality $4\kappa_p^2 < v < 4\kappa_{p+1}^2$ and does not coincide with any of the characteristic velocities $4\kappa_p^2$. Now, as $t \to +\infty$, $\xi_n \to +\infty$, if $n \leqslant p$, and $\xi_n \to -\infty$, if $n \geqslant p+1$. Obviously, also $A_n \to 0$ if $n \leqslant p$ and $A_n \to A_n^{(0)}$ if $n \geqslant p+1$, and the $A_n^{(0)}$ satisfy the system of equations

$$\sum_{m=p+1}^{N} \frac{A_m^{(0)}}{\kappa_n + \kappa_m} + 1 = 0 . \tag{6.3.14}$$

The solution of this system leads, obviously, to $K(x, x) =$ constant and $u(x)=0$. Thus, as $t \to \infty$, the solution, in the (x, t)-plane is concentrated near the straight lines $x - 4\kappa_n^2 t$. This is also true, as $t \to -\infty$.

Now let $v = 4\kappa_2^2$. As earlier, as $t \to \infty$, we have $\xi_n \to \infty$, $A_n \to 0$ if $n \leqslant p$ and $\xi_n \to -\infty$ when $n \geqslant p + 1$. The system (6.3.12) takes the form

$$\sum_{m=p}^{N} \frac{A_m(x, t)}{\kappa_n + \kappa_m} + 1 + A_p\, e^{2\kappa_p \xi_p} \delta_{np} = 0 . \qquad (6.3.15)$$

Similarly, as $t \to -\infty$, we have

$$\sum_{m=1}^{p} \frac{A_m(x, t)}{\kappa_n + \kappa_m} + 1 + A_p\, e^{2\kappa_p \xi_p} \delta_{np} = 0 . \qquad (6.3.16)$$

The solution of the systems depends only on ξ_p, i.e. it is constant along straight lines $x - 4\kappa_p^2 t = \text{const}$. Let us show that the systems (6.3.15) and (6.3.16) describe solitons.

To this end it is necessary to obtain two auxiliary formulae. Let $\kappa_1, \ldots, \kappa_n$ be N given numbers. Let us consider the rational function

$$a(\kappa) = \frac{Q(\kappa)}{P(\kappa)}, \quad Q(\kappa) = \prod_{n=1}^{N} (\kappa - \kappa_n), \quad P(\kappa) = \prod_{n=1}^{N} (\kappa + \kappa_n)$$

and expand it into simple fractions

$$a(\kappa) = 1 + \sum \frac{c_m}{\kappa - \kappa_m} . \qquad (6.3.17)$$

The quantities c_n coincide with the residues of the function $a(\kappa)$ at the points $\kappa = -\kappa_n$:

$$c_n = \frac{Q(-\kappa_n)}{P'(-\kappa_n)} = \frac{(-1)^N \prod\limits_{m} (\kappa_m + \kappa_n)}{(-1)^{N-1} \prod\limits_{m \neq n} (\kappa_m - \kappa_n)} = -a_n,$$

$$a_n = \frac{\prod\limits_{m} (\kappa_m + \kappa_n)}{\prod\limits_{m \neq n} (\kappa_m - \kappa_n)} . \qquad (6.3.18)$$

Substituting $\kappa = \kappa_k$ in (3.17), we obtain the formula

$$\sum \frac{a_m}{\kappa_k + \kappa_m} = 1 . \qquad (6.3.19)$$

Let us consider an arbitrary polynomial $H(\kappa)$ of degree $N - 1$ and write it in the form

$$H(\kappa) = \sum_k \frac{H(\kappa_k)}{P(\kappa_k)} \frac{Q(\kappa)}{\kappa - \kappa_k} a_k . \qquad (6.3.20)$$

Let $H(\kappa) = \dfrac{P(\kappa)}{\kappa + \kappa_p}$; then

$$\frac{P(\kappa)}{Q(\kappa)(\kappa + \kappa_p)} = \sum_k \frac{a_k}{(\kappa - \kappa_k)(\kappa_k + \kappa_p)}. \tag{6.3.21}$$

Substituting $\kappa = -\kappa_q$ in (6.3.21), we obtain

$$\sum_k \frac{a_q a_k}{(\kappa_q + \kappa_k)(\kappa_k + \kappa_p)} = \delta_{pq}, \tag{6.3.22}$$

which means that the matrix $\|a_n a_m/(\kappa_n + \kappa_m)\|$ is the inverse of the matrix $\|1/\kappa_n + \kappa_m\|$.

Applying now the formulae (6.3.19), (6.3.22) to the system (6.3.15), we obtain

$$A_m = -a_m\left(1 + \frac{a_p}{\kappa_m + \kappa_p} A_p\, e^{2\kappa_p \xi_p}\right),$$

and in particular,

$$A_p = -a_p\left(1 + \frac{a_p}{2\kappa_p} A_p\, e^{2\kappa_p \xi_p}\right). \tag{6.3.23}$$

Apart from this we also find

$$\frac{d}{dx}\sum A_m = -\frac{d}{dx} a_p A_p\, e^{2\kappa_p \xi_p}. \tag{6.3.24}$$

Finally, we obtain along the straight lines $x - v_p t = \text{constant}$

$$u_p(x) = -\frac{2\kappa_p^2}{\operatorname{ch}^2 2\kappa_p(x - v_p t - x_{0p}^+)}, \tag{6.3.25}$$

$$x_{0p}^+ = x_{0p} + \frac{1}{\kappa_p}\ln a_p = x_{0p} - \frac{1}{\kappa_p}\sum_{m=p+1}^{N} \ln\frac{\kappa_m + \kappa_n}{\kappa_m - \kappa_p}. \tag{6.3.26}$$

Analogously, as $t \to -\infty$, we find

$$u_p(x) = -\frac{2\kappa_p^2}{\operatorname{ch}^2 2\kappa_p(x - v_p t - x_0^-)}, \tag{6.3.27}$$

$$x_0^- = x_0 - \frac{1}{\kappa_p}\sum_{m=1}^{p-1} \ln\frac{\kappa_m + \kappa_p}{\kappa_p - \kappa_m}. \tag{6.3.28}$$

The formulae (6.3.25) and (6.3.27) show that an arbitrary non-reflecting initial condition, which has N eigenvalues, is decomposed into N noninteracting solitons as $t \to \pm\infty$. The amplitudes and velocities of the asymptotic solitons are determined by the conserved characteristic numbers and are identical as $t \to \pm\infty$. This allows to denote the exact solution of the KdV-equation, given by (6.3.10), as an N-soliton solution, and to consider it as describing the process of scattering of N-solitons on one another. During the scattering process, the solitons do not exchange energy, but only displacements of the coordinates of their centers occur. As $t \to -\infty$, the solitons are arranged in the order of decreasing velocities, and they are arranged in the inverse order, as $t \to +\infty$. The total displacement of the coordinate of the centre of the p-th soliton is

$$(\delta x_0)_p = x_{0p}^{+} - x_{0p}^{-} = \frac{1}{2\kappa_p}\left(\sum_{m=1}^{p-1} \ln \frac{\kappa_m + \kappa_p}{\kappa_p - \kappa_m} - \sum_{m=p+1}^{N} \ln \frac{\kappa_m + \kappa_p}{\kappa_m - \kappa_p}\right), \tag{6.3.29}$$

from which it follows that the total displacement of the centre of the soliton is equal to the algebraic sum of displacements during pair collisions with other solitons. The displacement of the p-th soliton when it meets the m-th soliton has the form

$$(\delta x_p)_m = \frac{1}{2\kappa_p} \operatorname{sgn}\{\kappa_p - \kappa_m\} \ln \frac{\kappa_p + \kappa_m}{\kappa_p - \kappa_m}. \tag{6.3.30}$$

The faster soliton is displaced forward and the slower one is displaced backward thus the solitons' behavior is like that of repelling particles.

Formula (6.3.27) shows that the total displacement of the soliton is independent of the details of the interaction. In particular, collisions are not assumed to be double in the usual sense (the latter meaning that during a collision of two solitons, the others are assumed to be sufficiently far away). From (6.3.27), it follows that the effects of collisions of three or more solitons are absent. This surprising circumstance cardinally distinguishes a soliton from usual particles considered in mechanics.

If the velocities of two colliding solitons are close, the displacement diverges logarithmically. Such solitons cannot draw closer together than $\delta x \sim \kappa^{-1}\ln(\kappa/\Delta\kappa)$; at this distance, the solitons exchange velocities and withdraw. For the displacements of solitons, the formula

$$\kappa_p(\delta\kappa_p)_m + \kappa_m(\delta\kappa_m)_p = 0\,, \tag{6.3.31}$$

which expresses the conservation of the centre of mass at a collision, is applicable.

If the initial condition in the KdV-equation is a potential well $[u(x) < 0]$ with the characteristic depth u_0 and the characteristic dimension l, and if $u_0 l^2 \gg 1$, then the behavior of the initial condition can be approximated in time to within terms of the order $\exp[-(1/u_0 l^2)^{1/2}]$ by an N-soliton solution. As $t \to \infty$, such an initial condition is decomposed into a great number of solitons ($N \sim \sqrt{u_0} l \gg 1$), the amplitudes of which can be evaluated by applying the WKB-approximation to the initial potential. For all the solitons, the quantities κ_n, with the considered accuracy, are solutions of the equation

$$\int_{x_1}^{x_2} \sqrt{|u(x)| - \kappa_n^2}\, dx = \pi \left(n + \frac{1}{2}\right). \tag{6.3.32}$$

The integration in (6.3.32) is carried out between the zeros of the radicand.

6.4 Complete Integrability of the KdV-Equation

The KdV-equation (6.1.1) can be written in the form

$$u_t = \frac{\partial}{\partial x}\frac{\delta H}{\delta u}, \quad H = \int_{-\infty}^{\infty} \left(u^3 + \frac{1}{2} u_x^2\right) dx\,. \tag{6.4.1}$$

Here the symbol $\delta H/\delta u$ denotes the functional derivative of the functional $H[u]$ with respect to the argument u. If $H = \int p(u, u\,, \dots u^{(n)})\, dx$, then

$$\frac{\delta H}{\delta u} = \frac{\partial p}{\partial u} - \frac{\partial}{\partial x}\frac{\partial p}{\partial u_x} + \dots + (-1)^n \frac{\partial^n}{\partial x^n}\frac{\partial}{\partial u^{(n)}}\,. \tag{6.4.2}$$

In the case under consideration, $|u| \to 0$, as $|x| \to \infty$, and hence, (6.4.1) can be rewritten in the form

$$\int G(x - x')\frac{\partial u(x')}{\partial t}\, dx' = \frac{\delta H}{\delta u}\,, \tag{6.4.3}$$

where $G(\xi) = -G(-\xi) = -1/2 \operatorname{sgn}\{\xi\}$.

The KdV-equation in the form (6.4.3) shows that this is a Hamiltonian dynamical system with a continuous infinity of degrees of freedom.

When investigating Hamiltonian systems, it is very important to find out their unique integrals of motion (i.e. those, which do not contain time explicitly). According to the Liouville theorem, a Hamiltonian system with N degrees of freedom, having N independent integrals of motion, such that the Poisson brackets between the latter ones are equal to zero, is completely integrable, i.e. it permits the introduction of action-angle variables. In terms of these variables (P_n, Q_n), the Hamiltonian depends only on the generalized momenta P, all generalized momenta are conserved, and all generalized coordinates are cyclic.

The KdV-equation possesses a continuous infinity of conservation laws; as was shown in Sect. 6.1, the function $a(k)$, $-\infty < k < \infty$ is conserved. In the present section, we show that under the zero conditions at infinity this is sufficient, and the KdV-equation is a completely integrable system.

Let us rewrite the equation $L\psi = \lambda\psi$ with $\lambda = k^2$ in the form

$$\frac{d^2\psi}{dx^2} - u(x)\,\psi - k^2\psi = 0 \tag{6.4.4}$$

and represent the Jost function ψ in the form

$$\psi = \exp\left[ikx + \int_x^\infty s(k, x')\,dx'\right].$$

For $s(k, x)$ we obtain the Riccati equation

$$-\frac{1}{2ik}(s_x + s^2 - u) = s\,. \tag{6.4.5}$$

Let $\mathrm{Im}\,\{k\} > 0$. We then have the relation

$$\int_{-\infty}^{\infty} s(k, x)\,dx = -\ln a(k)\,, \tag{6.4.6}$$

which can also be continued to the real axis $\mathrm{Im}\,\{k\} = 0$.

Assuming $u(x)$ to be infinitely differentiable, let us investigate the asymptotic behavior of $\ln a$ with respect to its expansion in powers of $1/k$ on the real axis. In view of (6.1.14) we have

$$\ln a(k) = \ln \sqrt{1 + |b_k|^2} + \mathrm{i}\arg\{a(k)\}\,.$$

Due to the relation $a^*(-k) = a(k)$, we have $\arg\{a(-k)\} = -\arg\{a(k)\}$. The function $u(k)$ being infinitely differentiable, $b(k)$ diminishes as $k \to \infty$ more rapidly than any power. Therefore $\ln a(k)$ can be expanded in odd powers of $1/k$:

$$\ln a(k) \cong \sum \frac{(-1)^n}{(2ik)^{2n+1}} I_n = \mathrm{i}\arg\{a(k)\}\,. \tag{6.4.7}$$

Let

$$s(k, x) = \sum \frac{1}{(2ik)^m} s_m(x)\,.$$

Obviously,

$$I_n = (-1)^n \int_{-\infty}^{\infty} s_{2n+1}(x)\,dx\,.$$

For $s_n(x)$ we have the recursion relation

$$s_{n+1}(x) = -\frac{d}{dx}s_n(x) - \sum_{k=1}^{n-1} s_{n-k}(x)\,s_k(x); \quad s_1 = u(x)\,.$$

Let us give explicitly the first few s_n's, calculated according to this formula:

$$s_2 = -u_x; \quad s_3 = -u^2 + u_{xx}; \quad s_4 = -u_{xxx} + 4uu_x\,;$$
$$s_5 = u_{xxxx} - 6uu_{xx} - 5u_x^2 + 2u^3\,.$$

We see that $s_2(x)$ and $s_4(x)$ are total derivatives. The same is true for all even n and can be proved by induction. Returning to the quantities I_n, we have

$$I_0 = \int_{-\infty}^{\infty} u(x)\,dx; \quad I_1 = \int_{-\infty}^{\infty} u^2(x)\,dx; \quad I_2 = \int_{-\infty}^{\infty}[2u^3(x) + u_x^2(x)]\,dx\,. \qquad (6.4.8)$$

Due to the conservation of a(k), all the I_n are independent of time. We have shown that the KdV-equation possesses a countable set of conservation laws, which are integrals over polynomials in the function $u(x)$ and a finite number of its derivatives. The Hamiltonian of the KdV-equation is $H = I_2/2$.

Assume now that we have transformed the KdV-equation from the variable u to some variables P_k and Q_k. Calculating the derivative of P_k with respect to time we find

$$\begin{aligned}\frac{\partial P_k}{\partial t} &= \int_{-\infty}^{\infty} \frac{\delta P_k}{\delta u(x)}\frac{\partial u(x)}{\partial t}\,dk = \int_{-\infty}^{\infty}\frac{\delta P_k}{\delta u(x)}\frac{\partial}{\partial x}\frac{\delta H}{\delta u(x)}\,dx \\ &= \int dx\,\frac{\delta P_k}{\delta u(x)}\frac{\partial}{\partial x}\left\{\int dk'\left(\frac{\delta H}{\delta P_k'}\frac{\delta P_k'}{\delta u(x)} + \frac{\delta H}{\delta Q_k'}\frac{\delta Q_k}{\delta u(x)}\right)\right\} \\ &= \int_{-\infty}^{\infty} dk'\left\{[P_k, P_{k'}]\frac{\delta H}{\delta P_k'} + [P_k, Q_{k'}]\frac{\delta H}{\delta Q_{k'}}\right\}. \end{aligned} \qquad (6.4.9)$$

Here the square brackets (Poisson brackets) $[S, T]$ for arbitrary functionals S and T of u have the form

$$[S, T] = \frac{1}{2}\int_{-\infty}^{\infty} dx\left\{\frac{\delta S}{\delta u(x)}\frac{\partial}{\partial x}\frac{\delta T}{\delta u(x)} - \frac{\delta T}{\delta u(x)}\frac{\partial}{\partial x}\frac{\delta S}{\delta u(x)}\right\}. \qquad (6.4.10)$$

Differentiating Q_k, we obtain

$$\frac{\partial Q_k}{\partial t} = \int_{\infty-}^{\infty} dk'\left\{[Q_k, P_{k'}]\frac{\delta H}{\delta P_{k'}} + [Q_k, Q_{k'}]\frac{\delta H}{\delta Q_{k'}}\right\}. \qquad (6.4.11)$$

If the relations

$$[P_k, P_{k'}] = [Q_k, Q_{k'}] = 0, \quad [P_k, Q_{k'}] = \delta(k - k')\,, \qquad (6.4.12)$$

are fulfilled, then (6.4.9) and (6.4.11) take the Hamiltonian form

$$\frac{\partial P_k}{\partial t} = -\frac{\delta H}{\delta Q_k}, \quad \frac{\delta Q_k}{\delta t} = \frac{\delta H}{\partial P_k}, \tag{6.4.13}$$

and the variables P_k and Q_k are canonically conjugate.

If, apart from the continuous variables P_k and Q_k, discrete variables P_n and Q_n exist, then these variables will be canonical, provided they satisfy the relations

$$\begin{aligned}
&[P_k, P_n] = [Q_k, P_n] = [P_k, Q_n] = [Q_k, Q_n] = 0\,, \\
&[P_n, P_{n'}] = [Q_n, Q_{n'}] = 0, \quad [P_n, Q_{n'}] = \delta_{nn'}\,.
\end{aligned} \tag{6.4.14}$$

The brackets in (6.4.14) are to be understood as in (6.4.10).

Let us now calculate the functional derivatives of the scattering data λ_n, c_n, $a(k)$ and $b(k)$ with respect to $u(x)$. Varying (6.4.4), we obtain

$$\frac{d^2}{dx^2}\frac{\delta\psi(x,k)}{\delta u(z)} + [k^2 - u(x)]\frac{\delta\psi(x,k)}{\delta u(z)} = \delta(x-z)\,\psi(x,k)\,. \tag{6.4.15}$$

Moreover, $\delta\psi(x,k)/\delta u(z) \to 0$; as $x \to +\infty$. Solving (6.4.15), we obtain

$$\frac{\delta\psi(x,k)}{\delta u(z)} = \begin{cases} 0 & \text{as} \quad x > z\,, \\ \dfrac{\psi(z,k)}{2ik}\,[\psi(z,k)\,\bar{\psi}(x,k) - \bar{\psi}(z,k)\,\psi(x,k)] & \text{as} \quad x < z\,. \end{cases} \tag{6.4.16}$$

As $x \to -\infty$

$$\frac{\delta\psi(x,k)}{\delta u(z)} = \mathrm{e}^{ikx}\frac{\delta a(k)}{\delta u(z)} + \mathrm{e}^{-ikx}\frac{\delta b(k)}{\delta u(z)}\,. \tag{6.4.17}$$

Letting $x \to -\infty$ in (6.4.16), comparing with (6.4.17) and using (6.2.2), we obtain

$$\begin{aligned}
\frac{\delta a(k)}{\delta u(z)} &= \frac{\psi(z,k)}{2ik}\,[\psi(z,k)\,b^*(k) - \bar{\psi}(z,k)\,a(k)] \\
&= -\frac{\bar{\varphi}(z,k)\,\psi(z,k)}{2ik}\,,
\end{aligned} \tag{6.4.18}$$

$$\begin{aligned}
\frac{\delta b(k)}{\delta u(z)} &= \frac{\psi(z,k)}{2ik}\,[\psi(z,k)\,a^*(k) - \bar{\psi}(z,k)\,b(k)] \\
&= -\frac{\varphi(z,k)\,\psi(z,k)}{2ik}\,.
\end{aligned} \tag{6.4.19}$$

Let us now introduce the quantities

$$P_k = \frac{2k}{\pi}\ln|a(k)|^2, \quad Q_k = \arg b_k, \quad k > 0\,. \tag{6.4.20}$$

From (6.4.18) and (6.4.19) we have

$$\frac{\delta P_k}{\delta u(z)} = \frac{1}{\pi} \operatorname{Im}\left\{\frac{1}{a(k)}\right\} \bar{\varphi}(z, k)\, \psi(z, k)\,,$$

$$\frac{\delta Q_k}{\delta u(z)} = \frac{1}{2} \operatorname{Re}\left\{\frac{1}{b(k)}\right\} \varphi(z, k)\, \psi(z, k)\,. \tag{6.4.21}$$

We evaluate the Poisson bracket between P_k and Q_k. To calculate the expression (6.4.10) we integrate between finite limits $-L < x < L$ and then let $L \to \infty$. The integrals can be evaluated with the help of the directly verifiable formula

$$\psi_1\varphi_1 \frac{\partial}{\partial x} \psi_2\varphi_2 - \psi_2\varphi_2 \frac{\partial}{\partial x} \psi_1\varphi_1$$
$$= \frac{1}{k_1^2 - k_2^2} \frac{\partial}{\partial x} [(\varphi_1\varphi_{2x} - \varphi_{1x}\varphi_2)(\psi_1\psi_{2x} - \psi_{1x}\psi_2)]\,. \tag{6.4.22}$$

Here ψ_1, φ_1 is an arbitrary pair of solutions of (6.4.4) with $k^2 = k_1^2$; ψ_2, φ_2 is an arbitrary pair of solutions of the same equation with $k^2 = k_2^2$. Evaluating the integrals and using the asymptotic expressions for the Jost function, we find that the quantities P_k and Q_k satisfy the conditions (6.4.12), i.e. they are canonical variables.

The function $b(k)$ can be completely reconstructed in terms of P_k and Q_k:

$$b(k) = (e^{\pi P_k/2k} - 1)^{1/2}\, e^{iQ_k}, \quad b(-k) = b^*(k), \quad k > 0\,, \tag{6.4.23}$$

While, the function $a(k)$ can be reconstructed in terms of P_k only to within the discrete spectrum. The function $a(k)$ can be represented by the formula

$$a(k) = \tilde{a}(k) \prod_n \frac{k - i\kappa_n}{k + i\kappa_n}\,, \tag{6.4.24}$$

where $\tilde{a}(k)$ is an analytic function in the upper halfplane, which does not have zeros there. On the real axis $-\infty < k < \infty$ we have $|a(k)| = |\tilde{a}(k)|$. The logarithm of the function $\tilde{a}(k)$ is also analytic in the upper halfplane. Therefore the argument of $\tilde{a}(k)$ can be reconstructed in terms of its modulus with the help of the Cauchy integral

$$\arg \tilde{a}(k) = -\frac{1}{\pi} \int_{-\infty}^{\infty} \frac{\ln |a(k')|}{k' - k}\, dk' = -\frac{1}{2} \int_0^{\infty} \frac{k P_{k'}\, dk'}{k'(k'^2 - k^2)}\,. \tag{6.4.25}$$

In the case of a discrete spectrum let us introduce a discrete set of variables

$$P_n = \kappa_n^2, \quad Q_n = -\ln c_n^2\,. \tag{6.4.26}$$

The functional derivative $\delta P_n/\delta u(x)$ can be found from the known formulae of perturbation theory

$$\frac{\delta P_n}{\delta u(x)} = \frac{\delta \kappa_n^2}{\delta u(x)} = -g_n^2(x)\,, \tag{6.4.27}$$

where $g_n(x) = \psi_n(x)/M_n$ is an eigenfunction of the discrete spectrum, which is normalized by the condition $\int_{-\infty}^{\infty} g_n^2 dx = 1$.

In order to find $\delta Q_n/\delta_n(x)$, let us apply the following reasoning. Let us truncate the potential $u(x)$ at the point $x = L$, i.e. we set $u(x) = 0$ when $x > L$. The Jost function is defined by the condition $\psi(L, k) = \exp(ikL)$, $\psi_x(L, k) = ik \exp(ikL)$ and by virtue of the theorem about the analytical dependence on the parameter, it is an entire analytic function in the whole k plane. Then, obviously, $c_n = b(i\kappa_n)$,

$$\frac{\delta Q_n}{\delta u(x)} = -\frac{2}{b(i\kappa_n)} \frac{\delta b(i\kappa_n)}{\delta u(x)} = \frac{\varphi(x, -i\kappa_n)\,\psi(x, i\kappa_n)}{\kappa_n}\,. \tag{6.4.28}$$

Evaluating the Poisson bracket between P_n and Q_n with the help of (6.4.27), (6.4.28), we find that they satisfy the condition (6.4.14) and are canonical. Now letting $L \to \infty$ we see that they are canonical in the general case as well.

The formulae (6.4.24–26) show that the function $a(k)$ is completely reconstructed by the set of "generalized momenta" P_k and P_n:

$$\arg a(k) = -\frac{1}{2}\int_0^{\infty} \frac{kP_{k'}dk'}{k'(k'^2 - k^2)} + \mathrm{Im}\left\{\sum_n \ln \frac{k - iP_n^{1/2}}{k + iP_n^{1/2}}\right\}, \tag{6.4.29}$$

i.e. the set of variables P_n, Q_n, P_k, Q_k is proved to be complete. Moreover, if H is an arbitrary functional of $a(k)$, then, in terms of these variables, the Hamilton equations are

$$\frac{\partial P_k}{\partial t} = -\frac{\delta H}{\delta Q_k} = 0, \quad \frac{\partial P_n}{\partial t} = -\frac{\delta H}{\delta Q_n} = 0\,, \tag{6.4.30}$$

from which it follows that $P_n = \text{const}$, $P_k = \text{const}$, and $a(k) = \text{const}$. For Q_n, Q_k, we have

$$Q_n(t) = Q_n(0) + \frac{\delta H}{\delta P_n}t\,,$$

$$Q_k(t) = Q_k(0) + \frac{\delta H}{\delta P_k}t\,. \tag{6.4.31}$$

Evaluating the derivatives (6.4.31), with the help of (6.4.29), we obtain a new proof of the basic formulae (6.1.22), which describe the dependence of the scattering data on time.

The formulae (6.4.30, 31) show that any dynamical system of the form $\partial u/\partial t = (\partial/\partial x)(\delta H/\delta u)$ where H is a functional of $a(k)$, is completely integrable, P_n, P_k

playing the role of action variables and Q_n, Q_k that of cyclic angular variables. In particular, this is true for any system of the form

$$\frac{\partial u}{\partial t} = \frac{\partial}{\partial x} \frac{\delta}{\delta u(x)} \sum_n \alpha_n(t) I_n ,$$

including the KdV-equation.

6.5 Shabat's Method

The inverse scattering method exposed above has an important defect: the potential $u(x)$ has to decrease as $x \to \infty$. Shabat [6.4] suggested another method of solving the KdV-equation, without this condition. Shabat's method is highly convenient also from the point of view of generalization. For any equation which admits a reasonable (L, A)-pair, the method allows one to obtain broad families of exact solutions, bypassing the difficult operation of solving the inverse scattering problem for the operator L.

Let us consider the linear integral equation

$$K(x, y) = F(x, y) + \int_x^\infty K(x, s) F(s, y) \, ds . \tag{6.5.1}$$

In this equation, each of the functions $K(x, y)$ and $F(x, y)$ can be regarded either as given or unknown. The equation (6.5.1) can be rewritten in the symbolic form

$$K = F + K{*}F . \tag{6.5.2}$$

Let the functions K and F depend on a parameter (for example on time t). Let us differentiate (6.5.1) with respect to this parameter

$$\begin{aligned} \frac{\partial K(x, y)}{\partial t} = \frac{\partial F(x, y)}{\partial t} &+ \int_x^\infty \frac{\partial K(x, s)}{\partial t} F(s, y) \, ds \\ &+ \int_x^\infty K(x, s) \frac{\partial F(s, y)}{\partial t} \, ds \end{aligned} \tag{6.5.3}$$

or in the symbolic form

$$D_t K = D_t F + D_t K{*}F + K{*}D_t F \quad (D_t = \partial/\partial t) . \tag{6.5.4}$$

Let us inquire about finding a more general class of operators which differentiate the convolution $K * F$. Here we do not demand that the operators which differentiate F and K, coincide. In the general case, we have

$$\tilde{D} K = DF + \tilde{D} L{*}F + K{*}DF . \tag{6.5.5}$$

We shall call the operators D and $\tilde{D}$ a pair of differentiation operators.

Let us consider the operator

$$D_n = \frac{\partial^n}{\partial x^n} + (-1)^{n+1}\frac{\partial^n}{\partial y^n} \tag{6.5.6}$$

and apply it to (6.5.1). We obtain

$$D_n K(x, y) = D_n F(x, y) + \int_x^\infty K(x, s)\, D_n F(s, y)\, ds$$
$$+ \frac{\partial^n}{\partial x^n}\int_x^\infty K(x, s)\, F(s, y)\, ds - \int_x^\infty K(x, s)\frac{\partial^n F(s, y)}{\partial s^n}\, ds\,. \tag{6.5.7}$$

Differentiating with respect to x and performing the n-fold integration by parts in the last term of the formula (6.5.7), we obtain

$$D_n K(x, y) = D_n F(x, y) + \int_x^\infty D_n K(x, s)\, F(s, y)\, ds$$
$$+ \int_x^\infty K(x, s)\, D_n F(s, y)\, ds + M_0 F(x, y)\,. \tag{6.5.8}$$

Here M_0 is a linear differential operator with respect to x, of order $N-2$, with variable coefficients, the latter constituting polynomials of partial derivatives of the kernel $K(x, y)$ with respect to x and y, taken at $x = y$. In (6.5.8), let us replace $F(x, y)$ by its expression (6.5.1). Then

$$M_0 F(x, y) = M_0 K(x, y) - \int_x^\infty M_0 K(x, s)\, F(s, y)\, ds + M_1 F\,,$$

where M_1 is a linear differential operator of order $N-3$ with respect to x. Let us repeat this process, replacing successively F by its expression (6.5.1). Finally, we obtain

$$M_0 F(x, y) = MK(x, y) - \int_x^\infty MK(x, s)\, F(s, y)\, ds\,,$$

$$M = M_0 + M_1 + \dots$$

It is now clear that (6.5.8) has the form (6.5.5), where

$$D = D_n, \quad \tilde{D} = D_n - M\,. \tag{6.5.9}$$

We proved that each operator D_n can be taken as the first term of the pair $(D, \tilde{D})$, the second term $\tilde{D}$ of the pair being then uniquely reconstructible. The operators D and $\tilde{D}$ coincide in the leading order.

Let us present the explicit expressions for the first three D, $\tilde{D}$:

$$D_1 = \tilde{D}_1 = \frac{\partial}{\partial x} + \frac{\partial}{\partial y},$$

$$D_2 = \frac{\partial^2}{\partial x^2} - \frac{\partial^2}{\partial y^2},$$

$$\tilde{D}_2 = \frac{\partial^2}{\partial x^2} - \frac{\partial^2}{\partial y^2} - u(x),$$

$$u(x) = -2\frac{d}{dx}K(x, x), \tag{6.5.10}$$

$$D_3 = \frac{\partial^3}{\partial x^3} + \frac{\partial^3}{\partial y^3},$$

$$\tilde{D}_3 = \frac{\partial^3}{\partial x^3} + \frac{\partial^3}{\partial y^3} - \frac{3}{4}\left(u\frac{\partial}{\partial x} + \frac{\partial}{\partial x}u\right) + \frac{3}{2}w,$$

$$w = \frac{\partial}{\partial x}\left[K(x, x)^2 + \frac{\partial}{\partial x}\left(\frac{\partial K}{\partial x} - \frac{\partial K}{\partial y}\right)\bigg|_{y=x}\right]. \tag{6.5.11}$$

In place of a pair $(D, \tilde{D})$, one may equally well choose the linear combinations

$$D = \alpha D_t + \sum_n \alpha_n(t)\, D_n = \alpha\, D_t + L_x^0 - L_y^{0+},$$

$$\tilde{D} = \alpha D_t + \sum_n \alpha_n(t)\, \tilde{D}_n = \alpha D_t + L_x - L_y^{0+}. \tag{6.5.12}$$

In the formulae (6.5.12)

$$L_x^0 = \sum \alpha_n(t)\frac{\partial^n}{\partial x^n}, \quad L_y^0 = \sum \alpha_n(t)\,(-1)^n\frac{\partial^n}{\partial y^n},$$

$$L_x = L_x^0 - M. \tag{6.5.13}$$

Now, let the function $F(x, y)$ satisfy the equation

$$DF = 0, \tag{6.5.14}$$

where the D-operator is of the form (6.5.12). Then from (6.5.5) it follows that

$$\tilde{D}K = \tilde{D}K * F. \tag{6.5.15}$$

If at the same time F satisfies the additional requirements of a unique solvability of (6.5.1) with respect to K, then it follows from (6.5.14) that

$$\tilde{D}K = 0. \tag{6.5.16}$$

Equation (6.5.14) is linear while (6.5.16) is nonlinear. Thus, the relation (6.5.1) associates with classes of nonlinear (with respect to K) equations (6.5.16), classes of linear (with respect to F) equations (6.5.14).

Let us now consider two operators

$$L^{(0)} = \sum_{n=1}^{N_1} \alpha_n \frac{\partial^n}{\partial x^n}, \quad A^{(0)} = \sum_{n=1}^{N_2} \beta_n \frac{\partial^n}{\partial x^n} \tag{6.5.17}$$

(α_n and β_n are arbitrary constants, for simplicity independent of time) and, use them construct the pair of differentiation operators

$$D^{(1)} = L_x^{(0)} - L_y^{(0)+} = \sum_{n=1}^{N_1} \alpha_n \left[\frac{\partial^n}{\partial x^n} - (-1)^n \frac{\partial^n}{\partial y^n}\right],$$

$$D^{(2)} = A_x^{(0)} - A_y^{(0)+} = \sum_{n=1}^{N_2} \beta_n \left[\frac{\partial^n}{\partial x^n} - (-1)^n \frac{\partial^n}{\partial y^n}\right],$$

$$\tilde{D}^{(1)} = L - L_y^{(0)+},$$

$$\tilde{D}^{(2)} = A - A_y^{(0)+}.$$

Here $L = L_x^{(0)} + \dots, A = A_x^{(0)} + \dots$

We require the function F to satisfy two equations

$$D^{(1)} F = 0, \tag{6.5.18}$$

$$(D_t + \mathrm{i} D^{(2)}) F = 0. \tag{6.5.19}$$

Differentiating (6.5.18) with respect to time and using the fact that the operators $D^{(1)}$ and $D^{(2)}$ commute, we see that the relation (6.5.18) is conserved in time and the equations (6.5.19) are compatible.

From (6.5.18, 19) it follows, see (6.5.16), that the function $K(x, y)$, satisfies the two equations

$$\tilde{D}^{(1)} K = 0, \tag{6.5.20}$$

$$(D_t + \mathrm{i} \tilde{D}^{(2)}) K = 0. \tag{6.5.21}$$

at all times. Differentiating (6.5.20) with respect to time and applying (6.5.21), we obtain

$$\left(\frac{\partial \tilde{D}^{(1)}}{\partial t} - \mathrm{i} [\tilde{D}^{(1)}, \tilde{D}^{(2)}]\right) K = 0,$$

whence

$$\left(\frac{\partial L}{\partial t} - \mathrm{i} [L, A]\right) K(x, y) = 0. \tag{6.5.22}$$

Since (6.5.22) is satisfied identically in y, we have

$$\frac{\partial L}{\partial t} = \mathrm{i}\,[L, A]\,. \tag{6.5.23}$$

Let us prescribe values $K(x, x)$ of the function $K(x, y)$ along the straight line $x = y$. Differentiating $K(x, x)$ with respect to x, we obtain the sum of partial derivatives $\partial K/\partial x + \partial K/\partial y$ at $y = x$. Therefore, independently of $K(x, x)$, one can prescribe only one partial derivative of first order. Continuing this reasoning, we see that, if the derivatives of order $N - 1$ are prescribed, then among the derivatives of Nth order, only one can be prescribed independently. Since $K(x, y)$ satisfies (6.5.20) which expresses the derivatives of Nth order in terms of the lower derivatives, exactly $N - 1$ independent combinations of the derivatives of $K(x, y)$ can be prescribed on the straight line $y = x$; we shall denote these combinations by $u_i(x)$ $(i = 1, \dots, N - 1)$.

The coefficients of the operators L and A are expressed in terms of the $u_i(x)$'s. Thus (6.5.23) constitutes a system of nonlinear partial differential equations governing the behavior of the functions $u_i(x)$ in time. This system automatically possesses an (L, A) pair; however, in order to solve it, it is not necessary to apply the methods of the inverse scattering problem for the operator L. Instead of this, let us note that if we require the additional condition $K(x, x + z) \to 0$ as $x \to \infty$, then the function $K(x, y)$ is reconstructed in the entire plane in terms of the given $u_i(x)$ by solving the Cauchy-Goursat problem for (6.5.20). This allows us to solve (6.5.23) according to the following scheme

$$\begin{aligned}&u_i(x)\,|_{t=0} \to K(x, y)\,|_{t=0} \xrightarrow{F=K-K*F} F(x, y)\,|_{t=0} \xrightarrow{F_t+D^{(2)}F=0}\\ &\xrightarrow{F_t+D^{(2)}F=0} F(x, y, t) \xrightarrow{K=F+K*F} K(x, y, t) \to u_i(x, t)\,.\end{aligned} \tag{6.5.24}$$

If $u_i(-\infty) = 0$, then the first two stages of the scheme, are equivalent to the solution of the direct scattering problem for the operator L, the fourth stage of the scheme is equivalent to the solution of the inverse problem for the operator L, while the third and fifth stages of the scheme are trivial. However, the scheme (6.5.24) has the advantage that it can be applied also if the $u_i(x)$ do not decrease, as $x \to -\infty$. Equation (6.5.1), in the second stage of the scheme, is solved with respect to F and in the fourth with respect to K.

For the KdV-equation the operator D_2 plays the role of $D^{(1)}$. The equation $D^{(1)}(F) = 0$ yields $F = F(x + y)$. The equation $\tilde{D}^{(1)}K = 0$ coincides with (6.2.6) obtained in Sect. 6.2. The operator $4D_3$ plays the role of the operator $D^{(2)}$. The equation $(D_t + \mathrm{i}D^{(2)})\ F = 0$ coincides with (6.3.2), and the equation $(D_t + \mathrm{i}\tilde{D}^{(2)})\,K = 0$ has the form

$$\frac{K(x, y)}{\partial t} + 4\left(\frac{\partial^3}{\partial x^3} + \frac{\partial^3}{\partial y^3}\right) K(x, y) - 3\left(u\frac{\partial}{\partial x} + \frac{\partial}{\partial x}u\right) K(x, y) = 0\,.$$

Inserting into the above, see (6.2.25),

$$K(x, y) = \int c(k)\, \psi(x, k)\, e^{iky}\, dk\,,$$

we arrive at (6.1.20).

It is clear from the above that a solution of (6.3.2) decreasing as $\xi \to \infty$, generates an exact solution of the KdV-equation, even if $F(\xi)$ does not decrease as $\xi \to -\infty$.

6.6 *N*-Soliton Solutions for the Equation of Nonlinear String

As an example of the application of the scheme presented above, let us consider the operators

$$\begin{aligned} D^{(1)} &= D_3 + s^2 D_1\,, \\ iD^{(2)} &= BD_2\,. \end{aligned} \tag{6.6.1}$$

From (6.5.10, 11), it follows that

$$L = \frac{d^3}{dx^3} - \frac{3}{4}\left(u\frac{d}{dx} + \frac{d}{dx}u\right) + \frac{3}{2}w + s^2\frac{d}{dx}, \tag{6.6.2}$$

$$A = B\left(\frac{d^2}{dx^2} - u\right). \tag{6.6.3}$$

The relation (6.5.23) leads to the system

$$\begin{aligned} u_t &= 2Bw_x\,, \\ w_t &= \frac{2}{3}B\left(-\frac{1}{4}u_{xxx} - s^2u_x + \frac{3}{2}uu_x\right), \end{aligned} \tag{6.6.4}$$

which is equivalent to the equation

$$u_{tt} = -\frac{4}{3}B^2\left(\frac{1}{4}u_{xxxx} + s^2u_{xx} - \frac{3}{2}(uu_x)_x\right). \tag{6.6.5}$$

When $B = i\sqrt{4/3}$, $s^2 = 1$, (6.6.5) has the form

$$u_{tt} = u_{xx} + \frac{1}{4}u_{xxxx} - \frac{3}{2}(uu_x)_x\,. \tag{6.6.6}$$

and describes the longitudinal long wavelength vibrations of a one-dimensional lattice, taking into account dispersion of sound. Setting $B = \sqrt{4/3}$, $s^2 = -1$, we obtain the equation

$$u_{tt} = u_{xx} - \frac{1}{4}u_{xxxx} + \frac{3}{2}(uu_x)_x\,. \tag{6.6.7}$$

Both these equations can be called the equations of a nonlinear string, because (6.6.7) describes the transverse vibrations of a nonlinear string, taking into account finite stiffness. We restrict ourselves in what follows to the consideration of the case (6.6.6).

In accordance with (6.5.18) and (6.5.19), $F(x, y, t)$ satisfies the equations

$$\frac{\partial^3 F}{\partial x^3} + \frac{\partial^3 F}{\partial y^3} + s^2 \left(\frac{\partial F}{\partial x} + \frac{\partial F}{\partial y}\right) = 0\,, \tag{6.6.8}$$

$$\frac{\partial F}{\partial t} + B \left(\frac{\partial^2 F}{\partial x^2} - \frac{\partial^2 F}{\partial y^2}\right) = 0\,. \tag{6.6.9}$$

Each solution of this system of equations, after finding $K(x, y)$ from (6.5.1), generates an exact solution of (6.6.5).

Let

$$F = -\, c^2(t)\, \mathrm{e}^{\lambda x + \lambda^* y}, \quad \mathrm{Re}\{\lambda\} < 0\,, \tag{6.6.10}$$

where λ satisfies the equation

$$\lambda^2 + \lambda^{*2} - |\lambda|^2 + s^2 = 0, \quad \lambda = -\,\eta + \mathrm{i}\nu, \quad \nu^2 = \frac{1}{3}(\eta^2 + s^2)\,. \tag{6.6.11}$$

Substituting (6.6.10) in (6.6.9), we obtain

$$c^2(t) = c^2(0)\, \mathrm{e}^{-\sqrt{4/3}\; 2\nu\eta t}\,.$$

Representing the solution of (6.5.1) in the form

$$K(x, y) = \varphi(x)\, \mathrm{e}^{\lambda^* y}\,,$$

we obtain, after simple calculations,

$$u(x, t) = -\,\frac{2\eta^2}{\cosh^2 \eta(x - vt - x_0)}\,,$$

$$u = 4\sqrt{\frac{2}{3}}\,\nu = -\,\frac{4\sqrt{2}}{3}\sqrt{\eta^2 + s^2}\,. \tag{6.6.12}$$

The solution (6.6.12) is a soliton.

In order to obtain N-soliton solutions, let us represent F in the form

$$F(x, y) = -\sum c_n^2(t)\, \mathrm{e}^{\lambda_n x + \lambda_n^* y}, \quad c_n^2(t) = c_n^2(0)\, \mathrm{e}^{-4\sqrt{2/3}\nu_n \eta_n t}\,.$$

Assuming $K(x, y)$ in the form

$$K(x, y) = \sum \varphi_n(x)\, \mathrm{e}^{\lambda_n y}\,,$$

we obtain for $\varphi_n(x)$ the system of equations

$$\varphi_n(x) + c_n^2 e^{-\lambda_n x} - c_n^2 \sum_m \varphi_m(x) \frac{e^{(\lambda_m^* + \lambda_n)x}}{\lambda_m^* + \lambda_n} = 0 . \tag{6.6.13}$$

The solution of this system can be obtained by the same method as the solution of the system (6.3.7):

$$u(x, t) = -2 \frac{d^2}{dx^2} \ln\Delta ,$$

where Δ is the determinant of the system (6.6.13).

In the problem under consideration the amplitude of a soliton determines only the modulus of its velocity; the sign of the velocity remains arbitrary. Therefore, as $t \to \pm\infty$, in general solitons run apart from each other in opposite directions.

The amplitudes and velocities of solitons during collisions are still conserved. Let us present the formula for the displacement of the center of a soliton with the eigenvalue $\lambda_1 = \eta_1 + i\nu_1$ after a collision with a soliton with the eigenvalue $\lambda_2 = \eta_2 + i\nu_2$:

$$|\delta| = \frac{1}{2\eta_1} \ln \frac{(\eta_1 - \eta_2)^2 + (\nu_1 - \nu_2)^2}{(\eta_1 + \eta_2)^2 + (\nu_1 - \nu_2)^2} . \tag{6.6.14}$$

Here the sign of ν is opposite to that of the velocity of the soliton and the sign of the displacement δ can be determined from the fact that the solitons repel each other during interaction. The derivation of (6.6.14) differs from that of (6.3.10) only by unimportant details.

6.7 The Toda Lattice

The method of the inverse problem can also be applied to discrete systems, for example, to the one described in Sect. 5.4, i.e. to the Toda lattice, the equations of which can be written in the form

$$\ddot{x}_n = e^{x_{n+1} - x_n} - e^{x_n - x_{n-1}} . \tag{6.7.1}$$

The operators L and A for this system have the form of infinite matrices

$$L_{nm} = i\sqrt{c_n}\, \delta_{n,m+1} - i\sqrt{c_m}\, \delta_{n+1,m} + v_n \delta_{nm} , \tag{6.7.2}$$

$$A_{nm} = \frac{1}{2} (\sqrt{c_n}\, \delta_{n,m+1} + \sqrt{c_m}\, \delta_{n+1,m}) , \tag{6.7.3}$$

where the subscripts n and m run through all integers, $c_n = \exp(x_n - x_{n-1})$ and $v_n = \dot{x}_n$. Carrying out direct calculations, we see that the system (6.7.1) is equivalent to the operator relation

$$\frac{\partial L}{\partial t} = \mathrm{i}[L, A]. \tag{6.7.4}$$

Let us consider the eigenvalue problem $L\psi = \lambda\psi$ where ψ is an infinite column of numbers

$$\mathrm{i}\sqrt{c_n}\,\psi_{n-1} - \mathrm{i}\sqrt{c_{n+1}}\,\psi_{n+1} + v_n\psi_n = \lambda\psi_n. \tag{6.7.5}$$

Conservation of the spectrum of the operator L in time is ensured by (6.7.4) and the hermiticity of L.

We shall consider an infinite lattice (6.7.1) with the natural boundary conditions

$$c_n \to 1, \quad v_n \to 0 \quad \text{as} \quad n \to \pm\infty. \tag{6.7.6}$$

Assume also that the sequences v_n, c_n converge to their limits rapidly enough so that the operator L has a finite number of nondegenerate discrete eigenvalues. Moreover, L has a two-fold degenerate continuous spectrum which fills the segment $-2 \leqslant \lambda \leqslant 2$.

Let us investigate in more detail the eigenvalue problem (6.7.5) for the operator L. Note, firstly, that if ψ_n is a solution of (6.7.5) with a real λ then $\tilde{\psi}_n = (-1)^n\psi_n^*$ is also a solution of (6.7.5), and secondly, that if $\psi_n^{(1)}$, $\psi_n^{(2)}$ are two arbitrary solutions of this system, then the quantity

$$w(\psi^{(1)}, \psi^{(2)}) = (-1)^{n+1}\sqrt{c_n}(\psi_n^{(1)}\psi_{n-1}^{(2)} - \psi_{n-1}^{(1)}\psi_n^{(2)}) \tag{6.7.7}$$

is independent of n.

Let us distinguish a special class of solutions (6.7.6) (Jost functions). Let $\psi_n(\xi)$, $\psi_n(\xi)$ be solutions of (6.7.5) with $\lambda = 2\sin\xi$ determined by the asymptotics

$$\psi_n(\xi) \to \mathrm{e}^{\mathrm{i}\xi n}, \quad n \to +\infty, \tag{6.7.8}$$

$$\varphi_n(\xi) \to \mathrm{e}^{\mathrm{i}\xi n}, \quad n \to -\infty, \tag{6.7.9}$$

then it can be shown that the function $\psi_n(\xi)$ can be analytically continued in the upper halfplane of the complex variable ξ and $\varphi_n(\xi)$ in the lower one. Furthermore, the function

$$\tilde{\varphi}_n(\xi) = (-1)^n\,\varphi_n^*(\xi^*) \tag{6.7.10}$$

is analytic when $\mathrm{Im}\{\xi\} > 0$.

The functions $\varphi_n(\xi)$, $\bar{\varphi}_n(\xi)$ constitute a complete set of linearly independent solutions of (6.7.5) with $\lambda = 2\sin\xi$. Therefore

$$\psi_n(\xi) = \alpha(\xi)\,\varphi_n(\xi) + \beta(\xi)\,\bar{\varphi}_n(\xi)\,. \tag{6.7.11}$$

Calculating $w(\psi, \bar{\varphi})$ in the limit $n \to \pm\infty$, we find that

$$|\alpha(\xi)|^2 - |\beta(\xi)|^2 = 1\,. \tag{6.7.12a}$$

It is also not difficult to see that

$$\alpha(\xi) = \frac{w[\psi(\xi), \bar{\varphi}(\xi)]}{2\cos\xi}\,. \tag{6.7.12b}$$

The relation (6.7.12b) and the analytical properties of the Jost functions show that $\alpha(\xi)$ is analytic when $\mathrm{Im}\{\xi\} > 0$.

Let us find the dependence of α and β on time. To this end we differentiate the equation $L\psi = \lambda\psi$ with respect to t, whence

$$(L - \lambda)\left(\frac{\partial\psi}{\partial t} + \mathrm{i}A\psi\right) = 0\,.$$

The condition of conservation in time of the definition of the Jost function $\psi_n(\xi)$ (6.7.8) yields

$$\frac{\partial\psi}{\partial t} + \mathrm{i}A\psi = \mathrm{i}\cos\xi\psi\,. \tag{6.7.13}$$

Now substituting (6.7.11) into (6.7.13) and letting $nk \to \infty$ we obtain

$$\frac{\partial}{\partial t}\alpha(\xi) = 0, \quad \frac{\partial}{\partial t}\beta(\xi) = 2\mathrm{i}\cos\xi\beta(\xi)\,,$$

$$\beta(\xi, t) = \beta(\xi, 0)\,\mathrm{e}^{2\mathrm{i}\cos\xi t}\,. \tag{6.7.14}$$

The points of the upper half ξ-plane at which $\alpha(\xi) = 0$ correspond to the discrete spectrum of the operator L. Due to the fact that $\lambda = 2\sin\xi$ is real, they lie on the straight lines $\mathrm{Re}\{\xi\} = \pm\pi/2$ (as a consequence of the obvious periodicity of the Jost functions in ξ with a period 2π, we restrict ourselves to the strip $-\pi \leqslant \mathrm{Re}\{\xi\} \leqslant \pi$ in the upper halfplane). Let us denote the zeros of $\alpha(\xi)$ by $\xi_k^\pm$;

$$\xi_k^\pm = \pm\frac{\pi}{2} + \mathrm{i}\eta_k^\pm\,.$$

Due to (6.7.12b), at these zeros

$$\psi(\xi) = c_\xi\bar{\varphi}(\xi)\,. \tag{6.7.15}$$

The dependence of c_ξ on time is also easily found:

$$c_{\xi\pm}(t) = c_{\xi\pm}(0)\,e^{\pm 2\sinh\eta t}\,. \qquad (6.7.16)$$

The set $\xi_k^\pm$, $c_\xi^\pm$, $\alpha(\xi)$, $\beta(\xi)$ form the "scattering data" for the operator L and, as will be shown below, determines completely the latter. This allows us to solve the Cauchy problem for the Toda chain using the usual procedure of the inverse scattering method (6.1.24), since the scattering data depend on time in a simple manner.

We proceed to the problem of reconstructing the matrix of the operator L from the scattering data. Let us note first of all, that due to the periodicity of $\psi_n(\xi)$ mentioned before, this function can be represented in the form of a Fourier series

$$\psi_n(\xi) = \sum_{|\infty}^{\infty} K_{nm} e^{i\xi m}\,. \qquad (6.7.17)$$

However, since $\psi_n(\xi)\exp(-i\xi n)$ is analytic in the upper halfplane and tends to the constant c_n as $\xi \to i\infty$, its Fourier transform (6.7.17) is truncated for $m < n$, i.e., for ψ_n, a triangular representation

$$\psi_n(\xi) = \sum_{m \leqslant n} K_{nm} e^{i\xi m}\,. \qquad (6.7.18)$$

is valid.

Quite analogously the existence of a triangular representation for $\varphi_n(\xi)$ is found; it is realized by the matrix A_{nm}, namely

$$\varphi_n(\xi) = \sum_{m > n} A_{nm} e^{i\xi m}\,. \qquad (6.7.19)$$

Substituting (6.7.18, 19) into (6.7.5) we obtain

$$\sqrt{c_n} = \frac{K_{n,n}}{K_{n-1,\,n-1}}\,; \quad v_n = i\left(\frac{K_{n,\,n+1}}{K_{n,n}} - \frac{K_{n-1,n}}{K_{n-1,\,n-1}}\right), \qquad (6.7.20)$$

$$\sqrt{c_n} = \frac{A_{n-1,\,n-1}}{A_{nn}} \quad v_n = i\left(\frac{A_{n+1,n}}{A_{n+1,\,n+1}} - \frac{A_{n,n-1}}{A_{n,n}}\right). \qquad (6.7.21)$$

Now it is obvious that K_{nn} and A_{nn} are real and positive. Moreover, it is clear that $A_{nn} = \gamma/K_{nn}$. The constant γ appearing here can be expressed in terms of $\alpha(i\infty)$. Substituting

$$\bar{\varphi}_n(\xi) = (-1)^n \sum_{m \leqslant n} A^*_{nm} e^{-i\xi m} \qquad (6.7.22)$$

and (6.7.18) into (6.7.12b) and letting $\xi \to i\infty$ we obtain

$$A_{nn} = \frac{\alpha(\mathrm{i}\,\infty)}{K_{nn}}. \tag{6.7.23}$$

Let us write (6.7.11) in the form

$$\frac{\psi_n(\xi)}{\alpha(\xi)} = \varphi_n(\xi) + \frac{\beta(\xi)}{\alpha(\xi)}\bar{\varphi}_n(\xi). \tag{6.7.24}$$

The left-hand side of (6.7.24) is a function of ξ, analytic in the strip $\mathrm{Im}\{\xi\} > 0$, $-\pi < \mathrm{Re}\{\xi\} < \pi$ except at the points at which $\alpha(\xi)$ vanishes and at which $\psi_n(\xi)/\alpha(\xi)$ has simple poles. When $\xi \to \mathrm{i}\infty$, it follows from (6.7.18) and (6.7.23) that $\psi_n(\xi)/\alpha(\xi)$ behaves as $\exp(\mathrm{i}\xi n)/A_{nn}$.

Let us multiply (6.7.24) by $\exp(-\mathrm{i}\xi m)$, $m \leqslant n$ and integrate the result with respect to ξ from $-\pi$ to π. When $m < n$, the integral on the left-hand side can be replaced by the integral along the closed contour, which consists of the segment of the real axis $-\pi = \xi < \pi$ and the straight lines $\mathrm{Im}\{\xi\} > 0$, $\mathrm{Re}\{\xi\} = \pm\pi/2$, the latter integral being expressible as the sum of the residues of the function $\psi_n(\xi)/\alpha(\xi)$. If $m = n$, then, from such a sum, the term

$$\lim_{n\to\infty} \int_{\mathrm{i}\eta-\pi}^{\mathrm{i}\eta+\pi} \frac{\psi_n \mathrm{e}^{-\mathrm{i}\xi n}}{\alpha(\xi)}\, d\xi = \frac{2\pi}{A_{nn}}$$

must be subtracted, Thus, in view of (6.7.15), we obtain

$$2\pi\mathrm{i} \sum_{\xi_k} \frac{c_\xi \bar{\varphi}_n(\xi_n)\,\mathrm{e}^{-\mathrm{i}m\xi_n}}{\alpha\xi(\xi_k)} - \frac{2\pi\delta_{nm}}{A_{nn}} = \int_{-\pi}^{\pi} \varphi_n(\xi)\,\mathrm{e}^{-\mathrm{i}\xi m}\, d\xi$$
$$+ \int_{-\pi}^{\pi} \frac{\beta(\xi)}{\alpha(\xi)} \bar{\varphi}_n(\xi)\,\mathrm{e}^{-\mathrm{i}\xi m}\, d\xi.$$

Substituting here the triangular representations (6.7.19,22) and also the representation for $\bar{\varphi}_n(\xi_k)$ which is obtained from (6.7.22) when $\xi = \xi_k$, we obtain the analog of the Marchenko equation for the matrix A_{nm} realizing the triangular representation (6.7.19)

$$A_{nm} - \frac{\delta_{nm}}{A_{nn}} + (-1)^n F_{n+m} A_{nn} + (-1)^n \sum_{m_1<n} A^*_{nm_1} F_{m_1+m} = 0, \tag{6.7.25}$$

where

$$F_n = -\mathrm{i} \sum_{\xi_k} \frac{c_{\xi_k} \mathrm{e}^{-\xi_k n}}{\alpha'(\xi_k)} + \frac{1}{2\pi} \int_{-\infty}^{\infty} \frac{\beta(\xi)}{\alpha(\xi)} \mathrm{e}^{-\mathrm{i}\xi n}\, d\xi. \tag{6.7.26}$$

Note that the constants c_ξ contained in (6.7.26) are real.

The system of equations (6.7.25) permits to find A_{nm} from the scattering data and constitutes the equation of the inverse spectral problem for the operator L. Now (6.7.25) constitutes, in fact, a system of linear equations. Introducing the notation $G_{nm} = A_{nn}A_{nm} (m < n)$, we obtain for G the system of equations

$$G_{nm} + (-1)^n F_{n+m} + (-1)^n \sum_{m_1<n} G^*_{nm_1} F_{m_1+n} = 0 \tag{6.7.27}$$

and the expression for A_{nn}

$$A_{nn} = [1 + (-1)^n F_{2n} + (-1)^n \sum_{m<n} G^*_{nm} F_{m+n}]^{-1/2} . \tag{6.7.28}$$

As for the continuum equations described above, the system (6.7.27) can be solved if $\beta(\xi) \equiv 0$. Such a solution is completely determined by prescribing the zeros ξ_k of the function $\alpha(\xi)$ and of the corresponding quantities c_{ξ_k}, and describes an N-soliton solution of the system (6.7.1). Let us consider the simplest situation, when $\alpha(\xi)$ has only one zero, which lies, for example, on the straight line $\mathrm{Re}\{\xi\} = \pi/2$: $\xi = \pi/2 + \mathrm{i}y$ The condition of periodicity of $\alpha(\xi)$ and the relation (6.7.12), which means that $|\alpha(\xi)| = 1$ when $\mathrm{Im}\{\xi\} = 0$ in the example under consideration, allows us to reconstruct $\alpha(\xi)$ uniquely:

$$\alpha(\xi) = \frac{\sin\left(\xi - \dfrac{\pi}{2} - \mathrm{i}\eta\right)}{\sin\left(\xi - \dfrac{\pi}{2} + \mathrm{i}\eta\right)} . \tag{6.7.28a}$$

The function F_n determined according to (6.7.26) has here the form

$$F_n = c \sinh(2\eta)\, \mathrm{e}^{-\mathrm{i}\pi\eta/2} \mathrm{e}^{\eta n} .$$

We seek a solution G_{nm} of the system (6.7.27) in the form

$$G_{nm} = G_n \mathrm{e}^{-\mathrm{i}\pi m/2} \mathrm{e}^{\eta m} .$$

By simple calculation, we find

$$G^*_n = -\gamma \frac{\mathrm{e}^{-\mathrm{i}\pi n/2} \mathrm{e}^{\eta n}}{1 + \dfrac{\gamma \mathrm{e}^{2\eta n}}{\mathrm{e}^{2\eta} - 1}} , \quad \gamma = c \sinh 2\eta ,$$

from which we immediately obtain

$$A_{nn} = \left(1 + \frac{\gamma \mathrm{e}^{2\eta n}}{1 + \dfrac{\gamma \mathrm{e}^{2\eta n}}{\mathrm{e}^{2\eta} - 1}}\right)^{-1/2} , \quad A_{n,n-1} = -\mathrm{i}\gamma \frac{\mathrm{e}^{\eta(2n-1)}}{1 + \dfrac{\gamma \mathrm{e}^{2\eta n}}{\mathrm{e}^{2\eta} - 1}} A_{nn} . \tag{6.7.29}$$

Using (6.7.21) and taking into account that $c_n = \exp(x_n - x_{n-1})$, we finally obtain

$$x_n = \ln \frac{\cosh \eta \left(n - x_0 - \dfrac{1}{2}\right)}{\cosh \eta \left(n - x_0 + \dfrac{1}{2}\right)} + \text{const},$$

$$v_n = \frac{\sinh^2\eta}{\cosh\eta\left(n - x_0 + \dfrac{1}{2}\right)\cosh\eta\left(n - x_0 - \dfrac{1}{2}\right)}, \tag{6.7.30}$$

where

$$x_0 = \frac{1}{2\eta}\ln\frac{1}{c\cosh\eta} \tag{6.7.31}$$

is the coordinate of the center of the soliton.

Substituting in (6.7.31) the time-dependence determined by (6.7.16), we find: $x_0(t) = x_0(0) - (\sinh\eta)/\eta$ Thus, the solution (6.7.30) is a single wave propagating through the chain (6.7.1) with velocity $v = -\sinh\eta/\eta$ without changing its form, i.e. it is a soliton. Analogous calculations show that to a zero of $a(\xi)$ on the straight line $\mathrm{Re}\{\xi\} = \pi/2$ corresponds a soliton which moves in the opposite direction.

In a more general case, when $a(\xi)$ has N zeros in the strip $-\pi/\mathrm{Re}\{\xi\} < \pi$ of the upper half-plane among which N_+ are located on the straight line $\mathrm{Re}\{\xi\} = \pi/2$, and N_- on the straight line $\mathrm{Re}\{\xi\} = -\pi/2$, it can be shown that the corresponding solution of the system (6.7.1) consists, asymptotically as $t \to \pm\infty$, of two sets of solitons, $N_-(N_+)$ being the number of solitons which have positive (negative) velocities, i.e. it describes an N-soliton collision. One can also see that the amplitudes (and hence, the velocities) of solitons appearing as $t \to + \to$, exactly coincide with the same quantities, as $t \to -\infty$; this is quite obvious from the point of view of the method under consideration, since these parameters of the solitons are determined by the eigenvalues of the operator L, which are independent of time. The whole effect of the collison of the solitons is, as it was in the cases of KdV- and nonlinear string equations, in changing the quantities $x_0(t)$ (6.7.31) with respect to any moment, for example, to $t = 0$.

Now we shall show a simple method of calculating the change of the above quantities. For definiteness let us consider the collision of two solitons moving in the positive direction. To such a solution there correspond two zeros of $a(\xi)$ on the straight line $\mathrm{Re}\{\xi\} = -\pi/2$ at points $\xi_1 = -\pi/2 + i\eta_1$, $\xi_2 = -\pi/2 + i\eta_2$. Let us set $\eta_2 < \eta_1$ Then, since the velocity of the second soliton is greater than that of the first, the first one is to the right of the second, as $t \to -\infty$ and as $t \to +\infty$ the sequence of the solitons is opposite. Let us consider in detail the behavior of the eigenfunction $\psi_n(\xi_2)$ of the operator L in this case. Let us denote the quantities $x_0(t)$ for the first and the second solitons by $x_0^{(1)}$ and $x_0^{(2)}$ respectively. As $t \to -\infty$, $x_0^{(2)}(t) \ll x_0^{(1)}(t)$. In the region $h \gg x_0^{(1)}$ the function $\psi_n(\xi_2)$ has the form $\psi_n(\xi_2) = \exp(i\xi_2\eta)$ When passing through the soliton, $\psi_n(\xi_2)$, according to (6.7.11), transforms into $a_1(\xi_2)\exp(i\xi_2\eta)$, where $a_1(\xi_2)$ as a consequence of (6.7.28a) has the form

$$a_1(\xi_2) = \frac{\sin(\xi_2 - \pi/2 - i\eta_1)}{\sin(\xi_2 - \pi/2 + i\eta_1)} = \frac{\sinh(\eta_1 - \eta_2)}{\sinh(\eta_1 + \eta_2)}.$$

For the second soliton, the function $\exp(i\xi_2 \eta)$ is an asymptote of its eigenfunction to the right of it. Therefore, in the region $n \ll x_0^{(2)} \psi_n(\xi_2)$ we have, see (6.7.15),

$$\psi_n(\xi_2) = c_2^{(-)} \frac{\sinh(\eta_1 - \eta_2)}{\sinh(\eta_1 + \eta_2)} \bar{\varphi}_n(\xi_2), \tag{6.7.32}$$

where $c_2^{(-)}$ is connected with $x_0^{(2)}(t)$ by means of (6.7.31). Now, let $t \to +\infty$. In this case $x_0^{(1)} \ll x_0^{(2)}$. When $n \ll x_0^{(1)}$ $\psi_n(\xi_2)$ due to (6.7.15), we have

$$\psi_n(\xi_2) = c_{\xi_2} \bar{\varphi}_n(\xi_2) \tag{6.7.33}$$

(that concerns also the case $t \to -\infty$). Taking the complex conjugate of the relation (6.7.11) and solving the resulting systems with respect to $\bar{\varphi}$, we find

$$\bar{\varphi}_n(\xi) = \alpha(\xi)\, \psi_n(\xi) - \beta^*(\xi^*)\, \psi_n(\xi). \tag{6.7.34}$$

Using (6.7.33,34), we obtain the expression for $\psi_n(\xi_2)$ in the region $x_0^{(1)} \ll n \ll x_0^{(2)}$

$$\psi_n(\xi_2) \simeq c_{\xi_2} \alpha_1(\xi_2)\, (-1)^n \, e^{-i\xi_2 n}.$$

But $(-1)_n \exp(i\xi_2 \eta)$ is the 'left' asymptote of the eigenfunction of the second soliton. Thus, when $n \gg x_0^{(2)}$ we must have

$$\psi_n(\xi_2) = c_{\xi_2} \alpha_1(\xi_2) \frac{1}{c_2^{(+)}} \psi_n(\xi_2), \tag{6.7.35}$$

where $c_2^{(+)}$ is connected with $x_2^{(2)}(t)|_{t\to+\infty}$ by the relation (6.7.31). Hence, from (6.7.35) it follows that

$$c_{\xi_2}|_{t\to+\infty} = c_2^{(+)} \frac{1}{\alpha_1(\xi_2)} = c_2^{(+)} \frac{\sinh(\eta_1 + \eta_2)}{\sinh(\eta_1 - \eta_2)}. \tag{6.7.36}$$

Comparing the relations (6.7.33) and (6.7.32) we obtain

$$c_{\xi_2}|_{t\to-\infty} = c_2^{(-)} \frac{\sinh(\eta_1 - \eta_2)}{\sinh(\eta_1 + \eta_2)}. \tag{6.7.37}$$

Further, excluding the time dependence from the relations (6.7.36, 37) and referrring them to $t = 0$, we obtain

$$c_2^{(+)}(0) = c_2^{(-)}(0) \frac{\sinh^2(\eta_1 - \eta_2)}{\sinh^2(\eta_1 + \eta_2)},$$

which, in veiw of (6.7.31) means that

$$\Delta x_0^{(2)} = x_0^{(2)+} - x_0^{(2)-} = \frac{1}{2\eta_2} \ln \frac{\sinh^2(\eta_1 + \eta_2)}{\sinh^2(\eta_1 - \eta_2)} . \tag{6.7.38}$$

Analogously for the slow soliton

$$\Delta x_0^{(1)} = -\frac{1}{2\eta_1} \ln \frac{\sinh^2(\eta_1 + \eta_2)}{\sinh^2(\eta_1 - \eta_2)} .$$

If now one of the solitons moves in the opposite direction (let this be the second soliton, for definiteness), then

$$\Delta x_0^{(1)} = \frac{1}{2\eta_1} \ln \frac{\sinh^2(\eta_1 + \eta_2)}{\sinh^2(\eta_1 - \eta_2)}, \quad \Delta x_0^{(2)} = -\frac{1}{2\eta_2} \ln \frac{\sinh^2(\eta_1 + \eta_2)}{\sinh^2(\eta_1 - \eta_2)} . \tag{6.7.39}$$

The method presented above can be applied also in the general case of N-soliton collison. It turns out then that as before only pair wise collisions take place, i.e. the displacement of any soliton is equal to the sum of the displacements, arising during the collisons of this soliton with all the others separately.

Note that the described procedure of the determination of the effect of collison of solitons can be used for all equations integrable with the help of the inverse scattering method.

6.8 Fermi-Pasta-Ulam Problem

In 1954, Fermi, Pasta, and Ulam performed a series of numerical experiments, their aim having been to clarify how "stochastization" and transition to the equipartition of energy in dynamical systems with a large number of degrees of freedom occur. As the object of the experiment, they took one-dimensional chains of nonlinear oscillators, which represented discrete models of a nonlinear string. The degree of nonlinearity and the number of oscillators in the chains were sufficiently large (in typical experiments, the chain consisted of 64 oscillators), so that the experimentalists hoped to see a rapid stochatization of the chain and the transition to the homogeneous distribution of energy among the degrees of freedom. Instead of this, they observed quasiperiodic exchange of energies between several initially excited modes and for sufficiently long computing time (up to several hundres periods of vibration) they could not reliably fix a tendency of a stochastic transition of energy to higher modes.

The problem of abnormally slow stochastisation of one-dimensional chains of nonlinear oscillators was named the Fermi-Pasta-Ulam problem. In subsequent years, many works have been devoted to this problem both analytical and containing results of numerical experiments.

The chain with quadratic nonlinearity, whose motion is described by the equations

$$\ddot{\xi}_n = \xi_{n+1} - 2\xi_n + \xi_{n-1} + (\xi_{n+1} - \xi_n)^2 - (\xi_n - \xi_{n-1})^2 . \tag{6.8.1}$$

demonstrated most clearly the "abnormal" behavior in the Fermi-Pasta-Ulam experiments. Regular quasi-periodic behavior of the chain was observed also in later and more detailed experiments.

Equation (6.6.5) is the continuous analog of the chain (6.8.1). As was shown in Sect. 6.6, (6.6.5) possesses an (L, A)-pair. Let us introduce the variables

$$v = -\frac{3}{4}u, \; \Phi_x = -\frac{3\sqrt{3}}{4\sqrt{2}}\,\mathrm{i}w .$$

In these variables (6.6.6) is transformed into the system

$$\begin{aligned} v_t &= \Phi_{xx} , \\ \Phi_t &= v + v^2 + \frac{1}{4} v_{xx} . \end{aligned} \tag{6.8.2}$$

The operator L in these variables has the form

$$L = \mathrm{i}\frac{\partial^3}{\partial x^3} + \mathrm{i}\left(v\frac{\partial}{\partial x} + \frac{\partial}{\partial x}v\right) + \mathrm{i}\frac{\partial}{\partial x} - \sqrt{\frac{4}{3}}\,\Phi_x . \tag{6.8.3}$$

Let us consider the system (6.8.2) under periodic boundary conditions on the segment $[0, l]$ and consider on the same segment the eigenvalue problem

$$L\psi = \lambda\psi \tag{6.8.4}$$

with periodic boundary conditions. Applying the reasoning carried out in Sect. 6.1, we see that all the eigenvalues are integrals of the system (6.8.2).

Observe that the system (6.8.2) can be written in the Hamiltonian form

$$\begin{aligned} u_t &= -\frac{\delta H}{\delta \Phi} . \\ \Phi_t &= \frac{\delta H}{\delta v} , \end{aligned} \tag{6.8.5}$$

where

$$H = \int_0^l \left(\frac{1}{2}v^2 + \frac{1}{2}\Phi_x^2 + \frac{1}{3}v^3 - \frac{1}{8}v_x^2\right) dx$$

is the Hamiltonian of the system. This allows us to determine the Poisson brackets for the functionals S and T of v and Φ:

$$[S, T] = \int_0^l \left(\frac{\delta S}{\delta v}\frac{\delta T}{\delta \Phi} - \frac{\delta S}{\delta \Phi}\frac{\delta T}{\delta u}\right) dx . \tag{6.8.6}$$

Let us show that the Poisson brackets between the eigenvalues λ_i of the operator L are equal to zero and in this sense λ_i are commuting integrals of the system (6.8.2). The variation $\delta\lambda$ of the eigenvalue λ, for infinitesimal perturbations δL of the operator L, has the form

$$\delta\lambda = \langle\psi|\delta L|\psi\rangle\,. \tag{6.8.7}$$

Here ψ is an eigenfunction of the operator L, which satisfies the equation

$$\mathrm{i}\psi_{xxx} + \mathrm{i}(2v+1)\,\psi_x + \mathrm{i}v_x\psi - \sqrt{\frac{4}{3}}\,\Phi_x\psi = \lambda\psi\,. \tag{6.8.8}$$

From (6.8.3), we have

$$\delta L = \mathrm{i}\left(\delta u\,\frac{d}{dx} + \frac{d}{dx}\,\delta u\right) - \sqrt{\frac{4}{3}}\,(\delta\Phi)_x\,. \tag{6.8.9}$$

Substituting (6.8.9) into (6.8.6), we obtain

$$\begin{aligned}\frac{\delta\lambda}{\delta u} &= \mathrm{i}(\psi^*\psi_x - \psi\psi^*_x)\,,\\ \frac{\delta\lambda}{\delta\Phi} &= \sqrt{\frac{4}{3}}\frac{d}{dx}\,|\psi|^2\,.\end{aligned} \tag{6.8.10}$$

Denoting, as in Sect. 6.1, the Wronskian of the functions χ_1 and χ_2 by

$$\{\chi_1, \chi_2\} = \chi_1\chi_{2x} - \chi_2\chi_{1x}\,,$$

we find from (6.8.6, 10), for the two eigenvalues λ_1 and λ_2

$$\begin{aligned}[\lambda_1, \lambda_2] &= \sqrt{\frac{4}{3}}\,\mathrm{i}\int_0^l \left[\{\psi_1^*, \psi_1\}\frac{d}{dx}\,|\psi_2|^2 - \{\psi_2^*, \psi_2\}\frac{d}{dx}\,|\psi_1|^2\right]dx\\ &= \sqrt{\frac{4}{3}}\,\mathrm{i}\int_0^l \frac{d}{dx}\left[\psi_1\psi_2\frac{d}{dx}\{\psi_2^*\,\psi_1^*\} - \psi_1^*\psi_2^*\frac{d}{dx}\{\psi_2\psi_1\}\right]dx\end{aligned} \tag{6.8.11}$$

Here ψ_1, ψ_2 are the corresponding eigenfunctions. From (6.8.8) it is then easy to obtain

$$\psi_1\psi_2 = \frac{\mathrm{i}}{\lambda_1 - \lambda_2}\left[\frac{d^2}{dx^2}\{\psi_2, \psi_1\} - \{\psi_{2x}, \psi_{1x}\} + (2u+1)\,\{\psi_2, \psi_1\}\right], \tag{6.8.12}$$

$$\frac{d}{dx}\{\psi_{2x}, \psi_{1x}\} = ux\,\{\psi_2, \psi_1\} + \mathrm{i}\sqrt{\frac{4}{3}}\,\Phi_x\{\psi_2, \psi_1\} + \mathrm{i}\,(\lambda_2\psi_2\psi_{1x} - \lambda_1\psi_1\psi_{2x})\,. \tag{6.8.13}$$

Substituting (6.8.12) into (6.8.11), and applying the relations (6.8.13), we obtain, after simple transformations,

$$[\lambda_1, \lambda_2] = \sqrt{\frac{4}{3}}\,\frac{\mathrm{i}}{2}\int_0^l \Big[\{\psi_2^*, \psi_1^*\}\frac{d}{dx}\psi_2\psi_1 - \{\psi_2, \psi_1\}\frac{d}{dx}\psi_2^*\psi_1^*\Big]\,dx = \frac{1}{2}[\lambda_1, \lambda_2]\,,$$

from which it follows that $[\lambda_1, \lambda_2] = 0$. All eigenfunctions of the periodic problem are commutating integrals of motion of the system (6.8.3).

An analogous result is true for the Toda chain.

The system (6.7.1) is obviously of Hamiltonian type. Its Hamiltonian is

$$H = \sum_n \left(\frac{v_n^2}{2} + \mathrm{e}^{x_n - x_{n-1}} - 1\right), \tag{6.8.14}$$

and the variables x_n, v_n are canonically conjugate. Consider now the Toda chain, which consists of N particles with periodic conditions, setting in the operator L (6.7.2): $c_{n+N} = c_n$, $v_{n+N} = v_n$ For the eigenvalue problem (6.7.5), we also consider the class of periodic functions, assuming $\psi_{n+N} = \psi_n$. Then the infinite matrix (6.7.2) is reduced to a Hermitian $N \times N$ matrix which has N independent eigenvalues λ_i, $i = 1, \dots, N$, constituting intergrals of the system (6.7.1).

Let us show that all the eigenvalues of the periodic problem commute with each other. To this end write, first of all, the Poisson bracket of some quantities, S and T

$$[S, T] = \sum_n \frac{\delta S}{\delta x_n}\frac{\delta T}{\delta v_n} - \frac{\delta S}{\delta v_n}\frac{\delta T}{\delta x_n} \tag{6.8.15}$$

in the variables v_n and $c_n = \exp(x_n - x_{n-1})$. Simple calculations yield for (6.8.15)

$$[S, T] = \sum_n c_n \left(\frac{\delta S}{\delta c_n}\frac{\delta T}{\delta v_n} - \frac{\delta S}{\delta v_n}\frac{\delta T}{\delta c_n}\right) - c_{n+1}\left(\frac{\delta S}{\delta c_{n+1}}\frac{\delta T}{\delta v_n} - \frac{\delta S}{\delta v_n}\frac{\delta T}{\delta c_{n+1}}\right). \tag{6.8.16}$$

Further, let us evaluate the functional derivatives $\delta\lambda_i/\delta c_n$, $\delta\lambda_i/\delta v_n$ using for this an analog of (6.8.7)

$$\delta\lambda_i = \langle\psi(i)|\delta L|\psi(i)\rangle = \sum_{n,m}\psi_n^*(i)\,L_{nm}\psi(i)\,, \tag{6.8.17}$$

where $\psi(i)$ is a normalized eigenfunction of the operator L, corresponding to the eigenvalue λ_i. Using the explicit form (6.7.2) of L, we obtain from (6.8.17)

$$\frac{\delta\lambda_l}{\delta v_n} = \psi_n^*(i)\,\psi_n(i)\,,$$

$$\frac{\delta\lambda_l}{\delta c_n} = \frac{\mathrm{i}}{2}\frac{1}{\sqrt{c_n}}[\psi_n^*(i)\,\psi_{n-1}(i) - \psi_{n-1}^*(i)\,\psi_n(i)]\,. \tag{6.8.18}$$

Moreover, note that for any pair of eigenfunctions $\psi(i)$ and $\psi(j)$ of the problem (6.7.5), whith the corresponding eigenvalues λ_l and λ_j, respectively, a directly verifiable relation

$$(\lambda_l - \lambda_j)\,\psi_n^*(j)\,\psi_n(i) = W_n[\psi^*(j), \psi(i)] - W_{n+1}[\psi^*(j), \psi(i)] \tag{6.8.19}$$

exists, where

$$W_n(\psi^*(j)\,\psi(i)) = \mathrm{i}\sqrt{c_n}[\psi_n^*(j)\,\psi_{n-1}(i) - \psi_{n-1}^*(j)\,\psi_n(i)]\,.$$

Calculating now the brackets $[\lambda_l, \lambda_j]$ (6.8.16) and using (6.8.18) and (6.8.19), we see that

$$[\lambda_l, \lambda_j] = \frac{1}{\lambda_l - \lambda_j}\sum_{n=1}^{N} |\,W_{n+1}[\psi^*(i), \psi(j)]\,|^2 - |\,W_n[\psi^*(i), \psi(j)\,|^2$$

$$= \frac{1}{\lambda_l - \lambda_j}\{|\,W_{N+1}[\psi^*(i), \psi(j)]\,|^2 - W_1[\psi^*(i), \psi(j)]\,|^2\}\,.$$

In view of periodic conditions $[\lambda_l, \lambda_j]$ vanishes.

Observe also that the Hamiltonian H of (6.8.14) is expressed in terms of the trace of the square of the operator L. For a chain, consisting of N particles,

$$H + N = \frac{1}{2}\,\mathrm{Sp}\{L^2\} = \frac{1}{2}\sum_{n=1}^{N}\lambda_n^2\,. \tag{6.8.20}$$

From the analysis of integrals of motion of the equations of the nonlinear string and the Toda chain, one can conclude that they are completely integrable. For the Toda chain, this follows directly from the Liouville theorem. Under periodic boundary conditions, the chain with N particles has exactly N commuting integrals λ_l. All these integrals can be chosen as action variables; the formula (6.8.17) represents a Hamiltonian, expressed in these variables. For a nonlinear string the existence of a countable set of commuting integrals λ_l is only a necessary condition of integrability. However, the equation of a nonlinear string can be represented as a continuum limit of the Toda chain (Sect. 5.4). Therefore, (6.6.5) is also a completely integrable system. This fact fully explains the results of the Fermi-Pasta-Ulam experiment. Since integratable systems do not become stochastic and perform quasi-periodic motions, it is the degree of deviation of a system from completely integrable continuous limit, that can lead to stochastization of the chain rather than degree of nonlinearity of the system.

For smooth initial conditions considered in Fermi-Pasta-Ulam experiments, this deviation had the order $\xi_{xx}/\xi \sim N^{-2}$ where N is the number of oscillators in the chain. When $N = 64$, this leads to the decrease of "effective nonlinearity" by $10^3 \div 10^4$ times and to the corresponding increase of time of stochastization up to values which are far in excess of those attained in numerical experiments. With an increasing number of oscillators in the chain, the corresponding increase of the time of stochastisation occurs.

From the above, it follows that caution must be exercised in estimating the time of stochastization of a one-dimensional dynamical system.

Among one-dimensional Hamiltonian systems there exist many completely integratable ones, both continuous and discrete, the problem of enumeration and classification of them being far from solved. When investigating a given Hamiltonian system, one has to take into account that the speed of its stochatization is determined by its "distance" from a "nearest" completely integrable system. Accidentally, this "distance" may be small (as it was with the chain (6.8.1)), which leads to an increasing time of stochastization.

6.9 Perspectives of the Method

The KdV-equation (6.1.1) and the two equations of a nonlinear string (6.6.6, 7) exhaust the physically interesting applications of the inverse scattering method, which uses scalar one-dimensional differential operators L, A. Apart from the case of the Toda chain, using scalar difference operators, one can integrate the differential-difference equation

$$\dot{f}_n = 2f_n(f_{n+1} - f_{n-1}) \, . \tag{6.9.1}$$

This equation occurs in the theory of plasma turbulence and describes in some cases the transfer of energy along the spectrum of turbulence. Just as the Toda chain in the continuum limit leads to the equation of a nonlinear string, equation (6.9.1) leads in the continuum limit to the KdV-equation. The operators L and A for (6.9.1) have the form

$$\begin{aligned} L_{nm} &= \sqrt{f_n}\,\delta_{n,m+1} + \sqrt{f_m}\,\delta_{n+1,m} \, , \\ A_{nm} &= \sqrt{f_n f_{n-1}}\,\delta_{n,m+2} - \sqrt{f_m f_{m-1}}\,\delta_{n+2,m} \, . \end{aligned} \tag{6.9.2}$$

Many more possibilities are contained in the matrix operators. Thus using the operator

$$L = \mathrm{i}\begin{bmatrix} 1 & 0 \\ 0 & -1 \end{bmatrix}\frac{\partial}{\partial x} + \begin{bmatrix} 0 & u^* \\ \pm u & 0 \end{bmatrix}, \tag{6.9.3}$$

one can integrate the equations

$$iu_t + \frac{1}{2}u_{xx} \mp |u|^2 u = 0 \tag{6.9.4}$$

(nonlinear Schrödinger equations). These equations occur in problems of nonlinear optics. The operator A for (6.9.4) has the form

$$A = \begin{bmatrix} 1 & 0 \\ 0 & -1 \end{bmatrix} \frac{\partial^2}{\partial x^2} - \mathrm{i}\begin{pmatrix} 0 & u^* \\ \pm u & 0 \end{pmatrix} \frac{\partial}{\partial x}$$
$$- \frac{\mathrm{i}}{2}\begin{pmatrix} 0 & u_x^* \\ u_x & 0 \end{pmatrix} \mp \frac{|u|^2}{2}\begin{pmatrix} 1 & 0 \\ 0 & -1 \end{pmatrix}. \tag{6.9.5}$$

Equations (6.9.4) have been studied almost to the same extent as the KdV-equation. For these, N-soliton solutions have been found, complete integrability, under certain conditions at infinity, has been proved, the asymptotic behavior at $t \to \pm\infty$ has been studied and the conservation laws have been obtained. As in the case of the KdV equation, the polynomial conservation laws generate a denumerable sequence of exactly integrable equations, (6.9.4) being the first nontrivial term of this sequence. The equation which can be reduced to the form

$$u_t + u_{xxx} \pm |u|^2 u_x = 0\,. \tag{6.9.6}$$

is the next term of the sequence. This equation, known as the "modified" Korteweg-de Vrise (KdV) equation is encountered in solid-state physics.

The special case of the operator L (6.9.3)

$$L = \mathrm{i}\begin{bmatrix} 1 & 0 \\ 0 & -1 \end{bmatrix} \frac{\partial}{\partial x} + \mathrm{i}\begin{bmatrix} 0 & u \\ u & 0 \end{bmatrix}, \tag{6.9.7}$$

where u is real, is of a particular interest. Apart from the usual sequence of integrable equations connected with the expansion of the coefficients $a(k)$ in powers of $1/k$ (the first term of this sequence is the modified KdV-equation), this operator generates a second sequence, connected with the expansion of $a(k)$ near $k = 0$. The integrable equation, applicable in nonlinear optics

$$u_{xt} = \sin u\,, \tag{6.9.8}$$

is the first in this sequence. Operator A for the equation (6.9.8) is an integral one. Vector generalization of (6.9.4,5) can also be easily obtained. Thus, with the help of the operator

$$L = \begin{bmatrix} 1 & 0 & 0 \\ 0 & -1 & 0 \\ 0 & 0 & -1 \end{bmatrix} \frac{\partial}{\partial x} + \begin{bmatrix} 0 & u_1^* & u_2^* \\ \pm u_1 & 0 & 0 \\ \pm u_2 & 0 & 0 \end{bmatrix} \tag{6.9.9}$$

the system appearing in nonlinear optics

$$
\mathrm{i}u_{1t} + \frac{1}{2}u_{1xx} \mp (|u_1|^2 + |u_2|^2)\,u_1 = 0\,,
$$

$$
\mathrm{i}u_{2t} + \frac{1}{2}u_{2xx} \mp (|u_1|^2 + |u_2|^2)\,u_2 = 0\,, \qquad (6.9.10)
$$

can be integrated,

Applying the operator

$$
L = \mathrm{i}\begin{bmatrix} 1 & 0 & 0 \\ 0 & 0 & 0 \\ 0 & 0 & -1 \end{bmatrix}\frac{\partial}{\partial x} + \begin{bmatrix} 0 & u_1 & u_2 \\ u_1^* & 0 & 0 \\ u_2^* & 0 & 0 \end{bmatrix}, \qquad (6.9.11)
$$

one can integrate the well-known (in non-linear optics) equations of generation of the second harmonic

$$
\frac{\partial u_1}{\partial t} + \frac{\partial u_1}{\partial x} = \mathrm{i}u_2u_1^*\,,
$$

$$
\frac{\partial u_2}{\partial t} - \frac{\partial u_2}{\partial x} = \mathrm{i}u_1^2\,, \qquad (6.9.12)
$$

In the special case of pure imaginary u_1, u_2, these equations can be transformed to the form

$$
\frac{\partial^2 \Phi}{\partial t^2} - \frac{\partial^2 \Phi}{\partial x^2} = \mathrm{e}^{-\Phi}\,,
$$

$$
u_1 = \mathrm{i}\mathrm{e}^{-\Phi}, \quad u_2 = -\mathrm{i}\left(\frac{\partial \Phi}{\partial t} + \frac{\partial \Phi}{\partial x}\right). \qquad (6.9.13)
$$

The equation

$$
\frac{\partial^2 \Phi}{\partial t^2} + \frac{\partial^2 \Phi}{\partial x^2} = \mathrm{e}^{-\Phi}
$$

is also integrated with the help of the operator (6.9.11)

The use of (4×4)-matrix operator with the principal term

$$
\mathrm{i}\begin{bmatrix} 1 & & & \\ & 1 & & \\ & & 0 & \\ & & & 0 \end{bmatrix}\frac{\partial}{\partial x}
$$

allows us to integrate the important "Sine-Gordon" equation

$$
\frac{\partial^2 u}{\partial t^2} - \frac{\partial^2 u}{\partial x^2} + \sin u = 0 \qquad (6.9.14)
$$

with respect to a generally complex function u and also the system of equations

$$\frac{\partial \hat{\rho}}{\partial t} + [H, \hat{\rho}] = 0, \quad \hat{\rho} = \begin{bmatrix} n_1 & \rho_{12} \\ \rho_{12}^* & n_2 \end{bmatrix},$$

$$\frac{\partial E}{\partial t} + \frac{\partial E}{\partial x} = \rho_{12}, \quad H = \begin{bmatrix} \omega_1 & \mathrm{i}E \\ -\mathrm{i}E^* & \omega_2 \end{bmatrix}, \tag{6.9.15}$$

which describes the propagation of light pulses in a resonating medium with infinite relaxation times.

In all the cases described above, construcing the equations of the inverse scattering problem does not present, in principle, any difficulties (though it can not be very simple technically): the equations of the inverse problem can be constructed by the method developed in Sect. 6.2.

Difficulties in constructing inverse scattering problem are encountered with the operator

$$L = \mathrm{i}\begin{bmatrix} a_1 & 0 & 0 \\ 0 & a_2 & 0 \\ 0 & 0 & a_3 \end{bmatrix}\frac{\partial}{\partial x} + \begin{bmatrix} 0 & u_2 & u_1 \\ \pm u_2^* & 0 & u_3 \\ u_1 & \pm u_3^* & 0 \end{bmatrix}, \tag{6.9.16}$$

which allows us to integrate the systems of equations

$$u_{1t} + v_1 u_{1x} = \mathrm{i}q u_2 u_3\,,$$
$$u_{2t} + v_2 u_{2x} = \mathrm{i}q u_1 u_3^*, \quad u_{3t} + v_3 u_{3x} = \mathrm{i}q u_1 u_2^* \tag{6.9.17}$$

$$u_{1t} + v_1 u_{1x} = \mathrm{i}q u_2^* u_3^*\,,$$
$$u_{2t} + v_2 u_{2x} = \mathrm{i}q u_1^* u_3^*, \quad u_{3t} + v_3 u_{3x} = \mathrm{i}q u_2^* u_3^*\,, \tag{6.9.18}$$

used to describe reasonance (decay and growth) interaction of three monochromatic waves in a nonlinear medium. The difficulties are connected with the fact that not all the Jost functions of the operator (6.9.16) are analystic, and are partially overcome by application of Shabat's method. For the complete solution of the inverse scattering problem for the operator (6.9.16), an essential improvement of traditionally known methods[1] is necessary. The use of matrix operators of higher dimensions permits to construct generalizations of the systems (6.9.17,18), also of physical interest.

All known one-dimensional integrable systems, which are interesting from the point of view of applications, are exhausted by the above list. Using the direct commutation of operators or applying Shabat's method, one can construct new integrable systems however, the complexity of these increases rapidly with the increase in the order of the operator L and the dimen-

[1]These difficulties were overcome by the works of Zakharov, Manakov, Shabat (USSR) and partially by Kaup (USA).

sion of the matrices it contains. Thus, the nth order operator with $N \times N$-matrices generates, in general, integrable systems for $(N^2 n - N)$ unknown functions. Undoubtedly, cases occur when these systems reduce to a smaller number of equations, there is still however no systematic way of determining such systems (among which physically interesting equations are most probable). For the already known integrable one-dimensional equations, there is an important problem of generalizing the method, to the case of nondecreasing initial conditions, as $X \to +\infty$; the the problem with periodic initial conditions mentioned in the previous section is the first of such a kind. The solution of the problem about the asymptotic behaviour as $t \to \infty$, of the initial condition, which decreases as $x \to +\infty$ and does not decrease, as $x \to -\infty$, for example tends to a constant (the problem of breakdown of the step junction as the initial condition) is of great interest.

The question about the possibility of extension to theories having more than one dimension presents a principal interest. In some cases, this can be done by very simple methods. Thus, if we assume, in Shabat's scheme, that the functions F and K depend on two parameters t and z and that F satisfies the equations

$$\frac{\partial F}{\partial z} + \mathrm{i} D_1 F = 0 \quad \frac{\partial F}{\partial t} + \mathrm{i} D_2 F = 0, \tag{6.9.19}$$

then instead of the conditions of compatibility (6.5.23), we obtain an integrable, according to the scheme (6.5.24), system

$$\frac{\partial L}{\partial t} - \frac{\partial A}{\partial z} = \mathrm{i}[L, A]\,. \tag{6.9.20}$$

The simplest system of the type (6.9.19) leads to the equation

$$\frac{\partial}{\partial x}(u_t - 6uu_x + u_{xxx}) = \pm \frac{\partial^2 u}{\partial z^2}, \tag{6.9.21}$$

which is a direct two-dimensional generalization of the KdV equation.

Other possibilities of two-dimensional and three-dimensional generalizations of integrable one-dimensional systems have not been sufficiently studied.

From the point of view of applications, the study of systems close to the above-described completely integrable ones, presents a great interest. The development of the corresponding perturbation theory should significantly extend the number of physical and mechanical problems; the method of inverse scattering problem appears to be useful for their solution.

6.10 Notes*

The method of the inverse scattering problem was discovered by Gardner, Green Kruskal, and Miura in 1967. In their first work [B4.21], an operator L was introduced, the behavior of scattering data in time was found and the principal scheme was suggested. The operator A was introduced by Lax [B4.36], who gave the modern form to carrying out the principal scheme. The inverse problem for the operator $-d^2/dx^2 + u(x)$ on the entire axis was solved by Kay and Moses [B4.30] and by Faddeev. N-soliton solutions were studied by Shabat [B4.56], Zakharov [B4.70], Hirota [B4.27] and Wadati and Toda [B4.66]. The fact that the contribution of multiple collisions is absent was proved for the first time in [B4.70]. The complete integrability of the KdV-equation was proved by Zakharov and Faddeev [B4.73]. Shabat's method is published in [B4.57]. This method was suggested independently by Gardner, Green Kruskal and Miura. The proof of the equivalence of both methods when $u \to 0$, as $|x| \to \infty$ is presented here for the first time.

The (L, A)-pair for a nonlinear string was found by Zakharov [B4.71].

The Toda lattice has been integrated by Manakov, and Sec. 6.7 is an account of his work [B4.43]. The proof of commuting of the eigenvalues for the string is presented in [B4.71] and that for the Toda lattice is presented in [B4.43].

Equation (6.9.4) has been integrated by Zakharov and Shabat [B.4.77], equation (6.9.1) has been integrated by Manakov [B4.43]. The modified KdV-equation (6.9.6) was considered by Wadati. Equation (6.9.5) has been solved by Takhtadzhyan [B4.59] and also by Ablowitz and others [B4.2]. The system of the nonlinear Schrodinger equations has been investigated by Manakov [B4.45].

The Since-Gordon-equation is completely solved in [B4.79], (L, A)-pairs for the equation of resonant interaction of three waves are found in [B4.74].

Supplement.[1] The number of works devoted to the inverse scattering method (ISM) and to a number of adjacent fields has increased considerably in recent years. It is significant that these investigations were joined by a number of well-known mathematicians, who use successfully the arsenal of modern mathematics, in particular, algebraic geometry, theory of group representations, etc. Below, without any pretension of completeness we will give reference to some basic works.

A detailed investigation of a wide class of equations, which are solved according to the ISM in 2×2-matrices (so-called AKNS-systems) was carried out by Ablowitz, Kaup, Newell and Segur [B4.3]. As a particular case, AKNS-systems contain many of equations, considered earlier, for example, KdV-, modified KdV-, Since-Gordon, nonlinear Schrodinger equations and others.

---

*The citations refer to the bibliography to be given in [1.12].

[1] Added by the author of the book.

The works [B4.34,60] contain a complete investigation within the scope of ISM, of the Since-Gordon equation and the Thirring model respectively. In the work of Zakharov and Shabat [B4.76], a method of dressing which gives a far reaching generalization of ISM is suggested. The method is considered in details in the survey [B4.72]. In particular, this method enables one to integrate certain equations with a number of independent variables higher than two. For example, the Kadomtsev-Petviashvily equation is such an equation. Note the work of Zakharov and Manakov [B4.75], in which the problem about three waves, which is integrated in terms of 3×3-matrices is investigated in detail. The three waves model appeared to be the first one with the presence of non-trivial interaction of solitons. Further development of concepts of the work [B4.75] in application to Einstein's equation is contained in [B4.6]. Calogero and Degasperis [B4.10] have suggested a variant of ISM, namely, the method of generalized Wronskians. This method also enables one to integrate equations with any number of independent variables.

Applicability of ISM to an important class of equations of field theory in two dimensions, namely, to the self-duality equations, which arise in the theory of Yang-Mills fields, is considered by Belavin and Zakharov [B4.5]. The equivalence of the nonlinear Schrodinger equation to Heisenberg's equation of ferromagnets is proved with the help of ISM in [B4.78].

Use of the theory of scattering restricted the possibilities of integration to the class of functions decreasing rapidly in the spatial variables. The periodic problem required essentially new concepts, and the first ones where invented by Novikov [B4.75] and Lax [B4.38]. Then this theory was developed in [B4.16, 17. 37, 47]. In [B4. 16–18], as well as in the works of Krichever [B4.32–34] and Manin [B4.42], the deep algebraic geometric nature of periodic and conditionally periodic solutions was understood.

When solving equations with the help of ISM, the method of formal variational calculus appeared to be useful. As regards this method and its applications, see [B4.22–25].

In [B4.15, 26, 35, 41, 48] possible applications of concepts of modern differential geometry in ISM were studied.

A number of investigations are devoted to the group-theoretical aspects of ISM, to the existence of infinite sets of conversation laws and to integrable equations. As regards these and some adjacent topics, see [B4.7, 9, 12–14, 31, 40, 68].

ISM seems to be closely connected with the prolongation structures method, suggested by Wahlquist and Estabrook [B4.67], though these connections are not quite clear. A further development of the method is given in [B4.10, 11, 51, 52].

As regards the connection between ISM, Bäcklund transformation and infinite set of integrals of motion, see, for example [B4.65].

Note surveys and review works on the subject, namely, [B4.1, 3, 4, 19, 20, 49, 55, 58, 62, 72].

A. Appendices

A.1 Summary of Fourier Transforms

In this context, Fourier transforms with respect to the spatial variables x and k play a major role. We summarize their principal features and write them in the form usually found in the physical literature. In order to perform the Fourier transform with respect to time, one has to carry out the substitution $x \to t$, $k \to -\omega$.

The Fourier transforms are understood in the sense of generalized functions [A.1]. In particular, the generalized function x^{-1} (and analogously k^{-1}) acts upon a test function $\varphi(x)$ according to the rule

$$[x^{-1}, \varphi(x)] = ⨍ \frac{\varphi(x)}{x}\, dx\,,$$

where the hatched integral sign denotes the principal value integral.

Some other notations are:

$\delta(x)$ is the Dirac δ-function;

$$\theta(x) = \begin{cases} 0 \text{ if } x < 0\,, \\ 1 \text{ if } x > 0\,; \end{cases}$$

$$\operatorname{sgn}\{x\} = \theta(x) - \theta(-x)\,;$$

$$\delta_\pm(x) = \frac{1}{2}\,\delta(x) \pm \frac{\mathrm{i}}{2\pi}\,x^{-1};$$

$P_n(x)$ is a polynomial of degree n;

$$B(k) = \begin{cases} 1 \quad \text{if} \quad |k| < \pi/a\,, \\ 0 \quad \text{if} \quad |k| > \pi/a\,; \end{cases}$$

$$\delta_B(x) = \frac{\sin(\pi x/a)}{\pi x}\,;$$

$$f(x) * u(x) = \int f(x - x')\, u(x')\, dx'\,;$$

$$f(k) * u(k) = \int f(k - k')\, u(k')\, dk'\,.$$

In the following table, A, b, α, β, are constants.

No	$f(x) = \frac{1}{2\pi} \int f(k)\, e^{ixk}\, dk$	$f(k) = \int f(x)\, e^{-ikx}\, dx$
1	$f(-x)$	$f(-k)$
2	$\overline{f(x)}$	$\overline{f(-k)}$
3	$f(Ax)$	$\lvert A \rvert^{-1} f(A^{-1}k)$
4	$f(x+b)$	$e^{ibk} f(k)$
5	$e^{i\beta x} f(x)$	$f(k-\beta)$
6	$f(x) * u(x)$	$f(k)\, u(k)$
7	$f(x) u(x)$	$\frac{1}{2\pi} f(k) * u(k)$
8	$\frac{d}{dx} f(x)$	$ikf(k)$
9	$P_n\left(\frac{d}{dx}\right) f(x)$	$P_n(ik) f(k)$
10	$xf(x)$	$\mathrm{i} \frac{d}{dk} f(k)$
11	$P_n(x) f(x)$	$P_n\left(i \frac{d}{dk}\right) f(k)$
12	1	$2\pi\delta(k)$
13	x	$2\pi i \frac{d}{dk} \delta(k)$
14	$P_n(x)$	$2\pi P_n\left(\mathrm{i} \frac{d}{dk}\right) \delta(k)$
15	$\delta(x)$	1
16	$\delta_B(x)$	$B(k)$
17	$\frac{d}{dx} \delta(x)$	ik

No	$f(x) = \frac{1}{2\pi} \int f(k)\, e^{ixk}\, dk$	$f(k) = \int f(x)\, e^{-ikx}\, dx$
18	$P_n\left(\frac{d}{dx}\right) \delta(x)$	$P_n(ik)$
19	$\delta(x+b)$	e^{ibk}
20	$e^{i\beta x}$	$2\pi\delta(k-\beta)$
21	$\sin \beta x$	$\pi i\, [\delta(k+\beta) - \delta(k-\beta)]$
22	$\cos \beta x$	$\pi[\delta(k+\beta) + \delta(k-\beta)]$
23	$\theta(x)$	$(ik)^{-1} + \pi\delta(k)$
24	$\mathrm{sgn}\, x$	$2(ik)^{-1}$
25	$\delta_+(x) = -(2\pi i x)^{-1} + \frac{1}{2}\delta(x)$	$\theta(k)$
26	$-(\pi i x)^{-1}$	$\mathrm{sgn}\, k$
27	$(1+x^2)^{-1}$	$\pi e^{-\lvert k \rvert}$
28	$e^{-\lvert x \rvert}$	$2(1+k^2)^{-1}$
29	$e^{-\alpha x^2}$, $Re\, \alpha > 0$	$\sqrt{\frac{\pi}{\alpha}} \exp\left(-\frac{k^2}{4\alpha}\right)$
30	x^{-2}	$-\pi \lvert k \rvert$
31	$\lvert x \rvert$	$-2k^{-2}$

When passing to three-dimensional Fourier transforms, one has to understand x and k as the vectors

$$x = x^\alpha e_\alpha, \quad k = k_\alpha e^\alpha$$

where e_α is a basis of an affine (oblique angle) coordinate system, and e^α is the reciprocal basis.

Then

$$f(x) = \frac{1}{(2\pi)^3} \int f(k)\, e^{ixk}\, dk, \quad f(k) = \int f(x)\, e^{-ik}\, dx\,,$$

where $xk = x^\alpha k_\alpha$. The transforms numbered (1)–(22) remain valid, if one makes the replacements

$$2\pi \to (2\pi)^3, \quad b \to b^\alpha, \quad \beta \to \beta_\alpha,$$

$$A \to A^\alpha_\beta, \quad |A^{-1}| \to |\mathrm{Det}(A^\alpha_\beta)|^{-1} \neq 0,$$

$$\frac{d}{dx} \to \frac{\partial}{\partial x^\alpha}, \quad \frac{d}{dk} \to \frac{\partial}{\partial k_\alpha},$$

and assumes that $B(k)$ is the characteristic function of the parallelepiped $-\pi/a \leqslant k \leqslant \pi/a$, v_0 is the volume of the elementary cell, which is constructed over the vectors e_α and

$$\delta_B(x) = \frac{1}{\pi^3 v_0} \prod_{\alpha=1}^{3} \frac{\sin(xe^\alpha)}{xe^\alpha}.$$

In the case of an isotropic medium, it is assumed that $r = |\boldsymbol{r}|$, $k = |\boldsymbol{k}|$. For functions which depend only on r and k, the Fourier trnasform is written in the form ($f(-r) = f(r)$ and $f(-k) = f(k)$)

$$f(r) = -\frac{\mathrm{i}}{4\pi^2 r} \int_{-\infty}^{\infty} kf(k)\, \mathrm{e}^{\mathrm{i}rk}\, dk,$$

$$f(k) = \frac{2\pi \mathrm{i}}{k} \int_{-\infty}^{\infty} rf(r)\, \mathrm{e}^{-\mathrm{i}kr}\, dr;$$

in the presence of poles it is necessary to carry out the corresponding regularization of the integrals.

A.2 Retarded Functions and Dispersion Relations

Let $f(t)$ be a generalized function, which does not have a singularity at the point $t_0 = 0$. Then it can be represented in the form

$$f(t) = f_+(t) + f_-(t), \tag{A.2.1}$$

$$f_+(t) = \theta(t) f(t), \quad f_-(t) = \theta(-t) f(t). \tag{A.2.2}$$

The generalized functions $f_+(t)$ and $f_-(t)$ have supports on the right and left half-axes, respectively. As for the Green's functions (Sect. 4.1), let us call the function $f_+(t)$ a retarded one and $f_-(t)$ an advanced one. Obviously, $\theta(t)$ and $\theta(-t)$ are the corresponding projection operators.

For the Fourier transforms of the functions $f_+(t)$ and $f_-(t)$, we have

$$f_+(\omega) = \delta_+(\omega) * f(\omega), \quad f_-(\omega) = \delta_-(\omega) * f(\omega). \tag{A.2.3}$$

Here, $\delta_{\pm}(\omega)$ are the Fourier transforms of $\theta(t)$ and $\theta(-t)$, divided by 2π. According to the formulae of the preceding section (after substituting $x \to t$, $k \to -\omega$)

$$\delta_{\pm}(\omega) = \frac{1}{2}\,\delta(\omega) \pm \frac{\mathrm{i}}{2\pi}\,\omega^{-1}\,, \tag{A.2.4}$$

where the generalized function ω^{-1} acts in the sense of a principal value.

We see that the generalized functions $\delta_{+}(\omega)$ and $\delta_{-}(\omega) = \delta_{+}(-\omega)$ are the kernels of the projection operators onto the corresponding sub-spaces and satisfy the identities

$$\delta_{+}(\omega)*\delta_{+}(\omega) = \delta_{+}(\omega), \quad \delta_{-}(\omega)*\delta_{-}(\omega) = \delta_{-}(\omega)\,, \tag{A.2.5}$$

$$\delta_{+}(\omega) + \delta_{-}(\omega) = \delta(\omega)\,. \tag{A.2.6}$$

The product of two retarded functions $f_{+}(t)$ and $g_{+}(t)$ (when it is defined) is obviously again a retarded function. But this means that the convolution $f_{+}(\omega)*g_{+}(\omega)$ is also the Fourier transform of a retarded function.

It is easy to prove that the same is valid for the convolution

$$f_{+}(t)*g_{+}(t) = \int_{0}^{t} f_{+}(t-t')\,g_{+}(t')\,dt'\,, \tag{A.2.7}$$

and hence, for the product $f_{+}(\omega)\,g_{+}(\omega)$.

Thus the retarded functions and their Fourier transforms form rings with respect to the operations of multiplication and convolution. This is also true for advanced functions.

For applications, it is important that retarded functions are closely related with the causality principle which requires a result to follow the cause. Let F be a linear operator with the kernel $f(t, t')$ and the function $q(t)$ in the relation $u(t) = Fq(t)$ be considered as a "cause" which generated the "result" $u(t)$. Then the causality principle requires that for any t_0, the condition $q(t) = 0$ with $t < t_0$ implies $u(t) = 0$ for $t < t_0$. It is easy to prove that, for a difference kernel $f(t-t')$, the causality principle is equivalent to the condition that $f(t) = 0$ when $t < 0$.

The following problem is important: what are the restrictions on the Fourier transform $f(\omega)$ of the kernel of an operator, so that the latter satisfyies the causality principle? This is equivalent to finding conditions to be satisfied by functions of the type $f_{+}(\omega)$. The formulae presented above enable us to solve this problem easily for the class of functions under consideration.

Let

$$\mathscr{H}[f(\omega)] = -\frac{1}{\pi}\,\omega^{-1}*f(\omega) = -\frac{1}{\pi}\,⨍\,\frac{f(\omega')}{\omega-\omega'}\,d\omega'\,. \tag{A.2.8}$$

The correspondence $f(\omega) \to \mathscr{H}[f(\omega)]$ is called the Hilbert transform. We leave it to the reader to prove that, in the t-representation, to the Hilbert transform there corresponds multiplication of $f(t)$ by i sgn$t$. From here, in particular, it follows that

$$\mathscr{H}[\mathscr{H}[f(\omega)]] = -f(\omega)\,. \tag{A.2.9}$$

We saw earlier that the relation

$$f(\omega) = \delta_+(\omega) * f(\omega) \tag{A.2.10}$$

is a necessary and sufficient condition for the function $f(\omega)$ to be the Fourier transform of a retarded function $f(t)$.

Taking into account (A.2.4) and (A.2.8), one can write (A.2.10) in the form

$$f(\omega) = -\mathrm{i}\mathscr{H}[f(\omega)]\,. \tag{A.2.11}$$

Setting $f(\omega) = f'(\omega) + \mathrm{i}f''(\omega)$ and taking the real and imaginary parts, we obtain

$$f''(\omega) = -\mathscr{H}[f'(\omega)]\,, \tag{A.2.12a}$$

$$f'(\omega) = \mathscr{H}(f''(\omega)]\,. \tag{A.2.12b}$$

This relationship between the real and imaginary parts of the function $f(\omega) = f_+(\omega)$ solves the problem formulated above and, in physical literature, is called a dispersion relation. The kernels of causality operators, which are sometimes called causality functions, must satisfy these relations.

In a number of cases, it is expedient to enlarge the space of retarded functions by including the generalized functions, concentrated at the point $t = 0$. As is well-known [A.1], these functions are representable in the form of finite linear combinations of $\delta(t)$ and its derivatives. In the ω representation, these correspond to polynomials.

The space thus enlarged contains, in particular, kernels of integro-differential operators, which satisfy the causality principle. Their action is reduced to convolution with the corresponding generalized retarded function (or to multiplication by its Fourier transform). However, note that products of the type $\delta(t)\,\theta(t)$ are not, generally speaking, defined. From this it follows that the enlarged space does not already form a ring with respect to multiplication in the t representation.

In order to write down the dispersion relations for the Fourier transform $f(\omega)$ of a generalized retarded function, one has first to separate out a polynomial from $f(\omega)$ (the so-called subtracted dispersion relations [A.2]). Indeed, the causality principle does not impose any restrictions on the real and imaginary parts of a polynomial.

An important property of the Fourier transform $f(\omega)$ of a (generalized) retarded function is its analyticity in the upper halfplane. In fact, the Fourier integral admits analytical continuation to the upper half ω-plane due to the rapid decreas of $\exp(i\omega t)$ as $t > 0$ (as a generalized function, $f(t)$ increases no faster than a polynomial as $t \to \infty$). The converse statement is also valid [A.2]: if the region of analyticity of $f(\omega)$ is the upper halfplane, than $f(t)$ is a retarded function. Thus, this property of analyticity is a characteristic one.

Let us consider some important examples.

Let us look for a solution $G^r(t)$ of the equation

$$\left(\frac{d^2}{dt^2} + \omega_0^2\right) G^r(t) = \delta(t)\,, \tag{A.2.13}$$

which satisfies the condition $G^r(t) = 0$ when $t < 0$. In other words, we want to construct a retarded Green's function $G^r(t) = G_+(t)$, which would be the kernel of the causality operator, which is inverse to the operator $d^2/dt^2 + \omega_0^2$.

In the ω-representation,

$$(\omega^2 - \omega_0^2)\, G^r(\omega) = -1\,. \tag{A.2.14}$$

The general solution of this equation has the form

$$G^r(\omega) = -\frac{1}{\omega^2 - \omega_0^2} + A\delta(\omega^2 - \omega_0^2)\,. \tag{A.2.15}$$

In order to obtain the causality solution $G^r(\omega) = G_+(\omega)$ we are interested in, the coefficient A must be chosen in such a way that the dispersion relations (A.2.12) would be satisfied. However, for the sake of convenience of further generalizations, we shall construct the solution in a different and more constructive way.

Let us introduce the generalized function $(\omega + i0)^{-1}$ (sometimes it is denoted also as $(\omega + i\varepsilon)^{-1}$), which occurs frequently in the physical literature and which acts on a test function $\varphi(\omega)$ according to the rule

$$\int \frac{\varphi(\omega)}{\omega + i0}\, d\omega = \lim_{\varepsilon \to +0} \int \frac{\varphi(\omega)}{\omega + i\varepsilon}\, d\omega\,. \tag{A.2.16}$$

Note that $\varphi(\omega)$ is here generally speaking, defined only for real values of ω and does not admit continuation to the complex plane. The computation of the integral by the usual method yields

$$\int \frac{\varphi(\omega)}{\omega + i0}\, d\omega = ⨍ \frac{\varphi(\omega)}{\omega}\, d\omega - \pi i \varphi(0)\,. \tag{A.2.17}$$

Taking into account (A.2.4) we find an important relation

$$\frac{1}{\omega + i0} = -2\pi i \delta_+(\omega)\,. \tag{A.2.18}$$

We see that $(\omega + \mathrm{i}0)^{-1}$ is a causality function, which satisfies the simplest equation

$$\omega \frac{1}{\omega + \mathrm{i}0} = 1\,. \tag{A.2.19}$$

Note that $(\omega + \mathrm{i}0)^{-1}$ can be continued to the upper halfplane of ω as ω^{-1} and that the real axis is the boundary of the analyticity region.

With the help of this function, we can write the causal solution of (A.2.14) in a convenient form. In fact, we have the identity

$$(\omega - \omega_0)(\omega + \omega_0)\left[\frac{1}{2\omega_0}\left(\frac{1}{\omega + \omega_0 + \mathrm{i}0} - \frac{1}{\omega - \omega_0 + \mathrm{i}0}\right)\right] = 1 \tag{A.2.20}$$

and hence,

$$G^{\mathrm{r}}(\omega) = \frac{1}{\omega_0^2 - (\omega + \mathrm{i}0)^2} \stackrel{\mathrm{def}}{=} \frac{1}{2\omega_0}\left(\frac{1}{\omega + \omega_0 + \mathrm{i}0} - \frac{1}{\omega - \omega_0 + \mathrm{i}0}\right). \tag{A.2.21}$$

Taking into account (A.2.4) and (A.2.18) one can represent $G^{\mathrm{r}}(\omega)$ in another form

$$\begin{aligned} G^{\mathrm{r}}(\omega) &= \frac{\pi \mathrm{i}}{\omega_0}\left[\delta_+(\omega - \omega_0) - \delta_+(\omega + \omega_0)\right] \\ &= \frac{1}{2\omega_0}\left(\frac{1}{\omega + \omega_0} - \frac{1}{\omega - \omega_0}\right) + \frac{\pi \mathrm{i}}{2\omega_0}\left[\delta(\omega - \omega_0) - \delta(\omega + \omega_0)\right]. \end{aligned} \tag{A.2.22}$$

It is easy to verify that the last expression satisfies the dispersion relations (A.2.12).

Using the expression (A.2.21) for the function $G^{\mathrm{r}}(\omega)$, let us write down the function $G^{\mathrm{r}}(t)$ in the form

$$G^{\mathrm{r}}(t) = \frac{1}{2\pi}\int \frac{\mathrm{e}^{-\mathrm{i}\omega t}}{\omega_0^2 - (\omega + \mathrm{i}0)^2}\,d\omega = \lim_{\varepsilon \to 0} \frac{1}{2\pi}\int \frac{\mathrm{e}^{-\mathrm{i}\omega t}}{\omega_0^2 - (\omega + \mathrm{i}\varepsilon)^2}\,d\omega\,. \tag{A.2.23}$$

We see that the rules of passing around the poles of the function $(\omega_0^2 - \omega^2)^{-1}$ are chosen such that with $t < 0$, i.e. before the source is switched on, the Green's function $G^{\mathrm{r}}(t)$ vanishes. Thus, all the poles are passed around from above.

Let us use this rule to determine the retarded solution G^{r} of the equation

$$LG = I\,, \tag{A.2.24}$$

where the operator L generates the dispersion relation $L(\omega, k) = 0$ with a real ω corresponding to a real k.

Then the retarded Green's function is written down in the form

$$G^{r}(x,t) = \lim_{\varepsilon\to 0} \frac{1}{(2\pi)^2} \iint \frac{e^{i(kx-\omega t)}}{L(k, \omega + i\varepsilon)} dk\, d\omega \quad (\varepsilon > 0)\,. \tag{A.2.25}$$

The expression for $L(k, \omega + i\varepsilon)$ can be expanded into a series in powers of ε, and restricting ourselves to terms linear in ε, we obtain for $G^{r}(x, t)$, the expression

$$G^{r}(x, t) = \lim_{\varepsilon\to 0} \frac{1}{(2\pi)^2} \iint \frac{e^{i(kx-\omega t)}}{L(\omega, k) + i\varepsilon \dfrac{\partial L}{\partial \omega}} dk\, d\omega\,, \tag{A.2.26}$$

which can be rewritten in the form

$$G^{r}(x, t) = \frac{1}{(2\pi)^2} \iint e^{i(kx-\omega t)} \left\{\frac{1}{L(\omega, k)} - i\pi\, \mathrm{sgn} \left\{\frac{\partial L}{\partial \omega}\right\} \delta(L(\omega, k))\right\} dk\, d\omega\,. \tag{A.2.27}$$

A.3 Expansion of Functions, Given at a Finite Number of Points, in Special Bases

We shall construct here a discrete analog of the expansion of functions defined in the boundary region considered in Sect 2.11. As bases, the sets of elementary solutions of the following equation will be used:

$$\sum_{n'} \Phi(n - n')\, u(n') = 0\,. \tag{A.3.1}$$

From the results of Chap. 2, it follows that, for a homogeneous chain with nearest-neighbor interaction this equation has $2N$ linearly independent solutions (here and below we assume that $a = 1$)

$$\varepsilon_0(n) = 1, \quad \varepsilon_1(n) = n, \quad \varepsilon_m(n) = e^{ik_m n}\,, \tag{A.3.2}$$

where $k_m (m = 2, 3 \ldots, 2N - 1)$ are the roots of the equation $\Phi(k) = 0$.

Let us first consider the general problem of the expansion of an arbitrary function $f(n)$ prescribed at $2N$ points of the region $\tilde{S}$ $(-N + 1 \leqslant n \leqslant N)$, in-terms of the complete set of the basis functions $\{\varepsilon_m\}$

$$f(n) = \sum_{m=0}^{2N-1} f^m \varepsilon_m(n)\,. \tag{A.3.3}$$

In order to find the coefficients f^m let us construct the reciprocal basis $\{\varepsilon^m\}$, which is defined by the condition of biorthogonality

$$\sum_{m\in S} \varepsilon^{m'}(n)\, \varepsilon_m(n) = \delta_m^{m'} \,. \tag{A.3.4}$$

In order to achieve this, let us complete the definitions of all the functions $\varepsilon_m(n)$ and $\varepsilon^{m'}(n)$ by equating them to zero outside $\tilde{S}$ and let us introduce their Fourier transforms $\varepsilon_m(k)$, $\varepsilon^{m'}(k)$. The conditions of biorthogonality take the form

$$\mathrm{i}\,\frac{d}{dk}\,\varepsilon^{m'}(k)\,|_{k=0} = \delta_1^{m'}, \quad \varepsilon^{m'}(k_m) = \delta_m^{m'} \quad (m \neq 1). \tag{A.3.5}$$

We leave it to the reader to verify that these conditions are satisfied by

$$\varepsilon^0(k) = \frac{\Phi(k)}{\Phi''(0)\,(1-\cos k)}, \quad \varepsilon^1(k) = (\mathrm{e}^{-\mathrm{i}k} - 1)\,\varepsilon^0(k)\,,$$

$$\varepsilon^m(k) = \frac{\Phi(k)}{\mathrm{i}\Phi'(k_m)\,\mathrm{e}^{\mathrm{i}k_m}(\mathrm{e}^{\mathrm{i}k} - \mathrm{e}^{-\mathrm{i}k_m})} \quad (m = 2,3,\,\dots\,,2N-1)\,. \tag{A.3.6}$$

Returning to the inverse Fourier transforms, we obtain

$$\varepsilon^0(n) = -\frac{1}{\Phi''(0)} \sum_{n'} \Phi(n')\,|n - n'|\,,$$

$$\varepsilon^1(n) = \varepsilon^0(n-1) - \varepsilon^0(n) = -\,c_0^{-1} \sum_{n'\geqslant n} \Phi(n')\,, \tag{A.3.7}$$

$$\varepsilon^m(n) = \frac{\mathrm{i}\mathrm{e}^{-\mathrm{i}k_m n}}{\Phi'(k_m)} \sum_{n'\geqslant n} \Phi(n')\,\mathrm{e}^{\mathrm{i}k_m n'} \quad (m = 2,3\,\dots\,,2N-1).$$

Let us show that the $\varepsilon^m(n)$ vanish outside $\tilde{S}$. For $\varepsilon^0(n)$, we have

$$\varepsilon^0(n) = -\frac{1}{2c_0}\left[\sum_{n'<n} \Phi(n')\,(n-n') + \sum_{n'>n} \Phi(n')\,(n'-n)\right]. \tag{A.3.8}$$

When $n \geqslant N$, the second term in the brackets disappears since $\Phi(n') = 0$, when $n' > N$, and the first term disappears also due to the relations, cf, (2.1.9) and (2.1.10),

$$\sum_{n'<n} \Phi(n')\,(n-n') = \sum_{n'} \Phi(n')\,n' = 0\,. \tag{A.3.9}$$

It can be proved in an anlogous way that $\varepsilon^0(n) = 0$, when $n \leqslant -N$. Vanishing of the function $\varepsilon^1(n)$ outside $\tilde{S}$ follows from its expression in terms of $\varepsilon^0(n)$.

For the remaining $\varepsilon^m(n)$, when $n \leqslant -N$

$$\sum_{n'\geqslant n} \Phi(n')\,\mathrm{e}^{\mathrm{i}k_m n'} = \sum_{n'} \Phi(n')\,\mathrm{e}^{\mathrm{i}k_m n'} = \Phi(k_m) = 0\,. \tag{A.3.10}$$

For $n > N$, the above sum vanishes since $\Phi(n') = 0$ for $n' > N$. This completes the proof.

Thus, the coefficients of the expansion (A.3.3) are determined by the expressions

$$f^m = \sum_{n \in S} \varepsilon^m(n) f(n) . \tag{A.3.11}$$

If one writes down the boundary value problems which were formulated in Sect. 2.6 for a finite chain, then these problems are reduced to solving a finite system of linear algebraic equations. The advantage of the biorthogonal basis selected here is connected with the possibility of a transition to the approximation with respect to the first roots.

In reality, in accordance with the reasoning carried out in Sect. 2.11, the case of a semi-bounded lattice is of primary interest. The coefficients of increasing exponents in the expansion (A.3.3) must be set equal to zero due to natural conditions, as $n \to \infty$. Under such an additional condition, one can use the complete basis $\{\varepsilon_m, \varepsilon^{n'}\}$. However, it is more justifiable and convenient to introduce directly a basis, which contains only decreasing (as $n \to \infty$) exponentials.

Let us, for definiteness, consider the second basic problem, i.e. assume that the displacement $u(n)$ satisfies the homogeneous equation (A.2.1), when $n > 0$, and the values of $u_s(n)$ are given in the boundary region $S(-N+1 \leqslant n \leqslant 0)$. It is obvious that, for the solution of the problem, it is sufficient to expand $u_s(n)$ in the basis

$$e_0(n) = 1, \quad e_m(n) = e^{ik_m n}, \quad \operatorname{Im}\{k_m\} > 0 . \tag{A.3.12}$$

The coefficients of the expansion and consequently $u(n)$ will be found if one constructs the reciprocal basis $\{\varepsilon_m\}$. In order to achieve this, it is possible to use the method considered above, but replacing $\Phi(k)$ by another generating function.

Using the identity

$$\cos k - \cos k_m = -\frac{1}{2} e^{ik_m}(e^{ik} - e^{ik_m})(e^{-ik} - e^{ik_m}) ,$$

let us rewrite (2.8.10) in the form

$$\Phi(k) = \Phi_+(k)\, \Phi_-(k), \quad \Phi_+(k) = \Phi_-(-k) ,$$

$$\Phi_-(k) = A \sum_{m=0}^{N-1} (e^{ik} - e^{-ik_m}), \quad A = \text{const}, \tag{A.3.13}$$

were $\Phi_\pm(k)$ has no roots inside the upper (lower) half-plane.

Let us take into account that the Fourier transform of $1(k)$ is $\delta(n)$ and that to multiplication by $\exp(ik)$ corresponds a shift by unity to the left, i.e. $\delta(n+1)$. Hence, $\Phi_-(n)$ vanishes outside the region $-N \leqslant n \leqslant 0$.

Choosing $\Phi_-(k)$ as a generating function we find analogously to the above the reciprocal basis

$$e^m(n) = \frac{\mathrm{i}e^{-\mathrm{i}k_m n}}{\Phi'_-(k_m)} \sum_{n' \geqslant n} \Phi_-(n')\, \mathrm{e}^{\mathrm{i}k_m n'}\,, \tag{A.3.14}$$

which is concentrated on S, as can easily be proved.

For solving of the first fundamental problem we should take as a basis on S cf, (2.11.28),

$$\tilde{e}_0(n) = n, \quad \tilde{e}_m(n) = \mathrm{e}^{\mathrm{i}k_m n}, \quad \mathrm{Im}\,\{k_m\} > 0\,. \tag{A.3.15}$$

The reciprocal basis $\{\tilde{\varepsilon}^n\}$ is an appropriate subset of elements of the basis $\{\varepsilon^m\}$.

References

1.1 E. Cosserat, F. Cosserat: *Theorie des Corps Deformables* (Hermann, Paris 1909)

1.2 C. Truesdell, R.A. Toupin: "The Classical Field Theories", in: *Principles of Classical Mechanics and Field Theory*, Encyclopedia of Physics, Vol. III/1 (Springer, Berlin, Göttingen, Heidelberg 1960) pp. 226–793

1.3 Cz. Woźniak: Dynamic models of certain bodies with discrete-continuous structure. Arch. Mech. Stosow. *21*, 707–724 (1969)

1.4 G.N. Savin, Iu. N. Nemish: Investigations on stress concentration in couple-stress elasticity (survey). Prikl. Mekh. *6*, 1–17 (1968) [in Russian]

1.5 A.A. Il'jushin, V.A. Lomakin: "Couple-Stress Theories in the Mechanics of Solid Deformable Bodies", in *Prochnost i plastichnost* (Strength and Plasticity), ed. by A. Il'jushin (Nauka, Moscow 1971) pp. 54–61 [in Russian]

1.6 M. Misicu: *Mecanica Mediilor Deformabile. Fundamentele Eleasticitătii Structurale* (Continuum Mechanics. Foundations of Elasticity) (Acad. R.S.R., Bucharest 1967) [in Rumanian]

1.7 W. Nowacki: *Theory of Micropolar Elasticity* (Wydawn. uczeln. politechn., Poznań 1970) [in Polish]
W. Nowacki: *Theory of Nonsymmetric Elasticity* (Zaklad narod. im Ossolinskich PAN, Wroclaw 1970) [in Polish]

1.8 E. Kröner (ed.): *Mechanics of Generalized Continua.* Proc. IUTAM Symposium on the Generalized Cosserat Continuum and the Continuum Theory of Dislocations with Applications, Freudenstadt and Stuttgart 1967 (Springer, Berlin, Heidelberg, New York 1968)

1.9 I.A. Kunin, V.G. Kosilova (eds): *Teoria Sred s Mikrostrukturoi* (Theory of Media with Microstructure. Bibliography) (Sib. Otdelenie AN SSSR, Institut teplofiziki, Novosibirsk 1976)

1.10 E. Kröner: On the physical reality of torque stresses in continuum mechanics. Int. J. Eng. Sci. *1*, 261–278 (1963)

1.11 I.A. Kunin: *Teoria Uprugikh Sred c Mikrostrukturoi* (Theory of Elastic Media with Microstructure) (Nauka, Moscow 1975)

1.12 I.A. Kunin: *Elastic Media with Microstructure II. Three-Dimensional Models, Springer Series* in Solid-State Sciences (Springer, Berlin, Heidelberg, New York, forthcoming)

1.13 D.G.B. Edelen: Irreversible thermodynamics of nonlocal systems. Int. J. Eng. Sci. *12*, 607–631 (1974)

Chapter 2

2.1 I.M. Gel'fand, G.E. Shilov: *Generalized Functions* (Academic, New York 1964)
2.2 Ia.I. Khurgin, V.P. Yakovlev: *Finitnii Funkcii v Fizike* (Localized Functions in Physics and Engineering) (Nauka, Moscow 1971)
2.3 C.E. Shannon: *The Mathematical Theory of Communication* (University of Illinois Press, Urbana 1949)
2.4 I.S. Gradshtein, I.M. Ryzhik: *Table of Integrals, Series and Products*, 4th ed. (Academic, New York 1965)
2.5 I.A. Kunin: "Methods of Tensor Analysis in the Theory of Dislocations", in J.A. Schouten: *Tensor Analysis for Physicists* (Nauka, Moscow 1965) pp. 374–443 (Supplement to the Russian translation, available in English from the U.S. Department of Commerce, FSTI, Springfield, VA 1965)
2.6. V.M. Agranovich, V.L. Ginzburg: *Kristallo-optika s Prostranstvennoi Disperiey i Teoria Exitonov* (Crystal-Optics with Space Dispersion and the Theory of Excitons) (Nauka, Moscow 1965)
2.7 M. Born, K. Huang: *Dynamical Theory of Crystal Lattices* (Clarendon Press, Oxford 1954)
2.8 J.M. Ziman: *Electrons and Phonons* (Oxford University Press, Oxford 1960)
2.9 B.I. Levin: *Distribution of Zeros of Entire Functions* (Am. Math. Soc., Providence, RI 1964)
2.10 M.G. Krein: Integral equations on semi-axis with kernels depending on the difference of arguments. Usp. Mat. Nauk. *13*, 3–120 (1958) [in Russian]

Chapter 3

3.1 D. Jackson: *Fourier Series and Orthogonal Polynomials* (The Mathematical Association of America, Oberlin 1941)
3.2 G.Ia. Lubarski: *Teoria Group i Primenenia v Fizike* (Groups Theory and its Application in Physics) (Fizmatgiz, Moscow 1958)
3.3 A.O. Gel'fond: *Calcul des Differences Finies* (Dunod, Paris 1963)
3.4 R.D. Mindlin: Micro-structure in linear elasticity. Arch. Rat. Mech. Anal. *16*, 51–78 (1964)
3.5 R.A. Toupin: Theories of elasticity with couple-stresses. Arch. Rat. Mech. Anal. *17*, 85–112 (1964)
3.6 H. Bremermann: *Distributions, Complex Variables, and Fourier Transforms* (Addison-Wesley, Reading, MA. 1965)
3.7 S. Timoshenko, D.H. Young, W. Weaver: *Vibration Problems in Engineering* (Wiley, New York 1974)

Chapter 4

4.1 E.T. Copson: *Asymptotic Expansions* (Cambridge University Press, Cambridge 1965)
4.2 A. Erdélyi: *Asymptotic Expansions* (Dover, New York 1956)
4.3 R.I. Elliott, J.A. Krumhansl, P.L. Leath: The theory and properties of randomly disordered crystals. Rev. Mod. Phys. *46*, 465 (1974)
4.4 D. Bohm: *Quantum Theory* (Prentice-Hall, Englewood Cliffs, NJ 1952)
4.5 A.M. Kosevich: *Osnovi Mekhaniki Kristallicheskoi Reshetki* (Fundamentals of Crystal Lattice Mechanics) (Nauka, Moscow 1972)
4.6 R.G. Newton: *Scattering Theory of Waves and Particles* (McGraw-Hill, New York 1966)
4.7 N.I. Muskhelishvili: *Singular Integral Equations* (Wolters-Noordhoff, Groningen 1972)

Chapter 5

5.1 D.J. Korteweg, G. deVries: On the change of form of long waves advancing in a rectangular channel, and on a new type of long stationary waves. Philos. Mag. *39*, 422–443 (1895)
5.2 V.I. Karpman: *Non-linear Waves in Dispersive Media* (Pergamon, Oxford 1975)
5.3 M.M. Vainberg: *Variational Methods for the Study of Non-linear Operators* (Holden-Day, San Francisco 1964)
5.4 I.M. Gel'fand, S.V. Fomin: *Calculus of Variation* (Prentice-Hall, Englewood Cliffs, NJ 1963)
5.5 M. Toda: Wave in nonlinear lattice. Prog. Theor. Phys. *45*, 174–200 (1970) [Supplement]
M. Toda: *Theory of Nonlinear Lattices*, Springer Series in Solid-State Sci., Vol. 20 (Springer, Berlin, Heidelberg, New York 1981)
5.6 N. Ooyama, N. Saito: On the stability of lattice solitons. Prog. Theor. Phys. *45*, 201–208 (1970) [Supplement]
5.7 N.J. Zabusky: Solitons and bound states of the time-independent Schrödinger equation. Phys. Rev. *168*, 124–128 (1968)
5.8 R. Hirota: Exact N-soliton solutions of the wave equation of long waves in shallow-water and in nonlinear lattices. J. Math. Phys. *14*, 810–814 (1973)
5.9 M.Y. Beran: *Statistical Continuum Theories* (Interscience, New York 1968)
5.10 V.E. Zakharov: Kinetic equation for solitons. Sov. Phys. JETP *33*, 538–541 (1971)
5.11 A. Seeger: „Zur Dynamik von Versetzungen in Gitterreihen mit verschiedener Gitterkonstanten"; Thesis, Technische Hochschule Stuttgart (1948–49)
5.12 A. Kochendorfer, A. Seeger: Theorie der Versetzungen in eindimensionalen Atomreihen. I. Periodisch angeordnete Versetzungen. Z.

Phys. *127*, 533–550 (1950)
5.13 A. Seeger, H. Donth, A. Kochendorfer: Theorie der Versetzungen in eindimensionalen Atomreihen. III. Versetzungen, Eigenbewegungen und ihre Wechselwirkung. Z. Phys. *134*, 173–193 (1953)
5.14 A. Seeger: "Solitons in Crystals", in *Continuum Models of Discrete Systems* (CMDS3), ed. by E. Kroner, K.H. Anthony (Univ. Waterloo Press 1980) pp. 253–327

Chapter 6

6.1 C.S. Gardner, J.M. Green, M. Kruskal, R.M. Miura: Method for solving the Korteweg-de Vries equation. Phys. Rev. Lett. *19,* 1095–1097 (1967)
6.2 V.A. Marchenko: *Spektralnaya Teoria Operatorov Sturma-Liouvillia* (Spectral Theory of Sturm-Liouville Operators) (Naukova Dumka, Kiev 1972)
6.3 I. Kay, V.E. Moses: Reflectionless transmission through dielectrics and scattering potentials. J. Appl. Phys. *27*, 1503–1508 (1956)
6.4 A.B. Shabat: On the Korteweg-de Vries equation. Sov Math. Dokl. *14*, 1266–1270 (1973)

Appendices

A.1 I.M. Gel'fand, G.E. Shilov: *Generalized Functions* (Academic, New York 1964)
A.2 H. Bremermann: *Distributions, Complex Variables, and Fourier Transforms* (Addison-Wesley, Reading, MA 1965)

Subject Index